Supramolecular Chemistry on Surfaces

Supramolecular Chemistry on Surfaces

2D Networks and 2D Structures

Edited by Neil R. Champness

WILEY-VCH

Editor

Prof. Neil R. Champness
University of Birmingham
School of Chemistry
Edgbaston
B15 2TT Birmingham
United Kingdom

Cover Image: Reproduced from "On-surface chemical reactions characterised by ultrahigh resolution scanning probe microscopy" A. Sweetman, N.R. Champness, and A. Saywell. *Chem. Soc. Rev.*, 2020, **49**, 4189–4202, with permission from The Royal Society of Chemistry.

Library of Congress Card No.: applied for

British Library Cataloguing-in-Publication Data
A catalogue record for this book is available from the British Library.

Bibliographic information published by the Deutsche Nationalbibliothek
The Deutsche Nationalbibliothek lists this publication in the Deutsche Nationalbibliografie; detailed bibliographic data are available on the Internet at <http://dnb.d-nb.de>.

Print ISBN: 978-3-527-34491-8
ePDF ISBN: 978-3-527-81668-2
ePub ISBN: 978-3-527-81670-5
oBook ISBN: 978-3-527-81669-9

Typesetting Straive, Chennai, India
Printing and Binding CPI Group (UK) Ltd, Croydon, CR0 4YY

Printed on acid-free paper

C085831_091221

Contents

Preface

The field of supramolecular chemistry has developed from its inception to now influence thinking, strategies, and application across the chemical and materials sciences. Whilst remarkable progress has been made in many fields, the need to interface supramolecular systems to the real world has spurred interest in performing supramolecular chemistry on surfaces. The goal of studying supramolecular self-assembly processes has in turn engendered new ideas, new concepts, and ultimately a new field of study.

Interestingly, this fresh research focus has brought together experts from many different backgrounds creating new interdisciplinary connections, notably between synthetic chemists and physicists. Surface-based supramolecular chemistry is a truly multidisciplinary field. Indeed, the field has rapidly developed and the original focus on hydrogen-bonded systems has been joined by the exploitation of other supramolecular interactions. Similarly, some researchers have moved towards using self-assembly processes that enable the formation of covalent bonds and hence robust chemical systems, such as nanoscale graphenes.

These studies rely on characterisation techniques, particularly scanning-probe microscopies, that enable molecular, and even submolecular, resolution. Not only do such approaches result in visually inspiring images they also allow appreciation of supramolecular structures with a level of detail that is rarely achievable in traditional supramolecular chemistry. In turn, this has led to the discovery of complex quasi-crystalline arrays and highly complex arrangements. These fascinating structures spark the imagination and move beyond much that has been achieved in supramolecular chemistry.

All these remarkable developments and new avenues of research have spurred increasing attention to how these systems may be exploited in devices operating at the single-molecule level. The interaction between the surface and those absorbed molecules allows direct interaction between molecular systems and the macroscopic world and has led to increasing interest in developing devices, particularly employing electronic properties. Thus, the field is developing from simple curiosity and structural fascination towards applications.

It is timely to evaluate progress in the field and to appreciate where the focus has been and where it is going. Hence, this collection surveys the field from the point of view of experts who have devoted their endeavours to develop this new area of science. I am grateful to all those authors for their excellent contributions and for so clearly expounding their vision of the research area. I hope that the chapters contained herein will inspire the many researchers in the field but also those who currently sit around the periphery of this activity whether chemist, physicist, or the next generation of scientist.

Birmingham
April 2021

Neil R. Champness

1

Two-Dimensional Supramolecular Chemistry on Surfaces

Neil R. Champness

University of Birmingham, School of Chemistry, Edgbaston, Birmingham B15 2TT, UK

Supramolecular chemistry represents one of the central themes of modern chemical sciences. Crossing traditional boundaries of chemistry, materials science, biology, and physics, the field of supramolecular chemistry affords opportunities to create new molecules and materials, with far reaching implications for many and diverse applications. The significance of supramolecular chemistry lies behind two Nobel Prizes, 1987 [1] and 2016 [2–4], and is now not only a field in its own right but is also a central underpinning theme in almost any area of chemistry. The primary principle of supramolecular chemistry is the use of non-covalent interactions to create and control self-assembled structures. A large range of interactions is available to the supramolecular chemist to influence and control self-assembly processes. From hydrogen bonds [5–7] and halogen bonds [8, 9] to π-interactions [10, 11], coordination bonds [12, 13] and the mechanical bond [2, 3, 14–16], interactions of different strengths and varying degrees of geometrical preferences are available to design and create structures. When in its infancy, supramolecular chemistry focussed predominantly on synthetic strategies in combination with understanding the fundamental properties of the non-covalent interactions employed. Over recent years, the field has developed to such an extent that it is now commonplace to focus effort towards applications and these range across a vast spectrum. Supramolecular chemistry is so wide-ranging that its relevance can be applied to diverse fields, from biology [17, 18] and medicine [18, 19] to new materials [20, 21] and energy-related applications [22, 23].

The origins of supramolecular chemistry lie in solution-based systems, using intermolecular interactions to create supermolecules. From these origins, supramolecular chemistry is now observed in most phases, notably in the solid-state, through crystal engineering [24, 25], in liquid crystals [26] and ionic liquids [27], and even in the gas phase [28]. It was only natural that supramolecular chemistry strategies would come to be applied to the two-dimensional (2D) environment of surfaces (Figure 1.1). This seemingly natural progression also raised a number of challenges to practitioners of the subject, not least in terms of appreciating this quite different

Supramolecular Chemistry on Surfaces: 2D Networks and 2D Structures, First Edition.
Edited by Neil R. Champness.

environment and perhaps most importantly the different techniques that are used to characterise and interpret surface-based molecular systems.

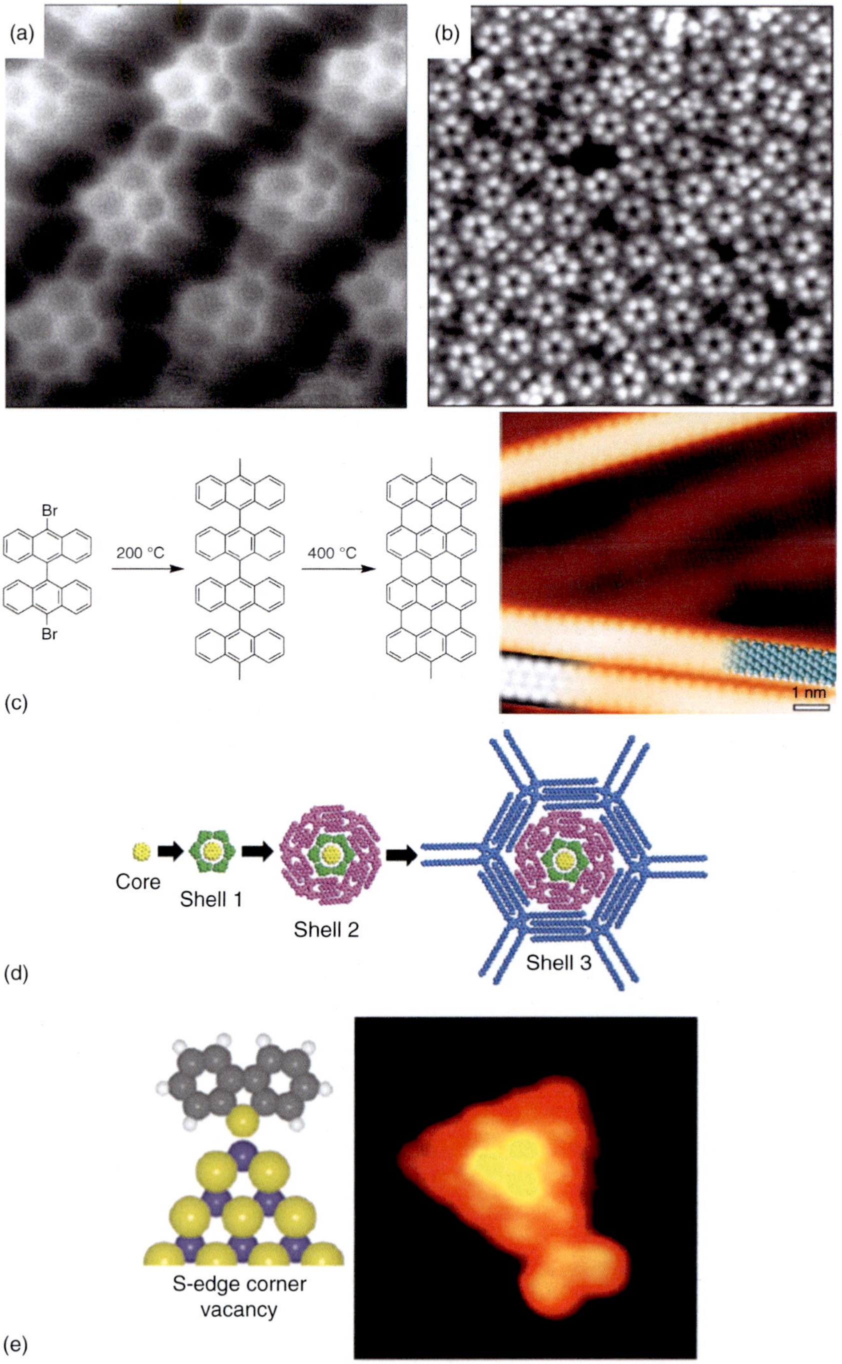

Figure 1.1 Examples of two-dimensional supramolecular chemistry on surfaces discussed within this volume. (a) ncAFM image of a hydrogen-bonded naphthalene-1,4:5,8-tetracarboxylic diimide (NTCDI) island on a Ag:Si(111) – $(\sqrt{3} \times \sqrt{3})$ R30° surface acquired at 77 K. The image reveals sub-molecular details of the self-assembled structure; (b) STM image of self-assembled arrays of ferrocene-carboxylic acid (FcCOOH); each bright feature represents a separate FcCOOH molecule, which then assemble into pentamers highly reminiscent of a Penrose tiling arrangement; (c) Surface-assisted C–C coupling reaction used to prepare straight graphene nanoribbons from bianthryl monomers, including a STM image of nanoribbon, following cyclodehydrogenation at 400 °C, with partly overlaid molecular model (right in blue) and a density-functional theory model (bottom left in grey); (d) Schematic representation of a strategy used to prepare a multicomponent system using a 'core–shell' approach. Each colour represents a different molecular building block; (e) Schematic representation and STM image showing dibenzothiophene bound to the corner vacancy of a S-edge-terminated MoS_2 nanocluster. Source: Images reproduced with permission as follows: (a) Sweetman et al. [29]; (b) reproduced with permission from Springer Nature from Wasio et al. [30]; (c) reproduced with permission from Springer Nature from Cai et al. [31]; (d) Mali et al. [32]; (e) reprinted and adapted with permission from Tuxen et al. [33]. Copyright (2010) American Chemical Society.

Whereas the techniques applied to characterising solution phase, or solid-state, supramolecular systems are common across synthetic chemistry, for example, NMR spectroscopy, mass spectrometry, and X-ray diffraction, characterisation of surface-bound molecules is a quite distinct domain. The most common approaches to characterising molecular species on surfaces are scanning probe microscopies (SPM). Specifically, techniques such as scanning-tunnelling microscopy (STM) [34] and atomic force microscopy (AFM) [35] represent the dominant characterisation methods used in the analysis of surface-based supramolecular systems. These imaging microscopies can be, and often are, supplemented by other approaches, such as X-ray photoelectron spectroscopy (XPS), but SPM approaches provide invaluable insight into specific molecular arrangements allowing determination of the geometric structure of organic molecules with molecular resolution. More recently, the development of noncontact atomic force microscopy (ncAFM) [36] allows the characterisation of supramolecular systems with sub-molecular resolution [37]. The use of SPM characterisation techniques in itself presents opportunities, which are rarely available to those working in other phases, not least because such microscopies function at the molecular, or even sub-molecular, level and as a result information, both structural and electronic, can be gathered for individual molecules and defined self-assembled arrays. In comparison, techniques such as NMR spectroscopy or X-ray diffraction rely upon the signal from comparatively large numbers of molecules. Thus, the characterisation of surface-based supramolecular systems can give a detailed picture of the structures and even transformations between different arrangements with a high degree of resolution. The complexities, challenges, and advantages of different SPM techniques are discussed in more detail by Sweetman, Champness, and Saywell in this volume.

A further aspect of the detailed imaging with molecular resolution is that this allows characterisation of structures that would prove extremely challenging by any other technique. Using SPM techniques allows ready identification of defects within supramolecular arrays but intriguingly allows the study of extended structures, which do not possess long-range order, with molecular resolution. This approach has been applied to the study of random, entropically stabilised, rhombus tilings [29, 38, 39], a molecular Penrose tile [30], quasicrystalline structures [40], and fascinating assemblies that exhibit the structure of Serpiński triangles [41]. The complex issues with studying and characterising quasicrystalline 2D arrays are discussed in detail by Kandel in this book.

The study of supramolecular chemistry on surfaces probably began with early studies of hydrogen-bonded assemblies [42–44] but has spread to employ other non-covalent interactions including coordination bonds [45, 46] and weaker van der Waals interactions [47–49]. The use of different intermolecular interactions is discussed throughout chapters in this volume. In particular, Mali, Teyssandier, Bilbao, and De Feyter discuss the use of hydrogen bonds and van der Waals interactions to create complex structures whereas the application of coordination bonds is presented by Lin and Liu. It will become clear to the reader that the choice of intermolecular interaction influences the choice of experimental conditions used, including deposition conditions, use of ultra-high vacuum (UHV) or studies at the solid–solution interface, and even the nature of the surface employed for surface self-assembly. The interactions between surface, substrate, solution, and self-assembled array are all important in determining the subtle energetic balance between different products [50].

These studies have now developed further to create covalently linked structures including nanographenes [51, 52] and covalent-organic frameworks (COFs) [53]. All of these strategies present their own distinct advantages, and disadvantages, but importantly represent a broad palette for researchers to employ and explore. Weaker interactions such as hydrogen bonds, van der Waals interactions, and even coordination bonds, form reversibly and therefore facilitate the formation of well organised, and relatively defect-free, supermolecule structures over comparatively large areas. Creating larger defect free structures can be more challenging using covalent bonds although the use of reversibly-formed bonds such as imines [54] has been developed to aid in this respect. Nanographenes, where carbon–carbon bonds are an absolute requirement, present quite different challenges but remarkable advances have been made in this area. In this volume Peng, Xing, and Wu discuss the use of intermolecular interactions to control on-surface reactions and Wang, Zhang, and Chi present developments in the field of on-surface reactions to create covalently bonded systems.

Another major challenge that requires thought when one considers surface-based supramolecular chemistry are the reaction environment and conditions. Firstly, it is typical to use a surface that is atomically flat or at least close to atomically flat. This rather stringent requirement facilitates the use of SPM characterisation and simultaneously controls the introduction of surface-based reactive sites to the self-assembly process. Even though atomically flat surfaces are commonly used, it would be a mistake to consider the surface as an innocent bystander in the self-assembly process. Indeed, adsorption between the surface and the molecules involved in self-assembly is essential to allow the formation of a surface-bound or surface-supported, supramolecular structure [50]. A range of surfaces are available to researchers investigating such systems but some are more common than others, notably highly-oriented pyrolytic graphite (HOPG) and Au(111). However, in some areas of study, the surface plays an integral role in the reaction process providing active sites, such as metal atoms, which catalyse the formation of a specific product [55].

The other major aspect that influences the self-assembly process is the experimental conditions of the experiment. SPM techniques can be used in both UHV conditions or at the interface between surface and solution. These quite different conditions present both advantages and disadvantages depending on the specific molecules and reaction processes being investigated. For example, studying molecules and self-assembled aggregates in UHV conditions can lead to higher resolution imaging, in part because lower temperatures (below the freezing point of solvents) can be accessed. Additionally, ncAFM imaging specifically requires UHV conditions. However, the introduction of molecules to the surface typically involves sublimation, and hence heating of the sample. Sublimation is not always possible and thermal degradation is a significant impediment for complex molecules. Milder electrospray deposition techniques have been developed [56] but the use of this approach is not yet widespread. In contrast, studies at the solution-solid interface directly image self-assembled structures in the presence of solvent. In terms of preparative conditions, this approach is quite straightforward, simply imaging at the interface between a drop of solvent containing the molecules of interest and the substrate. Although this approach offers many advantages the choice of solvent, which is limited by the requirements for imaging, can clearly influence the self-assembly process, potentially with solvent molecules interacting or even co-adsorbing with the target species. Although images tend to have lower resolution than UHV studies, this is not always the case and remarkable examples of molecular resolution with AFM have been reported [49].

Ultimately, the possibilities that arise from the various approaches to create supramolecular structures suggest the possibility of creating molecular level devices and the application of 2D materials. The advances in this area are illustrated in the chapter by Huang and Wee where they discuss the rapidly advancing field that studies 2D transition metal dichalcogenides and their potential integration with organic molecules for multifunctional flexible devices.

This book brings together perspectives from research leaders in the field. It can be seen that across the breadth of the subject, there are many fascinating examples of applying supramolecular chemistry to the development of surface-based arrays. Whether through the direct implementation of hydrogen bonds, coordination bonds, or well-designed van der Waals interactions, or through the controlled formation of covalently-bonded arrays, it is clear that strategies for creating 2D arrays on surfaces are well developed. A theme that commonly arises throughout the contributions is that of complexity. It is not a surprise that this subject has become prominent in the field of surface-based supramolecular arrays when one considers the specificity of the SPM characterisation techniques employed for characterisation. When one applies a technique that affords molecular resolution, allowing detailed appreciation of extended frameworks, their complexity becomes all the more apparent, drawing the attention of researchers and hence becoming a focus for investigation. Remarkable discoveries have been made across the field and in turn, spur new endeavours. An emerging aspect of the field is the implementation of synthetic strategies towards new applications with electronic properties of new structures receiving notable attention. However, other directions of research are also emerging at the solid–solution interface, for example, applying the chirality of surface arrays. Exploiting the interplay between surface-based arrays and solution chemistry promises to be of significance in applications ranging from sensing to the interface with biological processes.

In summary, as is common for new areas of science, the field now stands at a crossroads. The origins of the field have been based on developing an underpinning methodology for both synthesis and characterisation and an appreciation of the many factors that affect surface-based supramolecular assembly. Increasingly, there is a focus on developing these fascinating 2D materials for specific applications and for their incorporation into devices. I am confident that all the authors of the other chapters will agree that there is a promising and bright future for the area of 2D chemistry on surfaces.

References

1 Lehn, J.-M. (1988). *Angew. Chem. Int. Ed. Engl.* 27: 89–112.

2 Stoddart, J.F. (2017). *Angew. Chem. Int. Ed.* 56: 11094–11125.

3 Sauvage, J.-P. (2017). *Angew. Chem. Int. Ed.* 56: 11080–11093.

4 Feringa, B.L. (2017). *Angew. Chem. Int. Ed.* 56: 11060–11078.

5 Li, Z.T. and Wu, L.Z. (eds.) (2015). *Hydrogen Bonded Supramolecular Materials.* Heidelberg: Springer.

6 Prins, L.J., Reinhoudt, D.N., and Timmerman, P. (2001). *Angew. Chem. Int. Ed.* 40: 2382–2426.

7 Gonzalez-Rodriguez, D. and Schenning, A.P.H.J. (2011). *Chem. Mater.* 23: 310–325.

8 Metrangolo, P., Meyer, F., Pilati, T. et al. (2008). *Angew. Chem. Int. Ed.* 47: 6114–6127.

9 Gilday, L.C., Robinson, S.W., Barendt, T.A. et al. (2015). *Chem. Rev.* 115: 7118–7195.

10 Hunter, C.A. and Sanders, J.K.M. (1990). *J. Am. Chem. Soc.* 112: 5525–5534.

11 Bauzá, A., Deyà, P.M., and Frontera, A. (2015). Anion-π interactions in supramolecular chemistry and catalysis. In: *Noncovalent Forces, Challenges and Advances in Computational Chemistry and Physics*, vol. 19 (ed. S. Scheiner), 471–500. New York: Springer.

12 Chakrabarty, R., Mukherjee, P.S., and Stang, P.J. (2011). *Chem. Rev.* 111: 6810–6918.

13 Lanigan, N. and Wang, X. (2013). *Chem. Commun.* 49: 8133–8144.

14 Stoddart, J.F. (2009). *Chem. Soc. Rev.* 38: 1802–1820.

15 Bąk, K.M., Porfyrakis, K., Davis, J.J., and Beer, P.D. (2020). *Mater. Chem. Front.* 4: 1052–1073.

16 Mena-Hernando, S. and Pérez, E.M. (2019). *Chem. Soc. Rev.* 48: 5016–5032.
17 Peng, H.-Q., Niu, L.-Y., Chen, Y.-Z. et al. (2015). *Chem. Rev.* 115: 7502–7542.
18 Williams, G.T., Haynes, C.J.E., Fares, M. et al. (2021). *Chem. Soc. Rev.* 50: 2737–2763.
19 Smith, D.K. (2018). *Chem. Commun.* 54: 4743–4760.
20 Stupp, S.I. and Palmer, L.C. (2014). *Chem. Mater.* 26 (1): 507–518.
21 Amabilino, D.B., Smith, D.K., and Steed, J.W. (2017). *Chem. Soc. Rev.* 46: 2404–2420.
22 Griffin, S.L. and Champness, N.R. (2020). *Coord. Chem. Rev.* 414: 213295.
23 Cordova, K.E. and Yaghi, O.M. (2017). *Mater. Chem. Front.* 1: 1304–1309.
24 Desiraju, G.R. (2013). *J. Am. Chem. Soc.* 135: 9952–9967.
25 Aakeröy, C., Champness, N.R., and Janiak, C. (2010). *CrystEngComm* 12: 22–43.
26 Saez, I.M. and Goodby, J.W. (2005). *J. Mater. Chem.* 15: 26–40.
27 Hayes, R., Warr, G.G., and Atkin, R. (2015). *Chem. Rev.* 115: 6357–6426.
28 Schalley, C.A. (2000). *Int. J. Mass Spectrom.* 194: 11–39.
29 Blunt, M.O., Russell, J., Giménez-López, M.C. et al. (2008). *Science* 322: 1077–1081.
30 Wasio, N.A., Quardokus, R.C., Forrest, R.P. et al. (2014). *Nature* 507: 86–89.
31 Cai, J., Ruffieux, P., Jaafar, R. et al. (2010). *Nature* 466: 470–473.
32 Mali, K.S., Pearce, N., De Feyter, S., and Champness, N.R. (2017). *Chem. Soc. Rev.* 46: 2520–2542.
33 Tuxen, A., Kibsgaard, J., Gøbel, H. et al. (2010). *ACS Nano* 4: 4677–4682.
34 Binnig, G., Rohrer, H., Gerber, C., and Weibel, E. (1982). *Appl. Phys. Lett.* 40: 178–180.
35 Binnig, G., Quate, C.F., and Gerber, C. (1986). *Phys. Rev. Lett.* 56: 930–933.
36 Gross, L., Mohn, F., Moll, N. et al. (2009). *Science* 325: 1110–1114.
37 Sweetman, A.M., Jarvis, S.P., Sang, H. et al. (2014). *Nat. Commun.* 5: 3931.
38 Stannard, A., Russell, J.C., Blunt, M.O. et al. (2012). *Nat. Chem.* 4: 112–117.
39 Steeno, R., Minoia, A., Gimenez-Lopez, M.C. et al. (2021). *Chem. Commun.* 57: 1454–1457.
40 Urgel, J.I., Écija, D., Lyu, G. et al. (2016). *Nat. Chem.* 8: 657–662.
41 Shang, J., Wang, Y., Chen, M. et al. (2015). *Nat. Chem.* 7: 389–393.
42 Barth, J.V., Weckesser, J., Cai, C. et al. (2000). *Angew. Chem. Int. Ed.* 39: 1230–1234.
43 Griessl, S., Lackinger, M., Edelwirth, M. et al. (2002). *Single Mol.* 3: 25–31.
44 Theobald, J.A., Oxtoby, N.S., Phillips, M.A. et al. (2003). *Nature* 424: 1029–1031.
45 Stepanow, S., Lingenfelder, M., Dmitriev, A. et al. (2004). *Nat. Mater.* 3: 229–233.
46 Dong, L., Gao, Z., and Lin, N. (2016). *Prog. Surf. Sci.* 91: 101–135.
47 Elemans, J.A.A.W., Lei, S., and Feyter, S.D. (2009). *Angew. Chem. Int. Ed.* 48: 7298–7332.
48 Elemans, J.A.A.W., De Cat, I., Xu, H., and De Feyter, S. (2009). *Chem. Soc. Rev.* 38: 722–736.
49 Tobe, Y., Tahara, K., and De Feyter, S. (2021). *Chem. Commun.* 57: 962–977.
50 Song, W., Martsinovich, N., Heckl, W.M., and Lackinger, M. (2013). *J. Am. Chem. Soc.* 135: 14854–14862.

51 Narita, A., Wang, X.-Y., Feng, X., and Müllen, K. (2015). *Chem. Soc. Rev.* 44: 6616–6643.

52 Song, S., Su, J., Telychko, M. et al. (2021). *Chem. Soc. Rev.* 50: 3238–3262.

53 Jin, Y., Hu, Y., Ortiz, M. et al. (2020). *Chem. Soc. Rev.* 49: 4637–4666.

54 Xu, L., Zhou, X., Yu, Y. et al. (2013). *ACS Nano* 7: 8066–8073.

55 Judd, C., Haddow, S.L., Champness, N.R., and Saywell, A. (2017). *Sci. Rep.* 7: 14541.

56 Saywell, A., Magnano, G., Satterley, C.J. et al. (2010). *Nat. Commun.* 1: 75.

2

Characterisation and Interpretation of On-Surface Chemical Reactions Studied by Ultra-High-Resolution Scanning Probe Microscopy

Adam Sweetman[1], Neil R. Champness[2], and Alex Saywell[3]

[1] *University of Leeds, School of Physics and Astronomy, Leeds LS2 9JT, UK*
[2] *University of Birmingham, School of Chemistry, Edgbaston, Birmingham B15 2TT, UK*
[3] *University of Nottingham, School of Physics and Astronomy, Nottingham NG7 2RD, UK*

2.1 Introduction

The development of supramolecular chemistry on surfaces is reliant upon detailed characterisation at the molecular level. A variety of approaches have been employed to understand the detailed arrangement of molecules in self-assembled arrays but the dominant and typically most informative techniques are based upon scanning probe microscopy (SPM). In the main scanning-tunnelling microscopy (STM) [1] has been highly successful in establishing a detailed appreciation of the structure of supramolecular systems, often at the molecular level, but it can be helpful to supplement this approach with other techniques that allow an analysis of the chemical speciation or other structural features that STM cannot probe. For example, X-ray photoelectron spectroscopy (XPS) [2–5] allows investigation of the chemical composition of molecules within supramolecular arrays, and techniques such as X-ray standing wave (XSW) analysis [6] can probe the molecular conformations of adsorbed molecules. However, STM and atomic force microscopy (AFM) [7] are the most common techniques used to study surface-based supramolecular structures. Indeed, SPMs facilitate the characterisation of single molecules and assemblies of molecules, confined to a supporting substrate, on the molecular or sub-molecular level. The defining characteristic of all variants of SPM is the use of a probe to measure a specific probe–sample interaction over a grid of points, which is used to generate an 'image' of a well-defined spatial region of the surface; often resulting in resolution on the sub-Ångström level.

Conceptually, the probe is terminated with a single atom and it is the interaction between this atom and the molecule-substrate system which is measured. The origins of this probe–sample interaction determine the interpretation of the resulting image but commonly the information acquired provides a relatively simple pathway to understanding structural arrangements. Thus, the terminating atom at the apex of the probe is typically brought to within a few Ångström of the surface and, due to the strong distance dependence of the probe–surface interactions, the measured

Supramolecular Chemistry on Surfaces: 2D Networks and 2D Structures, First Edition.
Edited by Neil R. Champness.

interaction is dominated by the position and properties of the single terminating atom. In both STM [1] and AFM [7], the current flow between the probe and sample or the probe–sample interaction force, respectively, are measured. In the majority of applications, the probe is either formed from a metal wire, sharpened mechanically or etched chemically, or an etched silicon tip. The resolution obtainable can be further improved when the apex of the tip is functionalised with a well-defined terminating species, such as CO [8], providing a probe with a defined size and known intermolecular chemistry. It is the localised nature of the probe–sample interaction measured by the probe, which gives rise to the high spatial resolution which allows the study of molecule-substrate systems on the atomic and molecular level.

A noteworthy feature of all SPM techniques is that the acquired data directly corresponds to real-space measurements, which allow an image of the surface to be produced. This is distinct from techniques such as X-ray crystallography and low-energy electron diffraction (LEED) where ensemble reciprocal space measurements are converted to produce a real-space structure. Such images of the surface, particularly of molecule-substrate systems, may often offer what appears to be an easily accessible view of molecular structure and/or reaction processes. However, great care should always be taken when interpreting the data acquired; the acquired data provides a wealth of information on the electronic and chemical structure of the system under study which is distinct from, although often related to, the topography of the adsorbed molecules.

This chapter seeks to provide a background to SPM studies of molecule-substrate systems and how they can be employed to understand self-assembled structures and in particular surface-based reaction processes. The chapter will focus on the underlying theory and experimental considerations that are required to conduct and interpret the investigation of on-surface synthesis reactions using high-resolution SPM methodologies. However, the specific examples discussed in this chapter also provide the underpinning concepts that can be applied to related areas of surface-based molecular assembly such as those discussed in the other chapters within this volume. The chapter provides details of the basic premise of SPM studies for molecule-substrate systems, including an overview of the experimental conditions (Section 2.2), and provides an in-depth discussion of the technical aspects of performing noncontact atomic force microscopy (ncAFM) experiments (Section 2.3). The physical processes underlying the probe–molecule interaction will be used as a basis for discussion of image interpretation (Section 2.4), and in the final section (Section 2.5) examples of on-surface reactions investigated by SPM will be given; focusing specifically on the formation of graphene structures (including graphene nanoribbons – GNRs) and cyclisation reactions (e.g. Bergman cyclisation).

2.2 SPM Under UHV Conditions

Although there are many examples of the implementation of SPM in ambient, liquid, and even electrochemical environments, here we specifically focus on the ultra-high vacuum (UHV) studies conducted at cryogenic temperatures (e.g. <5 K – achievable using liquid helium). A UHV environment is usually a vital prerequisite for the formation of atomically flat and clean substrates. All SPM techniques work optimally, with regards to the characterisation of molecular species, when large areas ($>100\,nm^2$) of flat surface are accessible. These large atomically flat regions facilitate sub-molecular and atomic resolution, which is in itself a prerequisite for the characterisation of on-surface chemical reactions.

Sample preparation under UHV conditions allows contaminant-free surfaces to be produced (simply by limiting exposure to contaminant species), offers accurate temperature control for sample preparation (with specific temperatures required to form certain surface reconstructions), and facilitates the use of the cleaning procedures described in Section 2.3. Typically UHV chambers allow pressures down to $\sim 10^{-10}$ mbar, and lower, to be obtained. Cryogenic SPM systems also allow samples to be cooled to <5 K (inhibiting both molecular diffusion and the progress of chemical reactions – required to study intermediate states of on-surface reactions).

There is however a disconnect between the use of UHV and the environment in which industrial scale, or even lab-based, chemical reactions often take place: specifically with regards to the environment in which solution-phase reactions are performed. In general, it is not possible to introduce solvents into UHV (as the high vapour pressure of many solvents renders them incompatible with a UHV environment), meaning that reactions investigated by SPM under UHV are studied in the absence of solvents. In addition, studying such molecule-substrate systems under UHV, as opposed to ambient conditions, gives rise to several challenges (including the inherent technical difficulties of simply moving samples around in a UHV environment). Most notable is the issue of transferring the molecules to a surface held in UHV. In the simplest case, a crucible loaded with the molecules under study can be introduced to the UHV system with subsequent thermal evaporation used to produce a sub-monolayer to multi-layer film upon the substrate. However, in many cases, the molecules may be non-volatile or thermally labile and in such cases, one of a variety of alternative techniques has to be employed [9].

There are several benefits in utilising UHV-SPM compared to other characterisation techniques. The molecules to be studied do not have to be crystalline (as is the case for some diffraction-based techniques) and only very small quantities of material are required for study by SPM (compared to, for example, nuclear magnetic resonance [NMR]). Combined with the exceptionally high spatial resolution offered by SPM, the technique has recently gained importance as a characterisation technique that can provide 'real space' characterisation of molecule-substrate systems, which both complements and enhances the chemical and structural characterisation offered by ensemble averaging techniques.

2.2.1 On-Surface Reactions

An obvious consideration with regards to characterisation utilising SPM techniques is that the molecules investigated have to be studied on a supporting substrate; prohibiting the study of solvent confined systems. The operational mechanics of SPM lend themselves to the study of systems confined to a 2D substrate and provide an invaluable technique for investigating chemical reactions upon, a potentially reactive and/or catalytic [10], surface (see reviews [10–15] and references therein). As the systems to be studied are on a substrate, this precludes the use of transmission electron microscopy (TEM) which can also be used in principle to provide atomic-level resolution, but is generally unsuitable for the study of molecule-substrate systems due to the thickness of the substrates required.

The major benefit of characterisation via SPM is the level of spatial resolution achievable (vertical resolution of less than 5 pm and sub-angstrom lateral resolution is routine). This is based upon sensitive measurements of the probe–substrate/molecule interaction (vide infra). As the probe plays a vital part in the measurements, one needs to consider its shape, and its electronic and chemical properties, as these can potentially give rise to a variety of 'artefacts' (Section 2.3 discusses this in detail). An additional benefit of confining a chemical reaction to a 2D plane is the potential to control reactions via different methodologies to those available in solution [16]. The technique has also been shown to allow different stages during the progression of a chemical reaction to be studied (i.e. initial, final, and even intermediate states) [13].

Two main variants of SPM have commonly been employed to study on-surface reactions; STM and AFM. In particular, a specific variant of AFM, ncAFM, provides a sub-molecular resolution that allows characterisation of the spatial position of chemical groups within a molecule, as well as facilitating not only the observation of single chemical bonds [8] but providing a methodology to distinguish the bond order (i.e. single, double, or triple carbon–carbon bond species) [17]. It is important to note that the specific aspects of ncAFM (discussed in detail throughout this chapter) provide sub-molecular resolution, and therefore, sub-molecular resolution ncAFM is part of the family of SPM techniques, it is not simply a mode of operation that can be applied to other SPM systems and requires, at least in the current implementation, a specific experimental set-up.

The level of sub-molecular resolution provided by ncAFM can be used to complement traditional characterisation techniques (e.g. NMR, GLC, LEED) and, for example, allows a level of single-molecule characterisation, which can aid in the structural determination of completely new species (typified by the role of ncAFM in the characterisation of a planar, proton-poor compound in combination with computational studies [18]) as well as distinguishing between the structure of asphaltenes (polycyclic aromatic hydrocarbons within crude oil; whose structural analysis is a tremendous challenge for non-spatially resolved techniques) [19].

2.2.2 Characterisation of Molecule-Substrate Systems via STM

There are many 'flavours' of SPM all designated by a confusing menagerie of acronyms, including but not limited to STM, ncAFM, KPFM (Kelvin probe force microscopy), and SNOM (scanning near-field optical microscopy). The archetypal example of this set of methods is STM. In common with all SPM methodologies, STM works by scanning a probe across a surface, in this case with an applied bias (relative to the probe – which is usually defined as grounded). The conducting tip (usually metallic) is moved in a straight line across a conducting/semi-conducting surface and the interaction between the probe and the tip measured (in the case of STM, the measured quantity is the magnitude of the current flow due to electrons tunnelling between the probe and the surface, or vice versa). Details of the concepts underpinning STM are given in several excellent textbooks [20, 21], but in summary, the salient points are: (i) the substrate is biased relative to the probe (typically in the range ± 2 V), (ii) the resultant flow of electrons between the probe and the molecule/substrate is recorded, (iii) the magnitude of this tunnel-current (I) has an exponential dependence on the distance between the probe and the substrate/molecule, and (iv) the vertical probe position (z) can be varied in order to give a constant current as the probe is moved laterally across the surface (this feed-back mode is known as a constant-current operation) or (v) the vertical probe position is kept constant and the current is recorded at various lateral positions above the substrate/molecule, known as constant-height mode (see Figure 2.1).

An STM image is produced by obtaining a series of line scans (shown in Figure 2.1a), which are then combined to form a 2D image. In constant current mode, I is maintained at a fixed set-point, typically a few picoamperes, and the resultant image, therefore, shows the variation in z as the probe is scanned over the surface. In constant height mode, images will show the variation in I with tip position. It is important to note that the measured current, for a finite bias voltage, is proportional to the sum of the contributions for the local density of states (LDOS) from which tunnelling is possible [20, 21]; i.e. the measured current is related to the electronic structure of the molecule/substrate, and is not necessarily well correlated to the spatial position of the atomic nuclei. In this respect, the path of the probe in constant-current mode does not simply provide a topographic height but is better interpreted as a map of the LDOS. This issue manifests in the characterisation of molecules where molecular orbitals are often delocalised over the molecular species under study. Therefore, preventing the position of individual atoms, within similar chemical environments (e.g. conjugated aromatic carbons), from being resolved as they will often form part of the same feature observed within an STM image. However, in cases where electronic character is localised over specific chemical moieties, STM images may be compared (at least as an approximation) to the chemical structure of the molecule under study. An example of this is shown in Figure 2.1c where the structure of a brominated terfluorene molecule (α,ω-dibromoterfluorene [DBTF]) can be compared with a constant-current STM image [22]; features related to the peripheral Br atoms and central fluorene groups are visible. Such electronic structures are often compared with density functional theory (DFT) based simulations of STM images, which can help identify molecular structure and conformations [24].

While STM can provide sub-molecular resolution, it suffers, in common with all SPM techniques, with regards to the non-trivial interpretation of the acquired data. Although DFT studies used in conjunction with STM data often offer good agreement and provide a plausible interpretation of the results (in terms of a more complete appreciation of the expected LDOS), an overreliance on DFT can lead

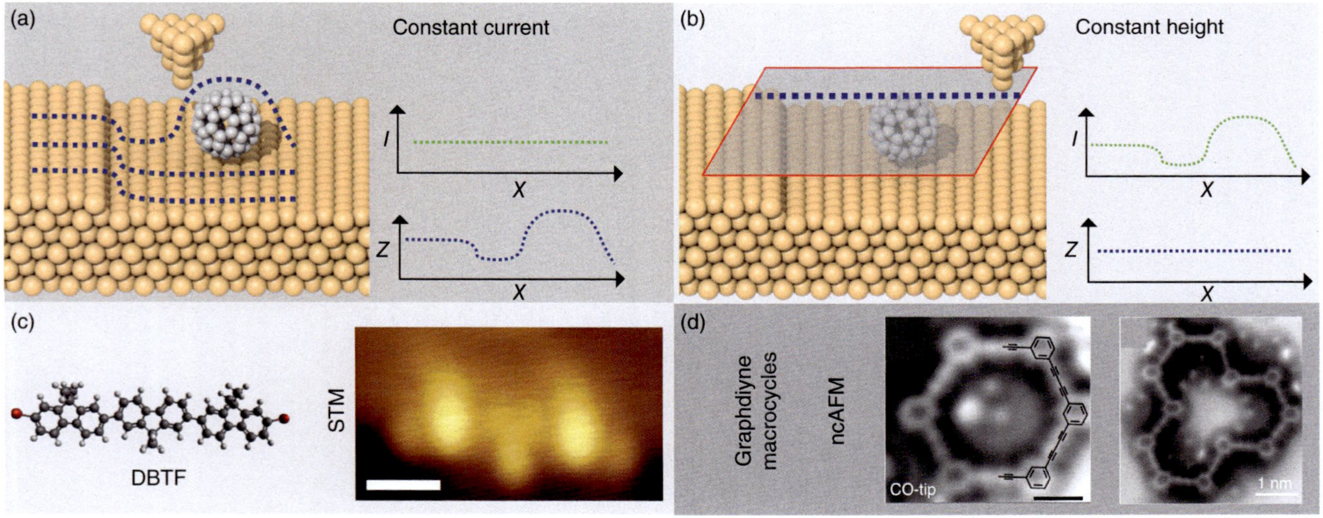

Figure 2.1 Outline of SPM image acquisition and examples of molecular characterisation. (a) Schematic showing image acquisition via a series of line profiles in constant current operation of STM. (b) Operation of STM in constant-height mode. (c) Example of STM characterisation of a single DBTF molecule via STM [22] (Scale bar: 1 nm, $V_{\text{Sample-bias}} = -0.4\,\text{V}$, $I_{\text{Set-point}} = 5.5\,\text{pA}$). (d) Example of characterisation of graphdiyne macrocycles via constant height ncAFM using a CO tip [23] (Scale bar: 0.6 nm). Source: STM image in (c) reproduced from Saywell et al. [22] with permission from John Wiley & Sons, Inc., Copyright 2012. Images in (d) reprinted with permission from Liu et al. [23]. Copyright 2018 American Chemical Society.

to potential pitfalls as calculating the energy and spatial distribution of molecular orbitals for surface adsorbed species can be challenging (specifically when taking into account hybridisation with electronic surface states). This is not to say that STM is not able to provide reliable and informative evidence with regards to the study of molecule-surface systems, but rather that it is best used in conjunction with complementary techniques to ensure robust characterisation of the structural, chemical, and electronic properties of the molecules under study.

2.2.3 ncAFM

Recently ncAFM has risen to prominence as a technique for characterisation of molecule-substrate systems on the sub-molecular level. The basic premise for data acquisition is the same as STM, with images formed line by line. In the case of ncAFM, the relevant interaction is the force between the probe and molecule substrate-system. The focus within this chapter is on the use of qPlus implementation of ncAFM (see Refs. [25, 26], and citations therein, for details of the technique). An excellent description of the underlying principles of ncAFM is given in Ref. [27], but to summarise: The probe, is affixed to a cantilever, which is oscillated at it's resonant frequency, and the interaction between the tip and sample results in a change in this resonant frequency. This shift in the resonant frequency, Δf, is the signal measured within ncAFM (in the same way I is recorded in STM). Feedback circuits are used to excite the cantilever at its resonance frequency and keep the oscillation amplitude constant (typically ~0.5 Å – as discussed by Giessibl [28]). Similar to STM constant current measurements, the z-height can be adjusted during scanning to keep Δf constant (constant Δf imaging), but within this chapter, we discuss constant height measurements where the z-height of the probe relative to the surface is kept fixed and the Δf signal is recorded as a function of probe position.

Example ncAFM images are shown in Figure 2.1d where the structure of two graphdiyne macrocycles are clearly resolved within the constant height ncAFM data [23]. By convention, bright features within the constant height ncAFM images correspond to a positive increase in Δf and are often simply interpreted as a topologically higher region of the molecule (however, as discussed in detail below, this is only the case if the nature of the interaction force between the 'higher' and 'lower' parts of the molecule is identical).

It is useful to briefly comment upon the nomenclature chosen for this particular version of AFM. The terminology ncAFM is used here, as opposed to dynamic force microscopy (DFM) or frequency modulated atomic force microscopy (FM-AFM); FM-AFM refers to the fact that the frequency shift, Δf, is the main observable. These terms are occasionally used interchangeably with ncAFM, and it is important to note that some confusion can arise with comparison to the so-called 'contact', 'intermittent contact', and 'tapping' modes of cantilever AFM. For the experiments discussed here, measurements are acquired in constant height operation and the 'non-contact' aspect refers to the fact that the method is distinct from the 'contact' modes of cantilever AFM.

ncAFM and STM techniques are often used in conjunction to characterise molecule-substrate systems. ncAFM provides greater lateral resolution, due in part to the shorter interaction range (Pauli-repulsion), and in principle offers a route towards chemical specificity. The image acquisition time for ncAFM is, however, significantly slower than STM, and so it is common practice to first characterise the molecule-substrate system using STM (of course, the initial STM characterisation can also provide important complimentary information on the electronic structure of the molecules). In addition, ncAFM images are predominantly acquired in constant height operation, which is not always compatible with non-planar molecules. The remainder of the chapter will focus on the application of the ncAFM technique and the interpretation of the data acquired for various molecule-substrate systems.

2.3 Practical Steps in Accomplishing Sub-Molecular Imaging

While the fundamental underlying physical principles of ultra-high resolution in ncAFM imaging of single molecules are relatively simple, and can be understood with reference to straightforward empirical models (see Section 2.4.1), the technical steps required to achieve it in practice are somewhat demanding and require a degree of specialist expertise to reproduce.

First, it should be emphasised that the substantial challenges of 'conventional' low temperature UHV SPM must be overcome. These include fundamentals such as the construction of high stability, vibrationally isolated, scan head, UHV generation, low levels of mechanical and electrical noise, and mounting of the instrument in a suitable low-temperature cryostat. Fortunately, many of these core challenges may now be routinely surmounted using commercially available systems, and so in this section, we only highlight those challenges specific to ultra-high resolution imaging of organic molecules with functionalised tips, assuming a fully functioning UHV SPM.

It should be noted that in principle sub-molecular resolution can be accomplished with a wide variety of sensors, including conventional silicon cantilevers [29, 30], and length extensional resonators such as the Kolibri sensor [31]. However, practically most of the literature on the topic has used the qPlus sensor [8, 25] implementation, and therefore in the following we will assume this is the setup under consideration.

2.3.1 Sample Preparation

Although in principle high resolution can be achieved on almost any atomically flat substrate [29, 30, 32] in practice, most imaging of organic molecules is done using single metal crystals with low index planes (e.g. Cu(111), Ag(111)) as a substrate. These are easily prepared in UHV via sputter/annealing cycles and allow for straightforward preparation of the tip (as described in Section 2.3.3) without the risk of creating an electrically insulating apex, as can be the case on semiconducting or insulating substrates. For the purposes of high-resolution imaging of organic molecules, there are a number of additional preparation steps that are worth covering in some detail.

2.3.1.1 Deposition of Organic Molecules at Low Temperature

Most small organic molecules have relatively low diffusion barriers on coinage metal surfaces, and readily aggregate into islands [33], or even reconstruct the surface [34], if deposited at room temperature. For the investigation of isolated molecules, it is therefore recommended that the deposition be performed at low temperatures. Practically, this is best done by direct deposition into the scan head itself. This is typically done using a thermal effusion (Knudsen) cell filled with a thermally purified powder (99% purity or better) of the required molecule, positioned to face one of the shutters of the SPM cryostat (note that large, or fragile, molecules, may require more sophisticated deposition techniques to be utilised [9]).

The cell is bought up to the required deposition temperature, and once a constant rate of deposition is measured (preferably calibrated previously using a quartz crystal micro balance [QCM] or similar), the shutter to the cryostat is opened for a short period of time. In order to prevent diffusion of the molecules on the surface, typical deposition times are on the order of less than a minute in order to prevent the substrate temperature exceeding $\sim$10 K. Depending on the exact molecule, and substrate, combination, this temperature requirement may be stricter, or more relaxed, and must be calibrated to each experimental setup. Generally, it is recommended to use higher fluxes, and shorter deposition times, in order to minimise the temperature increase, but practically the flux of a given molecule, in a particular microscope, must be calibrated on a case-by-case basis to give the required coverage. If the required coverage cannot be reached in the given time, multiple depositions can be performed, on the proviso that no single deposition exceeds the temperature threshold for diffusion.

2.3.1.2 CO Deposition

Although techniques vary, typically CO molecules are not deposited via direct (line of sight) deposition, but instead the UHV chamber is backfilled with CO gas up to pressure on the order of 10^{-8} mbar [8, 25], by admitting ultra-high purity gas via a UHV leak valve. Once the pressure has stabilised, the shutters of the cryostat are opened for a short period (as for the deposition of organic molecules). Although exposures are sometimes given in Langmuir, it should be noted that the pressure of a gas at the sample is often significantly lower than that read on the vacuum gauges, so coverages must again be calibrated on a system by system basis. CO deposition is typically performed after deposition of the organic molecules, as a diffusion barrier for CO is normally smaller than the diffusion barrier for the larger molecules, and so depositing the CO as a final step reduces the likelihood of inducing CO diffusion during deposition of organic molecules. It should be noted that this back-filling technique can result in a large quantity of CO gas being absorbed onto the cryostat shields themselves, and should the cryostat warm up, a large number of molecules will desorb from the shields, raising the pressure in the chamber, and inevitably contaminating any samples stored in the same chamber. Therefore, once CO gas has been dosed into the system, it is essential to keep the cryostat cold until the experiment is complete. The quantity of gas adsorbed may even be high enough to trip ion gauges and ion pumps, so particular care must be taken during warm-up after repeated CO deposition events at low temperature. A large number of passivating molecules/atoms have been shown to work for high resolution ncAFM of organic molecules (including CO, xenon, chlorine, bromine, and iodine [35]), but the overwhelming majority of imaging is performed either with CO, or xenon, mostly due to their ready availability, ease of deposition, and a large volume of work describing protocols for the manipulation. The former generally produces higher resolution images, but also produces a greater degree of distortion due to the flexibility of the apex. The exact choice of a passivating agent can also have a significant influence on the contrast due to its interaction with the short-range electrostatic field of the molecule [36, 37] (see Section 2.4.1.5).

2.3.1.3 Decoupling Layers

The use of a thin decoupling layer (e.g. 1–3 monolayers [ML] of NaCl or MgO) is a non-essential but often used step in sub-molecular contrast imaging. The absorption of molecules onto a thin insulating layer is used in STM studies to decouple the electronic structure by preventing hybridisation of molecular orbital's with the surface [38, 39], whereas for AFM studies the primary benefit is that the tip functionalisation may be performed more easily by picking the CO molecule up from the insulating layer, as the binding strength to the surface is dramatically reduced [8], a factor more important on reactive metals such as copper, compared to less reactive metals such as silver or gold. Generally, a low coverage (less than half a monolayer) is preferred, such that patches of clean metal remain for tip preparation. Growth of 2 ML thick islands of NaCl can be achieved by deposition onto the metal crystal outside of the scan head, with a sample temperature of around 293 K. Preparation techniques varies, but some groups report that deposition slightly below room temperature (e.g. 270 K) is preferential for the growth of 2 ML islands and helps prevent wetting of the surface by the NaCl. Particular care must be taken at this stage as deposition at too high a temperature will result in a complete wetting of the surface by 1 ML NaCl, which can be difficult to detect immediately via STM imaging.

2.3.2 Construction of the qPlus Sensor

Whilst the geometry of the qPlus sensor is well described [25], practically constructing a complete sensor from scratch requires a degree of experimental skill and preferably specialised equipment. Both the attachment of the tuning fork to a suitable base and the attachment of the metal tip to the end of the free tine of the fork must be done carefully such that the resonant frequency and Q factor of the cantilever are not compromised. In practice, this is best accomplished by using as small an amount of appropriate UHV compatible epoxy resin as possible, ensuring that no epoxy resin bridges the gap between the two tines of the tuning fork, and making the attached metal tip as small as possible. It is also highly recommended that the resonant frequency and Q factor of the sensors be tested ex situ using an external testing apparatus before introducing them into the UHV chamber, such that obviously compromised sensors can be excluded from use. More sophisticated strategies [40] may be employed to ensure a high degree of reproducibility between sensors.

2.3.3 Tip Preparation

Although, in principle, almost any material can be attached to the end of the qPlus sensor, in practice a typical STM tip material such as tungsten, or platinum-iridium, is normally used. qPlus sensors usually have the tip attached to the end of the tine of the tuning fork via UHV compatible epoxy resin, and, because these epoxies usually breakdown at high temperature, common STM tip treatment techniques such as heating via electron bombardment cannot easily be implemented. Practically, this means that sensors are often introduced into the scan head without further treatment, and are therefore prone to contamination from exposure to the atmosphere. Preparation of the tips in situ is essential before attempting high-resolution imaging and is most often accomplished via voltage pulsing, indentations into the surface, or field emission over the surface. During low-temperature operation, much of the contamination on the tip shaft is 'frozen out', and providing the apex of the probe can be coated in a layer of clean surface material, it is usually possible to obtain good resolution even with untreated tips. Nonetheless, ex situ preparation techniques such as focused ion beam (to improve the macroscopic radius of curvature) [41], or in situ techniques such as field ion microscopy (which can be used to clean the probe apex without heating) [40] are recommended in order to improve reproducibility between experiments.

2.3.3.1 Tip Functionalisation

The creation of a functionalised tip itself is an essential part of the imaging process in high-resolution ncAFM and requires some skill in the preparation of clean metal tips via STM methods in situ before it can be attempted. In the discussion below, we assume the functionalising molecule to be CO as this is the most commonly used approach (see schematic in Figure 2.2a–e).

First, a clean metal tip should be prepared by the controlled crashing of the tip into the metal surface, and good STM resolution should be achieved on both the CO molecule and the organic molecule to be imaged. It is also desirable, though not essential, that the frequency shift (Δf) of the tip during normal STM operation is relatively small as this indicates a relatively sharp tip apex (e.g. Δf less than −2 Hz at an STM setpoint of 2 pA and a gap voltage of 200 mV, with a 50 pm oscillation [41]). The pickup of the desired molecule is then accomplished by positioning the tip over the molecule, withdrawing the tip a few hundred picometres, and raising the bias above 2.5 V. Transfer of the molecule from the surface the tip is indicated by a jump in the tunnel current [44]. It should be noted that successful molecular manipulation is highly tip dependent, and these parameters are a guide only, and not every tip is capable of successful desorption.

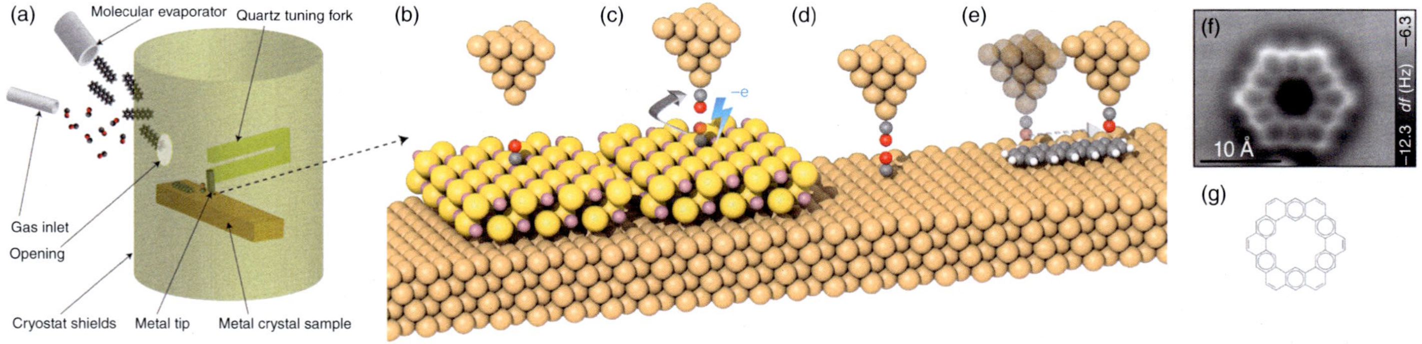

Figure 2.2 Overview of the setup and procedure for generating a suitable surface and tip for the high resolution imaging of absorbed organic molecules. (a) Cartoon showing the setup required for in situ deposition of organic molecules and gas molecules onto a cold metal sample (not to scale), as noted in the text typically gas and molecular deposition would be done sequentially, and not simultaneously as shown in the schematic. (b) Positioning of metal tip apex over CO molecule absorbed on a 2 ML NaCl/Cu(111) island. (c) Excitation of the CO molecule by tunnelling electrons causes the molecule to desorb from the surface and attach itself to the tip. (d) The CO functionalised tip is then characterised on a CO molecule absorbed on the clean Cu(111) surface. (e) An absorbed organic molecule is then imaged in constant height mode using the functionalised tip. (f) Constant height Δf image of kekulene molecule absorbed on Cu(111) [42]. (g) Clar annulenoid structure of kekulene molecule on same scale. Image in (f) reproduced with permission under a Creative Commons licence from Pozo et al. [42], copyright 2019. Further permissions related to the material excerpted should be directed to the ACS.

It is also possible to induce a transfer mechanically via a close approach of the tip at low biases, although this carries a higher risk of modifying the probe apex in an uncontrolled fashion. Although in principle CO can be picked up from the copper surface, in practice this can be hard to achieve due to the strong tip apex dependence. For this reason, the CO molecule is often picked up from the surface of a NaCl bilayer [8]. CO pickup on less reactive surfaces (e.g. Au or Ag) is often easier to achieve, whereas pickup from more reactive surfaces (e.g. Ir(111)) may be extremely difficult, or impossible to perform reproducibly [45]. However the transfer is accomplished, it is important to confirm the successful transfer has occurred by characterising the resulting tip apex on other nearby CO molecules. The characteristic contrast change for CO on the Cu(111) substrate is a switch from the imaging of CO molecules as depressions, to a characteristic 'sombrero' shape [44] during conventional STM. Characterisation of the CO terminated tip by ncAFM on metal adatoms can also provide important information on the precise orientation of the CO molecule at the end of the tip apex [25].

2.3.4 Practical Considerations for Imaging

In the standard implementation of qPlus ncAFM, performing large area scans, and preparing the tip, are best performed in STM constant current mode (assuming a conductive substrate), rather than in ncAFM feedback, due to its generally superior scan speed and stability. Typically a large area is imaged in STM (e.g. 50–100 nm^2), and then a small area of interest (e.g. 5–10 nm^2) is located and imaged at high resolution using conventional STM, before the tip is functionalised and ncAFM imaging in constant height mode can proceed. Due to the difficulty in establishing absolute tip-sample distances in SPM, normally a 'zero point' is established over the substrate in STM feedback at a low tunnel current setpoint (e.g. tip height = 0 nm at $V_{gap} = 0.2$ V, $I_t = 2$ pA), and the vertical position of the constant height images and spectroscopy curves is given relative to this position.

2.3.4.1 Drift and Creep

An essential element in achieving intramolecular contrast is operating in constant height mode. Because of the exceptionally strong dependence of the tip-sample force with distance [46], drift rates on the order of less than 0.5 pm/min are required, as a typical high-resolution image may take 30 minutes to acquire, resulting in a 15 pm variation in absolute tip-sample separation during this time. Practically, this can be accomplished by ensuring that the microscope is at thermal equilibrium before starting imaging and that no thermal fluctuations arise from variations in the temperature or air pressure of the local environment. It is also necessary to ensure that no residual piezoelectric creep is present. This is typically done by placing the tip in the required area and waiting for an extended period. Depending on the exact conditions, and scanner configuration, it may be necessary to wait for upwards of 12 hours before the creep becomes small enough to allow for stable imaging. This requirement is doubly necessary if complex force mapping or atomic manipulation protocols are to be used. If operation in more challenging environments is required (e.g. over a varied temperature range, or where time constraints are more pressing), some form of atom tracking and/or feedforward correction may be desirable [47]. Even in low drift conditions, for force mapping experiments over long periods, some form of post-processing image alignment may also be necessary. It should also be noted that any residual slope on the sample must be measured and corrected before constant height imaging begins, as even small errors in the measurement of the sample slope can produce visible effects during constant height imaging at a close approach.

2.3.4.2 Amplitude Calibration

The response of the cantilever in ncAFM is recorded as an electrical signal, and accurate calibration to convert this into a physical oscillation is mandatory. Due to the small oscillation amplitudes typically used during sub-molecular imaging, the normalised frequency shift method, commonly used in cantilever ncAFM, is generally not suitable. However, as most qPlus instruments combine STM and AFM techniques, an alternative normalised average tunnel current method introduced by Sugimoto et al. [48] may instead be used. In both these methods, the tip is positioned statically over the surface in feedback, and the z position of the tip monitored as the amplitude and feedback setpoint are systematically varied. By subsequently plotting the z position against the recorded oscillation amplitude (in volts as recorded by the detector), a conversion factor from volts to nanometres can then be established. Because of the variation between different sensors, each sensor must be calibrated separately.

2.3.4.3 Apparent Dissipation and Mechanical Coupling of the Sensor

Although the qPlus sensor can be electrically excited in principle, in practice most systems use a separate piezoelectric crystal to mechanically excite the oscillation. The mechanical coupling of this piezoelectric crystal to the cantilever can be highly variable, and depends both on the sensor construction, and the seating of the sensor onto the excitation piezo (in practice this is usually accomplished by mechanical clamping or magnetic attachment). Due to this non-ideal coupling, the mechanical transfer function between the excitation piezo and the cantilever is generally not flat with respect to frequency. This can result in significant 'Apparent dissipation' as described by Labuda et al. [49] and in the worst case even qualitative distortions in the frequency shift channel. In practice, this transfer function can only be rigorously checked after the tip has been approached to the surface by performing $\Delta f(\mathrm{V})$ curves (i.e. Kelvin parabolas) at a distance of several nanometres from the surface (so that no tunnelling occurs) and checking the parabolas are unperturbed, and the dissipation channel is flat throughout this measurement.

2.3.4.4 Crosstalk

Most low-temperature instruments now permit simultaneous measurement of STM and ncAFM channels. Depending on the exact instrumental setup, and construction of the sensor, it is possible for the tunnel current signal to capacitively couple to, and distort, the cantilever response channel [50]. Because it can be difficult to separate out crosstalk effects from genuine force interactions arising from the local current density [51], the behaviour of the ncAFM channels in the presence of significant current should be checked on a conductive substrate. The precise degree of crosstalk is highly dependent on the construction of the instrument and sensor, but practical steps that can reduce crosstalk include: the use of a high bandwidth tunnel current pre-amplifier, biasing the sample, grounding the tip, and using a separate wire for tunnel current collection rather than an integrated electrode. Whilst use of a separate wire for tunnel current collection is in principle preferred, attachment of the wire is in practice often difficult and can easily significantly modify the Q factor and resonant frequency of the sensor, and an integrated solution may be preferable if ncAFM measurement in the presence of high tunnel currents is not required.

2.3.4.5 Force Inversion

Whilst not a requirement for standard imaging, it is often desirable to extract quantitative force data from intramolecular resolution studies. If this is desired, a method must be selected for the inversion of the frequency shift to vertical tip sample force. Because the frequency shift is, to a first approximation, proportional to $-dF/dz$, this requires an integration of the complete frequency shift curve from a distance at which an unperturbed oscillation occurs, to the point of interest. Although several integration schemes exist, the most commonly used is that implemented by the Sader–Jarvis algorithm [52]. In principle, this technique can be used for arbitrary oscillation amplitude, whilst introducing uncertainties of less than 10%, but should be noted that recent work suggests there are limits to its applicability depending on the precise nature of the tip-sample interaction [43]. Additionally, it should also be noted that, in general, the *total* tip-sample force is not the quantity of interest, as this is dependent on the shape of the (generally uncontrolled) macroscopic tip apex. Instead, usually, the quantity of interest is the site-specific (short-range) force between the molecule and the last few atoms of the tip apex. In order to extract this quantity, the non-site-specific (long range) forces must be removed. This is best done by acquiring a $\Delta f(z)$ curve over the same height range away from the molecule (where no contrast is visible), subtracting this from the $\Delta f(z)$ curve taken on top of the molecule, to produce a 'short range' $\Delta f(z)$ curve, which can then be inverted using the Sader–Jarvis algorithm described above. Force inversion requires that the stiffness of the tuning fork be well known, and this quantity can depend on the precise geometry and construction of the individual sensor. Therefore, to avoid large systematic errors, the stiffness of the individual sensor used in the experiment should be independently verified.

2.4 Interpretation of Sub-Molecular Contrast at the Single Bond Level

At this juncture, we will take a step back from the discussion of the significant technical and instrumental discussion of Section 2.3, and turn our focus to the atomic scale physical/chemical forces present in the tip-sample junction. Assuming the technical hurdles outlined in Section 2.3 can be overcome and high-quality data obtained, the task then becomes understanding how best to interpret the collected data. In this section, we will outline the fundamental chemical and physical processes underpinning sub-molecular contrast in ncAFM, common methods used in interpreting the experimental images, and important subtleties that must be taken into account during this process.

2.4.1 Forces in the Tip-Sample Junction

Image contrast arises explicitly from the sum of all of the atomic scale forces that arise in the complete tip-sample junction (including contributions from the metallic tip, the functionalising molecule, the molecule under investigation, and the underlying substrate), and the mechanical response of the atoms in the junction to these forces. Therefore, we will elucidate the key components that result in the production of intramolecular contrast, and describe the state-of-the-art in their interpretation by comparison to different modelling techniques. In Section 2.4.1, we examine each of these forces in the junction in detail, but it is instructive to first examine a representative image (see Figure 2.2f–g) and cover, qualitatively, the key features. While examining Figure 2.2f, we first note that the images are the maps of frequency shift (Δf) – i.e. the change in the resonant frequency of the cantilever due to the tip-sample force. The relationship between force and frequency is complex (see Section 2.3.4.5), but to first order, we can assume that the frequency shift Δf is proportional to the inverse force gradient $-dF/dz$. Even more qualitatively, we can normally assume that more positive (bright) features correspond to an area of repulsive force, and more negative (darker) areas correspond to more attractive interactions.

2.4.1.1 Non-site Specific Interactions – The 'Background'

Inspecting the image further, the molecule sits within a uniform grey background. This corresponds to regions where the tip is far above the substrate such that the force on the tip is uniform (generally uniformly attractive dispersion and electrostatic forces), i.e. we see only *non-site specific* forces arising from the bulk interaction between the tip and substrate, which provide a background on which all the other interactions sit.

2.4.1.2 Local Dispersion Interactions – The 'Halo'

The net attractive interaction arising from the dispersion interaction between all of the atoms of the molecule, and the atoms at the probe apex results in an attractive 'well', which shows up as a dark (negative frequency shift) feature in constant height imaging, at close approach this results in a dark halo in which the geometric structure of the molecule sits.

2.4.1.3 Pauli Repulsion – The 'Carbon Backbone'

The onset of repulsive interactions due to the overlap of the electron orbitals of the tip apex and the molecule is the fundamental origin of intramolecular contrast in tip functionalised ncAFM. As a first-order approximation, the magnitude of the repulsive interaction can be considered proportional to the charge density in the space that the tip apex is attempting to probe. Because the total charge density closely mirrors the bonding structure of the molecule, the result is an image that closely mirrors the classic 'ball and stick' model used to describe molecular structure. As a repulsive interaction, these features typically show up as bright in constant height imaging (more positive frequency shift) which sit within the dark attractive halo of the dispersion interaction (see Section 2.4.1). However, it must be stressed that this is only a first-order approximation – as detailed in Section 2.4.2, there are significant differences between maps of the charge density of the molecular system and the ncAFM images, which arise from the subtle interplay of the different forces in the junction, the finite size of the terminating atom, and the response of the probe particle at the probe apex.

2.4.1.4 Chemical Bonding

The key mechanism behind tip functionalised ncAFM is that the addition of a passivating unit to the end of the tip apex inhibits chemical bonding between the tip and molecule, reducing the magnitude of the tip-sample force, and so prevents the molecule being picked up, or displaced, during the close approach of the tip. In principle, a chemical interaction does not strictly prohibit the possibility of obtaining intramolecular contrast, but in practice, strong chemical interactions are likely to result in either displacement of the molecule, or changes in the structure of the tip apex. For this reason, forces resulting from chemical bonding between tip and surface are assumed to be negligible in the following discussion.

2.4.1.5 Local Electrostatic Interactions

In ncAFM, there is often a net long-range electrostatic interaction between the bulk tip apex and the surface, which can be measured and used to determine the difference in workfunction between two materials [53]. However, for the purposes of intramolecular contrast, the quantity of interest is the *local* electrostatic field that arises over the molecule due to localised partial charges. How this local field affects the imaging depends on whether the probe apex itself also has a net charge or dipole, and it is trivial to note that accumulation of charge density (for example, due to bonding within the molecule) can also result in a localised electrostatic field, and so effects due to increased charge density and local electrostatics cannot be simply disentangled.

2.4.2 Response of the Probe Particle – Distortions in Imaging

Careful examination of the images shows that, although it is often easy to make a one-to-one comparison between the image and a structural model of the molecule, some features are somewhat distorted compared to the ball and stick model. In particular, some features appear more elongated than a simple structural model would suggest. Key to correctly interpreting these features is an understanding that due to the extremely close approach of the probe apex to the molecule, the probe can no longer be considered to be a 'weakly perturbing' interaction (as is often assumed in conventional STM for example). As such we cannot assume that the images produce a simple map of the unperturbed state of the molecule, but instead must explicitly consider the dynamic response of the complete system. Strictly a complete quantum mechanical treatment of the entire system (for example, via simulation of the complete junction using DFT [17]) is preferable, but this is prohibitively computationally expensive in most situations. Fortunately, an important simplifying assumption that can be made is that the attachment of the probe particle to the metal tip apex is the most mechanically flexible part of the system, and therefore almost all the relaxation that occurs in the junction will occur at this position. Moreover, it has been shown that a good qualitative understanding of these systems can be extracted from simulations that assume simple empirical potentials between the probe and the surface geometry, assuming that only the probe particle is able to move [54]. These simulations have the advantage of taking minutes to run on a desktop computer, as compared to weeks of computational time on a supercomputing cluster for the full quantum mechanical treatment.

2.4.2.1 Flexibility of Adsorbed CO

The vibrational modes of a CO molecule absorbed in an upright configuration on a metal surface are well-known [55], and further direct evidence of its flexibility in the context of SPM imaging comes from a combined experimental and theoretical study [56]. The flexibility of the CO probe was first explicitly described in the imaging of organic molecules as a 'enhancement' effect [17] and was supported by detailed DFT calculations showing how the tip apex geometry changed during simulated imaging. On this basis, Hapala et al. [54, 57] developed a simple model using empirical potentials to approximate the CO – molecule interaction to rapidly create simulated images (see Figure 2.3a). Moreover, the link between the frustrated vibrational modes of the CO molecule detected during inelastic tunnelling spectroscopy (IETS) and the deflection of the molecule during close approach in high-resolution ncAFM imaging is strongly supported by the apparent common origin of the imaging contrast for both techniques [54].

The consequence of the mechanical deformation that occurs within the probe particle junction requires explicit consideration – because of the low lateral stiffness of the CO molecule, the particle has the tendency to deflect sideways when approaching a ridge in the potential energy landscape (see Figure 2.3b). As a result, any ridge experienced by the probe particle will have the tendency to sharpen into a line connecting two centres. In the case of covalent bonding within a molecule, this ridge may be thought of as arising from the Pauli repulsion resulting from the charge density that lies within the bond itself, but it is essential to note that *any* ridge in the potential energy landscape will produce the same effect. As demonstrated in simulations using the probe particle model (PPM) where interactions can only arise from simple steric hindrance (mediated by the Lennard–Jones potential) similar ridges will arise even if there is no charge density in the space between two atoms.

Consequently, the critical question becomes how well the probe particle is able to penetrate the space between two atoms and to understand the physical phenomena responsible for the lack of penetration (a related concept is the solvent excluded

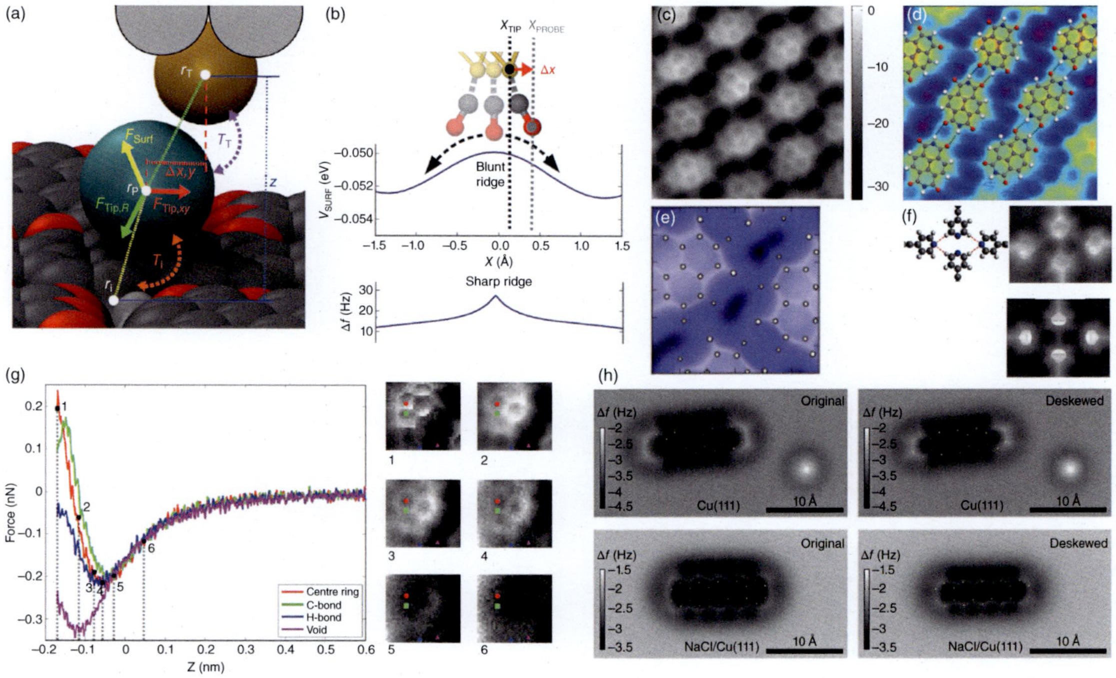

Figure 2.3 Key results in understanding intermolecular bond resolution in ncAFM. (a) Overview of the PPM probe particle and the forces acting upon it on a PTCDA island [57]. (b) Schematic showing the principal of the sharpening of features due to the deflection of the probe particle over a ridge in the potential energy landscape. (c) Constant height Δf image showing contrast in expected hydrogen bonding locations in a NTCDI molecular island absorbed on the Ag : Si(111) surface [58]. (d) Ball and stick model showing proposed geometries and location of expected hydrogen bonds overlaid on a high resolution image of the same type of molecular island. (e) Simulated image of NTCDI island using the PPM [57]. (f) Tetramer junction of bis(*para*-pyridyl)acetylene (BPPA) molecules on the Au(111) – ball and stick model (left) and experimental (top) and simulated (bottom) constant height images [59]. (g) Experimental 3D force map data showing $f(z)$ curves, and 2D XY force images over an NTCDI Island on Ag : Si(111) [58]. (h) Original (left) and 'de-skewed' (right) images of Pentacene on Cu(111) and 2 ML NaCl/Cu (111) [60]. Source: Images (a, b, and e) reprinted with permission from Hapala et al. [57]. Copyright (2014) by the American Physical Society. Images (c, d) reproduced with permission under a Creative Commons license from Sweetman et al. [58]. Image (f) reproduced with permission from Haemaelaeinen et al. [59]. Copyright (2014) by the American Physical Society. Image (g) reproduced with permission under a Creative Commons license from Sweetman et al. [58], Copyright 2014. Image (h) reprinted with permission from Neu et al. [60].Copyright (2014) by the American Physical Society.

volume of molecules when considering the penetration of water into protein structures). The case of hydrogen bonding between surface absorbed molecules has been particularly controversial, with initial reports directly assigning ridges in the position of expected hydrogen bonds to direct observation of the bonding itself [61].

Simulations using the PPM (see Figure 2.3e), and subsequent experimental results from molecular junctions where no hydrogen bonding is expected to occur [59] (see Figure 2.3f) reproduced similar contrast, despite not taking into account any hydrogen bonding effects, and other results confirmed that the hydrogen bonding contrast only arises in the repulsive regime where deflection and steric effects are key [58] (see Figure 2.3g). It now seems indisputable that atoms in close proximity to each other will appear to be linked *even if no bonding between them is present* due to simple steric hindrance effects resulting from the finite radius of the involved atoms. Steric hindrance arguments also suggest that much of the charge density in the intervening space due to any bonding is in fact already inaccessible to the probe due to the finite radius of the probe atom itself. In the case of hydrogen bonding, there remains no good theoretical justification for how hydrogen bonding can, even in principle, contribute to the contrast using accepted models for how intramolecular contrast arises. Whilst much discussion has focused on how the deflection of the probe significantly enhances this effect, it is important to note that even with a rigid probe steric hindrance effects will still result in a ridge between two atoms in close proximity, and therefore the deflection only serves to enhance an already existing effect. Recent reports using alternative functionalisations (resulting in stiffer tips) [62] have not provided any theoretical justification for how to distinguish these known steric hindrance effects from actual observation of hydrogen bonding nor an understanding of how the bonding itself can contribute to the observed contrast.

An interesting subtlety arises in the case of covalent bonding, as the same model reproduces the contrast in the positions of the expected covalent bonds without taking into account the accumulated charge density in the bond, and thus a question arises as to what extent the charge density between the atoms contributes to the overall contrast. Some confidence that the charge density associated with covalent bonds does affect the imaging can be taken from pronounced node-like features in the imaging of triple bonds [63, 64], which cannot be reproduced in models using simple Lennard–Jones potentials. Only by taking into account the full electrostatic force field produced by the total charge density calculated using DFT can these features be reproduced in simulation, and recently developed hybrid methodologies [65], offer a promising route towards simulating images with a firmer theoretical underpinning. Moreover, by collecting complete 3D force fields and estimating the lateral forces on the tip, it is possible to effectively correct for the apparent imaging distortion caused by the deflection without any apparent loss in resolution [60] (see Figure 2.3h).

2.4.2.2 Electrostatics

An additional subtlety arises from the short-range electrostatic forcefield above molecules with significant localised charges, and their interaction with the probe particle, which may have itself localised charges. The imaging of TOAT molecules absorbed on metal surfaces with CO and xenon functionalised tips was have shown to display significant, tip-dependent, distortions in the structure of the molecule at close approach, modelling of which necessitates consideration of the different local charges of the probe particle (see Figure 2.4) [67]. It is therefore clear that the flexibility of the junction can be strongly affected by a number of factors, which can only be elucidated via a combination of first-principles modelling and empirical potentials to allow for high-resolution simulations of the complete system, which are extremely difficult to achieve using ab initio simulation techniques alone.

2.4.2.3 Chemical Sensitivity

Prior to the development of intramolecular resolution imaging, it had been shown that ncAFM was, in principle, capable of achieving chemical sensitivity during atomic resolution imaging, by comparing the ratio of interactions between different atoms in mixed-phase semiconductor blends [69]. An open question, therefore, is whether this same sensitivity may be achieved during intramolecular imaging by similar methods, i.e. can the chemical identity of the atoms in a molecule be determined from a high-resolution image using CO terminated tips? Recent work has shown promise in this direction [41, 68] but has highlighted that severe complications arise as the most important interactions in chemical identification (i.e. the chemical interactions), are intrinsically suppressed by the use of a passivating unit on the tip apex. As such differences in the force interaction between atoms must be mediated by variations in the much weaker dispersion interaction. Only by direct comparison to high-quality ab initio simulations can attempts at this level of identification be made, and interpretation of the data is not straightforward, as any force difference arising from the difference in the chemical identity of the atom itself is strongly convoluted with geometric effects, and local charges, related to the structure of the molecule (see Figure 2.4). It is therefore still unclear whether this technique can be extended to provide a generalisable and unambiguous map of the chemical identity of the atoms within a molecule.

2.5 Characterising On-Surface Reactions with ncAFM

Following the introduction of the technical and theoretical aspects of the ncAFM technique, the remainder of this chapter will focus on several recent examples demonstrating how on-surface reactions can be characterised, and we will discuss the information about the progress of the reactions that may be obtained. It is apparent that individual molecules can be characterised using CO functionalised probes, but ncAFM can also be used to look at the initial, intermediate, and final structures of a reaction. All of the reactions discussed here are performed 'on-surface' and as such are confined to a supporting substrate. This on-surface synthesis approach has been seen to give rise to different reaction pathways to that observed in solution based-synthesis and potentially provides new options for influencing and controlling reactions [16].

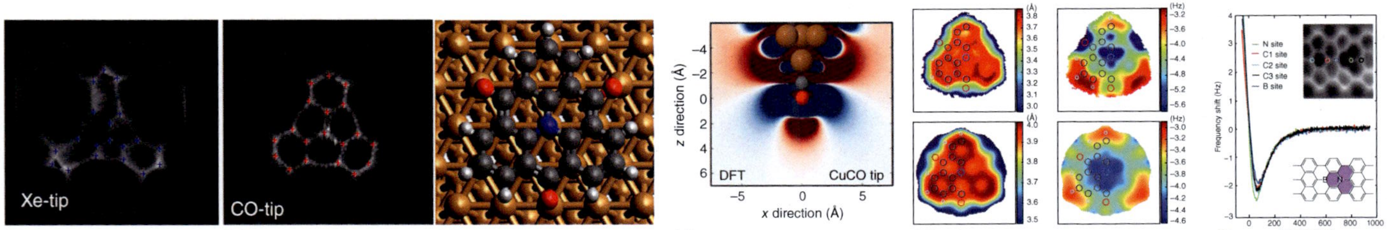

(a) (b) (c) (d) (e) (f)

Figure 2.4 Examples mapping the local electrostatic field, and chemical identity of atoms in small molecules. (a**)** TOAT molecule imaged with a xenon tip and (b) CO tip showing distortion due to the local electrostatic field of the molecule [37]. (c) Ball and stick model showing TOAT molecule adsorbed on Cu(111) surface [66]. (d) Calculated local electrostatic field for a CO molecule absorbed at the apex of a sharp metal tip [67]. (e) Experimental (top) and calculated (bottom) Z*and Δf^* maps of TOAT molecule highlighting local variation over chemically distinct atoms [66]. (f) Identification of substitution boron and nitrogen atoms in a surface synthesised graphene nanoribbon [68]. Source: Images (a) and (b) reproduced with permission under a Creative Commons license from Hapala et al. [37], copyright 2016. Images (c) and (e) reprinted with permission from van der Heijden et al. [66]. Copyright 2016 American Chemical Society. Image (d) reproduced with permission from Ellner et al. [67]. Image (f) reproduced with permission under a Creative Commons license from Kawai et al. [68], copyright 2018.

2.5.1 Practical Considerations for Characterising On-Surface Reactions

The acquisition time for SPM techniques are slow compared to the timescale of the reactions under study: the on-surface diffusion steps for the molecular species will be many orders of magnitude faster than the several minutes typically required to form a ncAFM image (although 'video rate' implementations of SPM have been demonstrated under ambient conditions; see Ref. [70] for an impressive example of the motion of a motor protein captured using cantilever AFM, high-speed ncAFM is technologically demanding and as yet has not been implemented under UHV conditions).

Therefore, in order to allow characterisation of on-surface reactions, studies are usually performed in discrete stages where: (i) Reactant molecules are deposited onto a substrate held at a low temperature, allowing the initial precursors to be characterised. (ii) Thermal energy is supplied to initiate an aspect of the reaction, i.e. formation of an intermediate/transition state, with the substrate then cooled to allow high-resolution image acquisition. (iii) Heating the substrate to a higher temperature then allows the reaction to progress to completion with the sample being subsequently cooled to allow product structures to be imaged and characterised.

In principle, it is possible to gain information about the chemical composition of the molecular species understudy from ncAFM characterisation (in a similar way to that demonstrated for distinguishing between mixtures of atomic species [71]). However, as discussed in Section 2.4.2.3, interpretation of chemical contrast is non-trivial, even when compared with a known reference material is possible, and therefore convolution with other probe–molecule interaction modalities must be carefully considered. More accessible, and of considerable importance, is the determination of the bond order of bonds between carbon atoms. Visually a variation in contrast (Δf) has been observed between single and triple carbon–carbon bonds, which has been attributed to a variation in charge density (which in turn gives rise to a variation in Pauli Repulsion between in probe and molecular and is displayed as the change in the measured Δf signal). Such an approach has been demonstrated in several cases (see for example Ref. [17]), although it is more feasible for planar molecules as convolution issues related to variation in molecular height are avoided.

It is also possible to estimate the adsorption height of the molecular species above the surface, but this requires calibration to a reference point [72] (such as XSW-based analysis) in order to obtain absolute adsorption heights relative to the substrate plane. In the absence of absolute height reference data, ncAFM can still be used to determine the relative adsorption heights across an adsorbed molecule.

For the remainder of the chapter, we focus on two case studies (i) the on-surface synthesis of GNRs, and (ii) the study of the cyclisation reactions. These demonstrate several aspects of the characterisation of on-surface synthesis including the ability to characterise the reaction progress at various stages.

2.5.2 Synthesis and Characterisation of Graphene Based Nanostructures

Graphene-based structures, including GNRs, have received much attention due to their electrical and optical adsorption properties as well as the potential for their applications within device structures (e.g. as molecular wires [73]). However, the atomically precise formation of GNRs is challenging, as is the characterisation of the resultant structures. Defects, in particular atomic-scale defects, could have important effects upon the desired characteristics of any device based upon GNRs. Therefore SPM techniques provide a method, which allows the reaction steps to be identified as well as facilitating structural characterisation of the resultant product.

The first on-surface synthesis of GNRs was shown by Cai et al. [74] and was based upon the use of dibromo functionalised bianthracene units (**1**) which were observed to couple together on Au(111) surface to form linear chains (see Figure 2.5a for mechanism). This type of covalent coupling reaction, based on Ullmann coupling (where iodine functionalised aryls are coupled over a copper catalyst), is a well-studied reaction and was first demonstrated in the on-surface synthesis of covalently coupled molecular architectures in the work by Grill et al. [79]. The reaction mechanism for the formation of the GNRs was postulated to be a two-step process involving: (i) C–C coupling giving rise to a non-planar product and (ii) a cyclodehydrogenation reaction forming the aromatic structure of the GNR. This assignment was based upon the distinct structural difference between the two structures, as shown within STM images (STM images showing GNRs following the cyclodehydrogenation step on Au(111) are shown in Figure 2.5b,c). However, the STM data (while supporting the proposed mechanism) does not show the sub-molecular resolution offered by ncAFM to allow the spatial position of the bonds to be observed.

In order to study the structure of the resultant GNRs, ncAFM has been employed allowing characterisation of the edge structure of GNRs which may be in either an armchair [80] or a zig-zag [76] motif. Characterisation of the structure may be obtained from a simple visual inspection of ncAFM images acquired above a GNR. The regions of brighter contrast (positive values of Δf) are assigned to the carbon backbone of the system. Figure 2.5d shows a ncAFM image of the middle section from an on-surface synthesised GNR where alternating columns of 3 and 2 fused benzene rings are visible. Such images allow ready characterisation of the edge structure – in the example here the so called 'armchair' edge structure is present. Similarly in Figure 2.5e the end of a GNR can be observed with the right-hand side being terminated by a 'zig-zag' structure. The exact termination of the edge-structure of the GNR (e.g. 'zig-zag' or 'armchair') has been observed to have an effect on the electronic states [73] in terms of the energy and delocalisation.

In order to form GNRs with specific edge structures, different precursor molecules are required (see Figure 2.5a,f for details of two alternative reactions – Refs. [76, 80]). By changing the precursor molecule, GNRs with zig-zag edges can be produced [76].

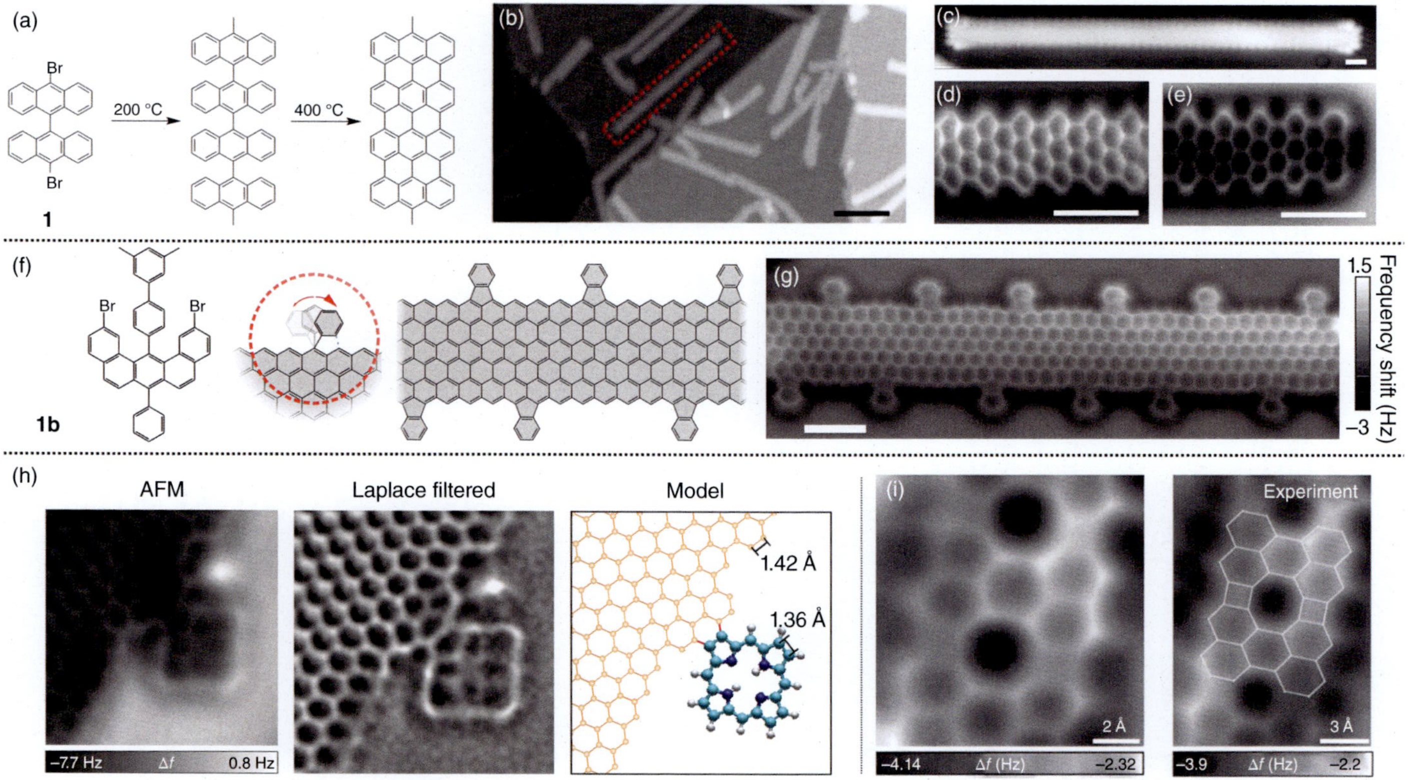

Figure 2.5 ncAFM characterisation of the synthesis of 2D graphene structures. (a) Schematic showing the two-step formation of graphene nanoribbons (GNRs) using a dibromo bianthracene precursor (**1**) progressing via C–C coupling giving rise to a non-planar product and cyclodehydrogenation [74]. (b, c) STM (Scale bars: 10 and 1 nm, respectively) and (d, e) ncAFM images of GNRs with armchair edge structure visible in ncAFM images [75] (Scale bars: 1 nm). (f) Scheme for the synthesis of functional GNRs, obtained via an alternative precursor species (**1b**); proposed structure [76] and (g) ncAFM images shown (Scale bar: 1 nm) [76]. (h) Characterisation of pyrrole units fused to a graphene sheet imaged via ncAFM [77]. (i) ncAFM characterisation of graphene-like nanoribbons periodically embedded with eight and four membered rings [78]. Source: Images (b–e) reproduced with permission under a Creative Commons license from Lit et al. [75], copyright 2013. Images (f and g) reprinted by permission from Ruffieux et al. [76], Copyright 2016. Image (h) reprinted by permission from He et al. [77], copyright 2017. Image (i) reproduced with permission under a Creative Commons license from Liu et al. [78], copyright 2017.

Figure 2.5f shows the precursor molecule (**1b**) employed by Ruffieux et al. to produce GNRs with functionalised zig-zag edges. The reaction is assumed to progress by a mechanism where the reactant molecules couple via an Ullmann-type reaction and subsequent cyclodehydrogenation. At the activation temperatures required to form the GNRs (573 K) [76] the external phenyl groups undergo a ring-closing reaction with the body of the GNR forming a fluoranthene-type sub-unit incorporating a five-membered ring. Structural characterisation of the resultant GNR (Figure 2.5g) clearly demonstrates that the additional phenyl group fuses to the edge with a cyclic motif and is not connected via a σ bond as in the precursor molecule (**1b**).

Additional functionalisation of graphene structures has been demonstrated by fusing tetrapyrroles (free-base porphyrins, 2H–P) to the edges of extended graphene structures [77]. In the work by He et al. a Ag(111) surface, partially covered with graphene structures, was exposed to 2H–P (with 2H–P found to adsorb as individual molecules on bare Ag(111) as well as at the edges of the graphene structures). A coupling reaction was initiated by annealing the sample at 620 K, with the resulting ncAFM characterisation indicating the precise nature of the bonding between 2H–P and the graphene structures (shown in Figure 2.5h).

As a final point, the bond resolving power of ncAFM can also facilitate the investigation and characterisation of local 'defects' within graphene structures. In the work by Liu et al. [78], four and eight-membered rings can be observed within the graphene structure. Such defects are important as they may be beneficial with regards to tuning the electronic properties of the structures (shown in Figure 2.5i). This example illustrates how ncAFM can be used to characterise the structure of carbon-containing planar structures.

2.5.3 Studying the Evolution of On-Surface Reaction

The previous examples demonstrate the use of ncAFM as a structural characterisation technique for reaction products (e.g. GNRs). ncAFM can also be used to study the evolution of on-surface chemical reactions. Cyclisation processes are a feature in many reaction pathways, with the conversion of neighbouring alkyne units to aromatic rings being common place. Work by Oteyza et al. [63] shows how the reactants and products of a cyclisation reaction involving enediynes can be studied using ncAFM. Such cyclisation reactions can result in the formation of a variety of products, and therefore detailed characterisation via ensemble techniques can be non-trivial. ncAFM, therefore, offers a route towards single-molecule characterisation of the reaction pathway. The reactant molecule 2-bis((2-ethynylphenyl)ethynyl)benzene was deposited on a Ag(100) surface held at room temperature and subsequently imaged at $\leq$7 K. Initial characterisation of the reactants and products was carried out using STM (with the sample annealed to $T > 90\,°C$ to initiate the reaction). High-resolution ncAFM images allow the reactant molecular features to be assigned to the aromatic rings and alkyne groups. Variation in contrast at the position of the triple bond can be observed within the ncAFM images as an enhanced value of Δf (similar contrast is not visible within the STM image).

Following heating of the surface ($T > 90\,°C$) three reactant products, all displaying distinct and different appearances within STM images were identified (in the ratio $(51 \pm 7\%) : (28 \pm 5\%) : (7 \pm 3\%)$). From ncAFM characterisation of the product structures, the newly formed six-, five-, and four-membered rings can be seen within the product molecules and allow the structures of these species to be inferred. From this knowledge of the structure of the reactants and products, Oteyza et al. were able to propose the thermal reaction pathways, supported by DFT based calculations [63].

A similar reaction system detailing cyclisation was the subject of a study by Riss et al. [81] where step-wise bimolecular enediyne coupling was observed on a Ag(100) surface. Here the temporal evolution of the reaction was observed (focusing on the conversion of intermediate structures to products as a function of reaction temperature). In order to identify the transient intermediates present within a multi-step reaction, thermal cycling and quenching of the system were employed. As in the previous study, the chemical structure of the precursor molecule was characterised. Next, the surface was repeatedly heated and cooled for 60 minute periods, steadily increasing the maximum temperature reached for each repetition. At each step, the relative abundance of various molecular components was determined.

The molecule, 1,2-bis(2-ethynyl phenyl)ethyne **2** (Figure 2.6a), was deposited on to Ag(100) substrate and characterised at ~4 K using ncAFM. This approach enabled identification of the rings and triple bond within the structure (similar to that seen in Figure 2.1d) of the reactant molecule with both cis and trans forms being observed. In order to track the progress of the on-surface reaction as a function of temperature, the sample was systematically heated to a range of temperatures between 290 and 460 K. In each case, the sample was heated then allowed to cool to room temperature (~1 hour for each cycle – see Figure 2.6b) with the sample then cooled to ~4 K for characterisation of the products.

The dimerisation of **2** was studied and it was noted that, following initial coupling, the dimer progressed through various degrees of cyclisation. The fully cyclised dimer product is shown in Figure 2.6a (**3**) and the relative abundance of the uncyclised, (**i**) half-cyclised, (**ii**) fully cyclised, and (**3**) dimers were characterised for each temperature cycle (Figure 2.6c). The relative ratio of the abundance of the respective species gradually shifts towards molecules with a higher degree of cyclisation following each anneal step. This suggests that the partially cyclised molecules are transient intermediates on the way to the fully cyclised dimer species. The total number of molecules was not observed to significantly decrease over the reaction; suggesting that desorption or trapping of intermediate species at surface sites (e.g. step-edges or defects) does not play an important role. In concert with DFT calculations, it is possible to obtain details of the reaction mechanism (by incorporating details of the structure of the known transition states).

The previous examples of on-surface synthesis have demonstrated how the structure of reactant, intermediate, and product molecular species can be determined via ncAFM. In particular, the cyclisation reactions that lead to new structural features that (e.g. six-, five-, and four-membered rings) are readily characterised by ncAFM. It is also possible to obtain information about the bond order (which on a simplistic level can be related to the bond number, i.e. a triple C—C has a bond number of 3,

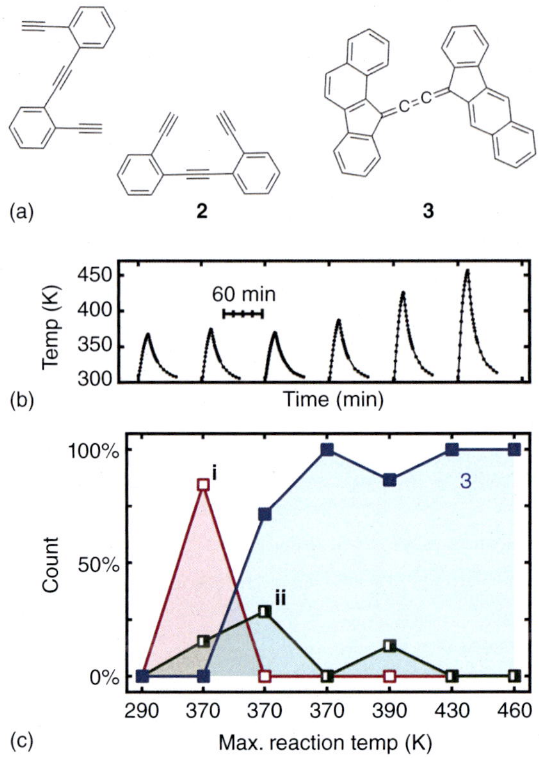

Figure 2.6 Details of on-surface cyclisation reactions where products and reactants are characterised by ncAFM [81]. (a) Reactant and product species for the cross-coupling and cyclisation of 1,2-bis(2-ethynylphenyl)ethyne on Ag(100). (b) Graph showing the systematic heating of Ag(100) surface with adsorbed **2** to a range of temperatures between 290 and 460 K. (c) Details of the relative abundance of the experimentally observed species (**3** being the product, with **(i)** and **(ii)** being intermediates) during a series of annealing steps (temperature profile shown in (b)). Source: Images (a–c) reprinted by permission from A. Riss, A. P. Paz, S. Wickenburg, H.-Z. Tsai, D. G. De Oteyza,A. J. Bradley, M. M. Ugeda, P. Gorman, H. S. Jung, M. F. Crommie, A. Rubio and F. R. Fischer, Nat. Chem., 2016, 8, 678–683, copyright 2016.

while C—H has 1). This information can be invaluable in determining the structure of a molecular species via ncAFM characterisation. Schuler et al. showed that it is possible to reversibly induce the Bergman cyclisation of a diyne with a 10-membered ring using the probe to perform atomic manipulation between the two structures (see Figure 2.7a) [82]; resulting in an increase/decrease in bond order within the intramolecular carbon–carbon bonds within the molecule.

A bromine functionalised anthracene molecule (9,10-dibromoanthracene – DBA, **4**) was deposited onto a bilayer NaCl film on Cu(111). Following the initial characterisation of the structure of DBA by ncAFM (see Figure 2.7b) the probe was positioned above the bright (positive Δf) features corresponding to the position of the Br substituents. Both Br atoms were removed via injection of electrons (achieved by increasing the bias applied to the sample for a short time – A so called 'bias pulse', a technique first used to induce local on-surface chemical bond-breaking by Hla et al. [84]). The first Br is removed using a bias pulse at 1.6 V. In the following images, the feature corresponding to Br has been removed and, interestingly, the contrast of the central ring has changed indicating a reduced adsorption height on the debrominated side of the molecule. The second Br can be removed using a higher voltage pulse (3.3 V – 10 seconds of picoampere current) with the probe withdrawn from the molecule so as to limit the current flow (a high current and elevated bias may induce additional unwanted bond cleavage). This doubly debrominated diradical species was then imaged (see Figure 2.7b – Diradical **5**). It should be noted

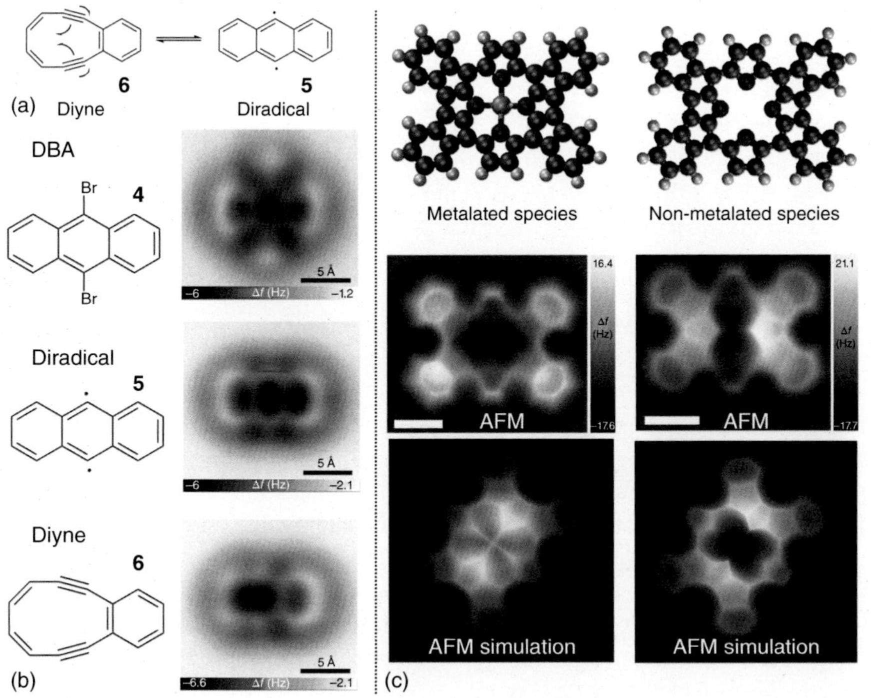

Figure 2.7 Examples of characterisation of bond order and chemical structure for carbon containing species. (a) Scheme showing the reversible Bergman cyclisation of the cyclic diyne 3,4-benzocyclodeca-3,7,9-triene-1,5-diyne (**6**) to generate the 9,10-didehydroanthracene diradical (**5**). (b) Structures and AFM imaging of the starting material, reaction intermediates and product formed by successive STM debromination and subsequent cyclisation. ncAFM images of the molecules were acquired on NaCl (2 ML)/Cu(111) using a CO tip [82]. (c) Chemical structure, ncAFM images and simulated ncAFM images of metalated and non-metalated planarised porphyrin monomers on Au(111) [83]. (Scale bars: 0.5 nm). Source: SPM images in (b) reprinted by permission from Schuler et al. [82], copyright 2016. Images in (c) from Cirera et al. [83]. Copyright 2019 American Chemical Society.

that the radical species is stabilised by the NaCl film (formation of the species was also observed on the Cu(111) surface but conversion to the diyne was not initiated indicating a strong interaction between the diradical and the Cu surface).

At this point applying additional bias pulses (1.7 V) above the diradical species results in conversion to a cyclised structure (Figure 2.7b – Diyne **6**). ncAFM images of the resulting product reveal that the structure consists of fused 6- and 10-membered rings, which suggests the formation of diyne **6** by homolytic cleavage of the C—C bond shared by two fused benzene rings. The authors state that this diyne was often created directly from the mono-radical without first observing the diradical.

Visual analysis reveals the presence of triple bonds (observed to appear with a distinctive elongation perpendicular to the bond direction, as previously found for alkynes imaged by CO tips [63, 64]). The detailed analysis provides further information about the bond order of the structure. The values for the Pauling bond order, *b*, for systems with a carbon-backbone can be calculated using:

$$b = -\frac{1}{p_1} \ln \left(\frac{l - p_2}{p_3} \right)$$

where $p_1 = 0.6852$, $p_2 = 1.0979$, $p_3 = 0.4397$ (obtained from fitting the lengths of the carbon–carbon bond within ethane, ethylene, and acetylene [82]). Therefore by measuring the length (l) of the bonds obtained from the ncAFM images and comparing them to the simulated ncAFM images based on DFT calculated structures (thereby taking the flexibility of the CO functionalised tip into account) it is possible to assign the experimentally observed features to structural models. It is important to note that due to the flexibility of the tip the apparent length of the bonds from the experimentally acquired images have to be compared to a simulated model of apparent distances, not the actual bond lengths as calculated by DFT. This is due to the fact that the tip flexibility results in a distortion of the apparent inter-atomic distances when the probe-sample distance is small (see Section 2.4.2.1) [17]. This means that although the approach provides some information on variation in bond order within a molecule, a direct measurement is not possible.

The success of comparison between simulated ncAFM images (based on the flexible probe model – Section 2.4.2.1) and experimental data in terms of structural determination is illustrated in the work by Cirera et al. [83]. A fluorinated free-base porphyrin (5,10,15,20-tetrakis-(4-fluorophenyl)porphyrin, 2H-4FTPP) was studied on a Au(111) surface and was observed to undergo a variety of thermally induced chemical transformations with the structures being characterised by ncAFM. In this system, the 2H-4FTPP species was observed to undergo a ring-closure reaction when heated; annealing at 500 K resulted in the formation of a planar species, with self-metallation (incorporation of Au adatoms into the core of the porphyrin macrocycle) occurring at 575 K. Structures of the non-metallated and metallated planarised porphyrin species are shown in Figure 2.7c.

The structure of the reaction products could be determined by comparison with simulated ncAFM images based upon the PPM (see Section 2.4.2). Figure 2.7c shows the ncAFM data for two of the observed planarised species (metallated and non-metallated) alongside the simulated ncAFM images based upon molecular models (structures obtained from DFT calculations). While there is a qualitative agreement between the simulated and experimentally obtained data, both showing a clear distinction between the metalated and non-metalated species, and a good description of the backbone of the molecule and the expected symmetry of the core, there are nonetheless significant differences in the observed contrast particularly around the metal core, highlighting the need for further progress towards a complete understanding of the contrast formation mechanism for more complex interaction.

2.6 Conclusions

This chapter has demonstrated the effectiveness of using SPM, and particularly ncAFM, as tools for imaging and characterisation of on-surface chemical reactions.

It is apparent from the systems discussed that ncAFM provides remarkable images of systems with intramolecular resolution but, importantly, these experiments also provide detailed chemical information for the systems studied. The level of detail is significant, allowing discrimination of individual atoms and/or chemical bonds. Such levels of information are highly unusual in chemistry and although other techniques may provide similar information (e.g. single-crystal X-ray diffraction) ncAFM is unparalleled in operating at the single-molecule level.

However, we have also sought to demonstrate the experimental challenges and limitations of performing such measurements. Indeed, it should always be recognised that ncAFM, and other SPM measurements, probe molecules adsorbed on a substrate and that the interaction between the molecule of interest and the substrate influences the properties observed. We have also discussed the importance of using UHV, low-temperature conditions to perform ncAFM imaging, and the factors associated with the choice of tip and probe molecule. We emphasise the importance of detailed studies that seek to confirm the exact nature of features observed in ncAFM images and how misassignment of features is a significant problem in the field.

It is apparent that SPM, and in particular ncAFM, is a remarkable advance that allows the identification of individual molecules and even reaction processes. These approaches will continue to grow in importance in the development of our understanding of single-molecule processes and we fully anticipate that ncAFM will become a tool used across the chemical sciences.

Acknowledgements

Alex Saywell acknowledges support via a Royal Society University Research Fellowship. NRC acknowledges financial support from the UK Engineering and Physical Sciences Research Council (EP/S002995/1). Adam Sweetman acknowledges support via Royal Society University Research Fellowship and via the ERC Grant 3DMOSHBOND. The authors also acknowledge Shashank S. Harivyasi for his assistance in generating the 3D models used in Figures 2.1 and 2.3.

References

1 Binnig, G., Rohrer, H., Gerber, C., and Weibel, E. (1982). *Appl. Phys. Lett.* 40: 178–180.
2 Yang, B., Björk, J., Lin, H. et al. (2015). *J. Am. Chem. Soc.* 137: 4904–4907.
3 She, Z., Yao, Z., Ménard, H. et al. (2019). *Nanoscale* 11: 13773–13782.
4 Saywell, A., Magnano, G., Satterley, C.J. et al. (2010). *Nat. Commun.* 1: 1–8.
5 Handrup, K., Richards, V.J., Weston, M. et al. (2013). *J. Chem. Phys.* 139: 154708.
6 Judd, C.J., Junqueira, F.L., Haddow, S.L. et al. (2020). *Commun. Chem.* 3: 1–8.
7 Binnig, G., Quate, C.F., and Gerber, C. (1986). *Phys. Rev. Lett.* 56: 930–933.
8 Gross, L., Mohn, F., Moll, N. et al. (2009). *Science* 325: 1110–1114.
9 Grill, L. (2010). *J. Phys. Condens. Matter* 22: 84023.

10 Vang, R.T., Lauritsen, J.V., Lægsgaard, E., and Besenbacher, F. (2008). *Chem. Soc. Rev.* 37: 2191.

11 Lindner, R. and Kühnle, A. (2015). *ChemPhysChem* 16: 1582–1592.

12 Shen, Q., Gao, H.-Y., and Fuchs, H. (2017). *Nano Today* 13: 77–96.

13 Fan, Q., Gottfried, J.M., and Zhu, J. (2015). *Acc. Chem. Res.* 48: 2484–2494.

14 Dong, L., Liu, P.N., and Lin, N. (2015). *Acc. Chem. Res.* 48: 2765–2774.

15 El Garah, M., MacLeod, J.M., and Rosei, F. (2013). *Surf. Sci.* 613: 6–14.

16 Clair, S. and Oteyza, D.G. (2019). *Chem. Rev.* 119: 4717–4776.

17 Gross, L., Mohn, F., Moll, N. et al. (2012). *Science* 337: 1326–1329.

18 Hanssen, K.Ø., Schuler, B., Williams, A.J. et al. (2012). *Angew. Chem. Int. Ed.* 51: 12238–12241.

19 Schuler, B., Meyer, G., Peña, D. et al. (2015). *J. Am. Chem. Soc.* 137: 9870–9876.

20 Wiesendanger, R. (2010). *Scanning Probe Microscopy and Spectroscopy: Methods and Applications*. Cambridge England; New York: Cambridge University Press.

21 Chen, C.J. (2007). *Introduction to Scanning Tunneling Microscopy*, 2e. Oxford University Press.

22 Saywell, A., Schwarz, J., Hecht, S., and Grill, L. (2012). *Angew. Chem. Int. Ed.* 51: 5096–5100.

23 Liu, M., Li, S., Zhou, J. et al. (2018). *ACS Nano* 12: 12612–12618.

24 Saywell, A., Greń, W., Franc, G. et al. (2014). *J. Phys. Chem. C* 118: 1719–1728.

25 Giessibl, F.J. (2019). *Rev. Sci. Instrum.* 90: 11101.

26 Giessibl, F.J. (2003). *Rev. Mod. Phys.* 75: 949–983.

27 Gross, L., Schuler, B., Pavliček, N. et al. (2018). *Angew. Chem. Int. Ed.* 57: 3888–3908.

28 Giessibl, F.J. (1998). *Appl. Phys. Lett.* 73: 3956–3958.

29 Iwata, K., Yamazaki, S., Mutombo, P. et al. (2015). *Nat. Commun.* 6: 7766.

30 Moreno, C., Stetsovych, O., Shimizu, T.K., and Custance, O. (2015). *Nano Lett.* 15: 2257–2262.

31 Berger, J., Kosmider, K., Stetsovych, O. et al. (2016). *J. Phys. Chem. C* 120 (38): 21955–21961.

32 Sweetman, A., Jarvis, S., Rahe, P. et al. (2014). *Phys. Rev. B* 90: 165425.

33 Barth, J.V. (2007). *Annu. Rev. Phys. Chem.* 58: 375–407.

34 Pai, W.W., Hsu, C.L., Lin, M.C. et al. (2004). *Phys. Rev. B* 69: 1–7.

35 Mohn, F., Schuler, B., Gross, L., and Meyer, G. (2013). *Appl. Phys. Lett.* 102: 73104–73109.

36 van der Lit, J., Di Cicco, F., Hapala, P. et al. (2016). *Phys. Rev. Lett.* 116: 96102.

37 Hapala, P., Svec, M., Stetsovych, O. et al. (2016). *Nat. Commun.* 7: 11560.

38 Repp, J., Meyer, G., Stojković, S. et al. (2005). *Phys. Rev. Lett.* 94: 26803.

39 Shin, H.J., Jung, J., Motobayashi, K. et al. (2010). *Nat. Mater.* 9: 442–447.

40 Labidi, H., Kupsta, M., Huff, T. et al. (2015). *Ultramicroscopy* 158 (1–5).

41 Kawai, S., Foster, A.S., Björkman, T. et al. (2016). *Nat. Commun.* 7: 11559.

42 Pozo, I., Majzik, Z., Pavliček, N. et al. (2019). *J. Am. Chem. Soc.* 141: 15488–15493.

43 Sader, J.E., Hughes, B.D., Huber, F., and Giessibl, F.J. (2018). *Nat. Nanotechnol.* 13: 1088–1091.

44 Bartels, L., Meyer, G., and Rieder, K.-H. (1997). *Appl. Phys. Lett.* 71: 213.
45 Boneschanscher, M.P., Van Der Lit, J., Sun, Z. et al. (2012). *ACS Nano* 6: 10216–10221.
46 Ternes, M., González, C., Lutz, C.P. et al. (2011). *Phys. Rev. Lett.* 106: 16802.
47 Abe, M., Sugimoto, Y., Custance, O. et al. (2005). *Appl. Phys. Lett.* 87: 173503.
48 Sugimoto, Y. and Onoda, J. (2019). *Appl. Phys. Lett.* 115: 173104.
49 Labuda, A., Miyahara, Y., Cockins, L., and Grütter, P.H. (2011). *Phys. Rev. B* 84: 125433.
50 Majzik, Z., Setvín, M., Bettac, A. et al. (2012). *Beilstein J. Nanotechnol.* 3: 249–259.
51 Weymouth, A., Wutscher, T., Welker, J. et al. (2011). *Phys. Rev. Lett.* 106: 1–4.
52 Sader, J.E. and Jarvis, S.P. (2004). *Appl. Phys. Lett.* 84: 1801.
53 Sadewasser, S. and Glatzel, T. (2018). *Kelvin Probe Force Microscopy from Single Charge Detection to Device Characterization*. Springer.
54 Hapala, P., Temirov, R., and Tautz, F.S. (2014). *Phys. Rev. Lett.* 113: 226101.
55 Lauhon, L. and Ho, W. (1999). *Phys. Rev. B* 60: R8525–R8528.
56 Sun, Z., Boneschanscher, M.P., Swart, I. et al. (2011). *Phys. Rev. Lett.* 106: 46104.
57 Hapala, P., Kichin, G., Wagner, C. et al. (2014). *Phys. Rev. B* 90: 085421.
58 Sweetman, A.M., Jarvis, S.P., Sang, H. et al. (2014). *Nat. Commun.* 5: 3931.
59 Haemaelaeinen, S.K., Heijden, N.V.D., Lit, J.V.D. et al. (2014). *Phys. Rev. Lett.* 113: 186102.
60 Neu, M., Moll, N., Gross, L. et al. (2014). *Phys. Rev. B* 89: 205407.
61 Zhang, J., Chen, P., Yuan, B. et al. (2013). *Science* 342: 611–614.
62 Mönig, H., Amirjalayer, S., Timmer, A. et al. (2018). *Nat. Nanotechnol.* 13: 371–375.
63 de Oteyza, D.G., Gorman, P., Chen, Y.-C. et al. (2013). *Science* 340: 1434–1437.
64 Moll, N., Schuler, B., Kawai, S. et al. (2014). *Nano Lett.* 14: 6127.
65 Ellner, M., Pou, P., and Perez, R. (2019). *ACS Nano* 13: 786.
66 van der Heijden, N.J., Hapala, P., Rombouts, J.A. et al. (2016). *ACS Nano* 10: 8517.
67 Ellner, M., Pavliček, N., Pou, P. et al. (2016). *Nano Lett.* 16: 1974.
68 Kawai, S., Nakatsuka, S., Hatakeyama, T. et al. (2018). *Sci. Adv.* 4: eaar7181.
69 Sugimoto, Y., Pou, P., Abe, M. et al. (2007). *Nature* 446: 64–67.
70 Kodera, N., Yamamoto, D., Ishikawa, R., and Ando, T. (2010). *Nature* 468: 72.
71 Custance, O., Perez, R., and Morita, S. (2009). *Nat. Nanotechnol.* 4: 803–810.
72 Schuler, B., Liu, W., Tkatchenko, A. et al. (2013). *Phys. Rev. Lett.* 111: 106103.
73 Koch, M., Ample, F., Joachim, C., and Grill, L. (2012). *Nat. Nanotechnol.* 7: 713–717.
74 Cai, J., Ruffieux, P., Jaafar, R. et al. (2010). *Nature* 466: 470–473.
75 Lit, J., Boneschanscher, M.P., Vanmaekelbergh, D. et al. (2013). *Nat. Commun.* 4: 2023.
76 Ruffieux, P., Wang, S., Yang, B. et al. (2016). *Nature* 531: 489–492.
77 He, Y., Garnica, M., Bischoff, F. et al. (2017). *Nat. Chem.* 9: 33–38.
78 Liu, M., Liu, M., She, L. et al. (2017). *Nat. Commun.* 8: 14924.
79 Grill, L., Dyer, M., Lafferentz, L. et al. (2007). *Nat. Nanotechnol.* 2: 687–691.

80 Dienel, T., Kawai, S., Söde, H. et al. (2015). *Nano Lett.* 15: 5185–5190.
81 Riss, A., Paz, A.P., Wickenburg, S. et al. (2016). *Nat. Chem.* 8: 678–683.
82 Schuler, B., Fatayer, S., Mohn, F. et al. (2016). *Nat. Chem.* 8: 220–224.
83 Cirera, B., Torre, B., Moreno, D. et al. (2019). *Chem. Mater.* 31: 3248–3256.
84 Hla, S.W., Bartels, L., Meyer, G., and Rieder, K.H. (2000). *Phys. Rev. Lett.* 85: 2777–2780.

3

Complexity in Two-Dimensional Multicomponent Assembly

Kunal S. Mali, Joan Teyssandier, Nerea Bilbao, and Steven De Feyter

KU Leuven, Division of Molecular Imaging and Photonics, Department of Chemistry, Celestijnenlaan, 200F, Leuven B-3001, Belgium

3.1 Introduction

One of the strongest inspirations behind the development of early science was Mother Nature. The beauty and complexity of natural systems have always fascinated researchers and non-researchers alike. A number of remarkable features exhibited by natural systems, ranging from the strength of spider silk to the light-harvesting ability of plant leaves, arise from sophisticated, multicomponent architectures where biological molecules are segregated into discrete compartments with nanoscale precision. It is widely acknowledged that strategies which allow the fabrication of self-assembled materials that rival the sophisticated systems found in nature may enable future generation technologies capable of addressing societal challenges in diverse areas ranging from renewable energy to advanced medicine.

Molecular self-assembly, which constitutes a defining concept of supramolecular chemistry, offers a convenient way of making complex systems held together via non-covalent bonds. *It is a process in which components, either separate or linked, spontaneously form ordered aggregates* via *specific local interactions* [1]. Despite being relatively straightforward, the self-assembly strategy is often trial-and-error-based. Furthermore, man-made systems rarely exhibit the high level of complexity observed in natural systems. Notwithstanding these limitations, there is increasing interest in designing multicomponent self-assembled structures where chemically distinct molecular building blocks are precisely positioned with respect to each other in an ordered matrix. Such multicomponent systems are expected to exhibit functions that cannot be obtained from single-component systems [2].

Multicomponent supramolecular systems have been studied in diverse range of environments namely, in the solution phase [3], in the solid-state [4], and at solid interfaces [5]. In this chapter, we recapitulate the progress made in the design and construction of crystalline multicomponent supramolecular networks formed at solid interfaces and characterised using scanning tunnelling microscopy (STM). The networks formed at the air–solid, solution–solid, and vacuum–solid interface are

Supramolecular Chemistry on Surfaces: 2D Networks and 2D Structures, First Edition.
Edited by Neil R. Champness.

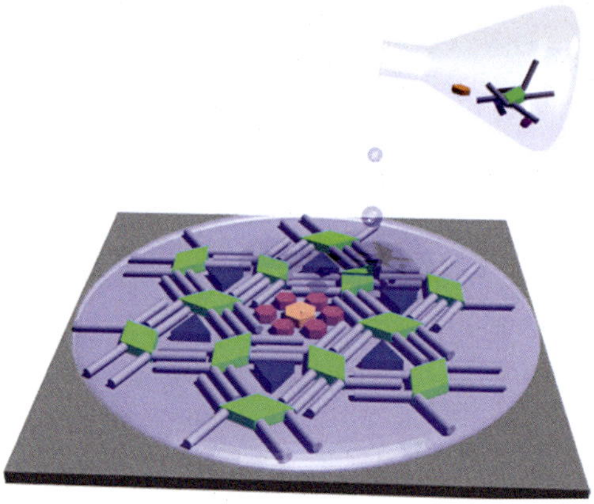

Scheme 3.1 On-surface fabrication of multicomponent self-assembled systems.

described. The fabrication of complex multicomponent architectures has remained one of the dominant themes in the 'on-surface' supramolecular chemistry and is popularly referred to as two-dimensional (2D) crystal engineering. 2D crystal engineering is the design and fabrication of single-molecule thick crystalline layers of (multicomponent) organic and metal–organic building blocks physisorbed on solid surfaces. This research is largely driven by potential applications in emerging fields such as organic- and molecular-scale electronics wherein precise positioning of molecular building blocks is highly desired to obtain predictable nanoscale morphologies and in turn, efficient performance. Although no breakthrough applications appear imminent, significant advances have been made in the recent past in controlling multicomponent assembly on surfaces [6].

Similar to supramolecular chemistry in solution and the solid-state, 2D crystal engineering employs a variety of molecular recognition processes where molecules recognise other molecules or ions and bind to each other via non-covalent interactions such as hydrogen and halogen bonds, van der Waals forces, ion–dipole, dipole–dipole, and π–π stacking interactions. These non-covalent interactions are often pre-programmed into the (supra)molecular design via the installation of specific structural features such as functional groups, alkyl chain substitution patterns and chiral centres during the synthesis step. When multiple building blocks are involved in self-assembly, identical molecules typically exhibit self-selection provided such information is 'encoded' in the (supra)molecular design via hierarchical design (Scheme 3.1).

Successful formation of multicomponent networks on surfaces often requires a good understanding of and precise control over intermolecular and interfacial interactions. Formation of a long-range ordered multicomponent crystalline network upon mixing of two or more building blocks at a solid interface is often a non-trivial process. When more than one building block is brought at a solid interface, the following three outcomes are possible. (i) Phase separation: In this case, both the building blocks adsorb on the surface however, each molecule forms an independent domain of its own within which it adopts a structure that is identical to that formed in a single component system. (ii) Random mixing: Here one component is randomly distributed within the crystalline matrix formed by the other component without significantly disturbing the unit cell of the former. Such mixing occurs when the building blocks have identical unit cell parameters, except for the unusual case of 2D solid solutions [7]. (iii) Co-crystallisation: In this case, the different components form a crystalline matrix with a defined surface stoichiometry giving rise to a new multicomponent 2D network. Typically, co-crystallisation occurs as a result of strong intermolecular interactions such as hydrogen bonding or space-filling constraints as observed for host–guest systems. A fourth possibility, often encountered when molecules self-assemble at the solution–solid interface, is that of preferential adsorption of one component over the other wherein the adsorption energy per unit area of one of the building blocks is higher than that of other constituents. In general, multicomponent self-assembly on solid surfaces strongly depends on both the symmetry and the quantitative similarity of the unit cells as well as the compatibility of the functional groups of the constituent building blocks.

Multicomponent self-assembly experiments carried out at the vacuum–solid interface differ significantly from those at the solution–solid interface. The latter represents a more dynamic environment where molecules can diffuse in-plane as well as out-of-plane whereas there is little or no dynamics present in the case of the former. As a result of the dynamic environment that promotes the diffusion of molecules, a number of systems exhibit dynamic re-configuration at the solution–solid interface wherein one building block causes a change in the adsorption behaviour of the other (vide infra). The manner in which the building blocks are brought in contact with the surface of the solid is also different for the two approaches. Experiments at the solution–solid interface can be carried out by pre-mixing the building blocks in solution in appropriate ratios prior to deposition on the surface. Alternatively, they can also be added to the surface in a sequential fashion. At the vacuum solid interface, sequential sublimation is employed most often while simultaneously annealing the substrate at varying temperatures to ensure enough dynamics.

Multicomponent self-assembled systems studied at solid interfaces can be classified into two categories; those based on host–guest interactions, and those that are not based on host–guest interactions. A majority of 2D multicomponent systems are based on host–guest interactions in which one molecule forms an open porous network with well-defined voids that are then occupied by one or more (type of) guest molecules. A minor class of multicomponent systems is based on non-host–guest type systems, in which none of the components involved forms an open porous network. Multicomponent network formation is driven by specific interactions between the assembling components that favour co-crystallisation over phase separation [8]. These two categories can further be divided into sub-categories depending on the number of different building blocks involved in the self-assembly as follows in the remainder of the chapter.

3.2 Two-Component Self-Assembled Systems

The examples presented in this chapter only concern crystalline networks adsorbed on a solid surface. These networks consist of a long-range ordered, periodic arrangement of two different molecules in a crystalline lattice. They are often referred to as 2D co-crystals in analogy with bulk co-crystals. Similar to that in bulk systems, co-crystallisation on a solid surface is a challenging task as it involves recognition between two different molecules and thus co-crystals are formed only when hetero-molecular interactions prevail over homo-molecular interactions. In other words, the formation of co-crystals is favoured only when the resulting structure is lower in energy compared to the native structures formed by individual components. Sections 3.2.1 and 3.2.2 describe selected few examples of bicomponent systems based on host–guest as well as non-host–guest systems.

3.2.1 Two-Component Systems: Host–Guest Architectures

The most commonly encountered example of bicomponent systems consist of host–guest architectures wherein one component forms the host network whereas the other acts as a guest and occupies the voids formed in the host network [9]. In most cases, the molecule that forms the host network is capable of forming an open porous structure via self-assembly thanks to appropriately installed design elements on the molecular backbone such as hydrogen bonding functional groups or long alkyl chains. Alternatively, the host network can also be obtained using intrinsically porous building blocks. Intrinsic porosity is inherent to the chemical structure and thus the building blocks exhibit permanent covalent cavities. The host molecule typically consists of a covalent (macro)cycle and can usually form host–guest complexes in solution as well as on the solid surface. A special class of host networks that exhibit covalent cavities is that of two-dimensional surface covalent organic frameworks (2D-sCOFs) which have also been recently used for guest capture.

3.2.1.1 Host Networks from Intrinsically Porous Building Blocks

The vast majority of host–guest systems studied in solution, including the seminal work on crown ethers and cryptands by Pedersen, Cram, and Lehn, involve a persistent macrocycle. Having intrinsically porous building blocks allows for specific guest binding with 1 : 1 stoichiometry as well as the direct comparison between recognition events occurring in solution and those on surface. On-surface host–guest chemistry using intrinsically porous building blocks however has some inherent drawbacks. For instance, the design of macrocycles with cavities of controlled dimensions requires a cumbersome synthetic process that needs to be repeated every time a new guest species is targeted. Moreover, not all porous building blocks are able to form organised arrays on surfaces, which imply that the macrocycles need to be functionalised further with additional groups in order to increase interactions with the surface and/or to promote self-assembly through intermolecular interactions. Despite these limitations, a few examples of guest binding, where the specific recognition occurs through a combination of non-covalent interactions and size and shape matching, have been reported on surfaces. Smaller cycles typically form coordination complexes with ions whereas larger macrocycles can trap aromatic molecules [10] or could be used for the growth of metal nanoparticles [11].

Inspired by the solution-phase host–guest chemistry, most of the early examples of ion binding on surfaces used crown ether derivatives. The most commonly known crown ethers are cyclic oligomers of ethylene oxide. The cavity size of crown ethers can be varied by changing the number of oligomers involved in a ring formation. Depending on the size of the cavity, the crown ethers show different binding affinity to typical cations. For example, 18-crown-6 has a high affinity for potassium ions whereas 12-crown-4 binds strongly with lithium ions. In fact, one of the early examples of on-surface binding of cations by a crown ether was reported for dibenzo-18-crown-6 (DB18C6) monolayers adsorbed on Au(111) surface characterised using electrochemical scanning tunnelling microscopy (EC-STM). DB18C6 forms densely packed stable monolayers at positive substrate potentials. In EC-STM images, the cavities of the macrocycle appear empty. The addition of KI solution changed the contrast of the cavities, which appear bright, indicating the formation of an inclusion complex between DB18C6 and K^+ ion. The addition of KI solution was also found to alter the packing arrangement of DB18C6 also highlighting the importance of spacing between the host molecules to allow this binding [12].

Although the ion binding chemistry is inspired by solution-phase studies, there exist some differences when such experiments are conducted at a solid interface. The positioning of molecules with respect to each other, the proximity of the ion-bound sites and the nature of the substrate dictates the efficiency of binding. For example, a 15-crown-5-ether-substituted cobalt(II) phthalocyanine, a molecule bearing crown ether units at the four vertices of the phthalocyanine core (Figure 3.1a), forms ordered monolayers on Au(111) surface. STM images revealed that diagonally opposite crown ether cavities appear occupied after the addition of Ca^{2+} to the solution (Figure 3.1b). Electrostatic repulsion between the filled crown ether cavities prohibits the occupation of all four cavities by Ca^{2+} ions (Figure 3.1c) [15]. Guest binding is also critically dependent on the interactions with the substrate lattice. Although the unit cell did not change upon the addition of Ca^{2+} solution, the molecules were found to be rotated by 10°. This rotation allows two of the crown ether units to move atop gold hollow sites, favouring the accommodation of Ca^{2+} while the empty crown ether units occupy the bridge sites. In fact, no guest binding was observed when Au(100) was used as a substrate. This is because if the crown ether moieties occupy the hollow sites, the central Co^{2+} ion of the phthalocyanine located at an on-top site, which is energetically unfavourable. The registry with the surface is therefore critical since the binding depends on the distance between the guest ion and the Au lattice [13]. Some studies also report the capture of anions within macrocycles with electropositive cavities. Typically, the presence of C–H groups can stabilise them via hydrogen bonding [16].

Host–guest chemistry with intrinsically porous building blocks is of course not limited to ion complexation. Hosts with larger cavities can accommodate larger guest molecules using various types of interactions. Calixarenes, which are vase-shaped molecules obtained via hydroxyalkylation of phenols with aldehydes, were one of the first intrinsically porous host molecules to be employed for on-surface guest encapsulation. Self-assembled arrays formed by calix[8]arene on Au(111) surface could be used for capturing C_{60} molecules [17]. Most often the guest

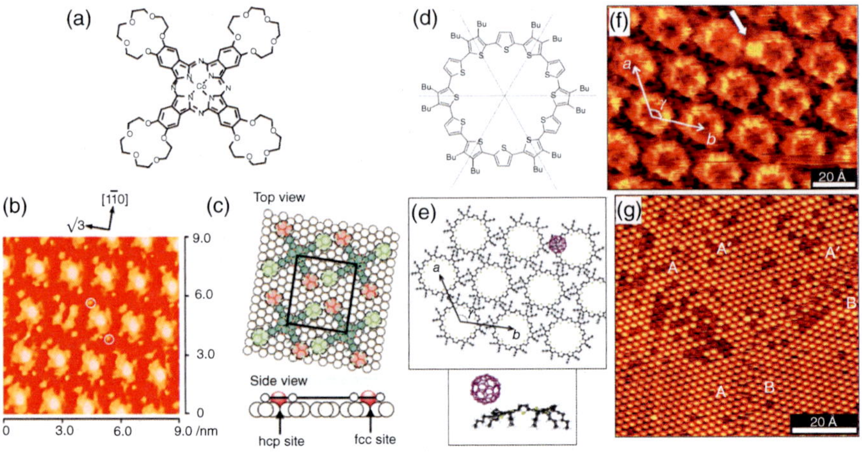

Figure 3.1 Host–guest chemistry using intrinsically porous building blocks. (a) Molecular structure of the cobalt phthalocyanine molecule bearing four crown ether units. (b) STM image showing the complexation between Ca^{2+} ions and two out of four crown ether units (white circles). (c) Molecular models (top and side views) showing the proposed host–guest complexation. The crown ether units with green colour bind with the Ca^{2+} ions whereas the others remain empty. (d) Molecular structure of the cyclothiophene macrocycle. (f) STM image showing the capture of C_{60} (white arrow) on the rim of the macrocycle. (e) Calculated model of a closely packed monolayer of the cyclothiophene macrocycle with a hexagonal arrangement together with the side view of the calculated energy minimum conformation of the complex. (g) A large scale STM image showing the host–guest complexation between cyclothiophene macrocycle and C_{60}. Source: (a, c) Yoshimoto et al. [13]. (b) Reproduced from Yoshimoto et al. [13] with permission from the American Chemical Society. (d, e) Mena-Osteritz et al. [14]. (f) Reproduced from Mena-Osteritz et al. [14] with permission from Wiley-VCH.

capture results as a consequence of size and shape complementarity between the host cavity and the guest moiety [18]. There are exceptions, however, where the guest molecules prefer to occupy the rim of the macrocycle instead of the central cavity in view of specific interactions. For example, C_{60} molecules were found to preferentially occupy the rim of an alkylated cyclothiophene derivative (Figure 3.1d,e) instead of the central cavity due to a combination of π–π and donor–acceptor interactions. Moreover, C_{60} was always found to complex in a 1 : 1 ratio with the macrocycle. This maximum ratio is a consequence of the donor–acceptor interactions that make the opposite end of the rim electron-deficient after binding of the first C_{60} molecule (Figure 3.1f,g). The electron-deficient 1 : 1 complex is thus unable to bind a second C_{60} molecule which is also electron-deficient [14].

In all the examples described above, the macrocycles were deposited from the solution phase. Most such macrocycles, being relatively high molecular weight compounds, are not suitable for deposition using sublimation methods as they tend to decompose before reaching the sublimation temperature. The development of electrospray ionisation has allowed the deposition of high molecular weight compounds and their imaging under ultra-high vacuum (UHV) conditions. Thanks to this method, ion complexation with crown ethers, which was mostly studied at the solution–solid interface, can also be investigated in a UHV environment. In contrast to ambient conditions, where the macrocycles form long-range ordered arrays, deposition using electrospray methods typically provides a relatively disordered arrangement of molecules and/or isolated cycles. Complexation between different cations and dibenzo-24-crown-8 was studied under UHV conditions on the copper surface where a combination of STM and DFT calculations indicate the occupation of the crown ether cavities by Na^+, Cs^+, and protons. The location of the cations inside the cavity was found to depend on their size and their interaction with the host cavity. The size of the central cavity was found to scale with the size of the ions [19]. Another class of high molecular weight multicomponent systems studied under UHV conditions consists of host–guest systems involving macrocycles and typical organic molecular guests. A notable example is the co-assembly of giant cyclic porphyrin nanorings with C_{60} via host–guest interactions on Au(111). The porphyrin nanorings that vary in diameter between 4.7 and 21 nm form stacks up to four layers high upon adsorption on the surface. The adsorption of C_{60} in the cavities of these nanorings was found to depend on the number of layers in the stack [20].

3.2.1.2 Host Networks from Self-Assembly of Building Blocks

In view of the challenges associated with the synthesis of macrocycles with tuneable dimensions, most of the host–guest systems studied on solid surfaces are based on non-porous building blocks. In contrast to the previous class of host networks, the host cavity in this case is not intrinsic to the molecule and is not designed and implemented in the synthetic step but is rather obtained via supramolecular design and assembly. Typically, the intermolecular interactions between the building blocks lead to the formation of a regular array of shallow nanowells on the surface. The size and shape of these nanowells can be precisely tuned by a careful choice of design elements. The tunability of the nanowells dimensions is highlighted by the concept of isoreticular self-assembled networks [9], meaning that molecules with different lengths and assembled through the same supramolecular synthons shall give rise to networks of identical symmetry but different periodicities. In reality, however, it is more challenging to obtain nanoporous networks with large cavities due to the propensity of molecular systems to undergo close-packing. Creating an open porous network with large void spaces at a solid interface is energetically expensive due to the lower adsorption enthalpy per unit area of the resultant porous network. However, despite this limitation, nanowells with a diameter up to 7.5 nm have been obtained via the self-assembly strategy [21].

Besides offering a variety of nanoporous networks with different symmetries and strength in a relatively straightforward fashion, the non-covalent nature of the self-assembled cavities also allows monitoring of the dynamic phenomena occurring during the guest binding process. In some cases, an otherwise densely packed network undergoes structural changes to accommodate the guest molecule. Furthermore, since the outcome of supramolecular assemblies is dependent on the experimental variables such as temperature, electric field, light irradiation, etc., stimuli-responsive host–guest systems can thus be directly observed by scanning probe methods. In the following sections “Hydrogen-Bonded Host Networks” and

"Host Networks Sustained by van der Waals Interactions Between Alkyl Chains", we survey the different types of bicomponent host–guest assemblies. The systems are classified based on the dominant intermolecular interactions involved in the formation of the host network.

Hydrogen-Bonded Host Networks Hydrogen-bonded host networks are the most frequently encountered motifs in multicomponent systems due to the strength and directional nature of hydrogen bonds. One of the archetypal examples of host–guest systems is the one based on hydrogen-bonded host network obtained from self-assembly of benzene-1,3,5-tricarboxylic acid (trimesic acid [TMA], Figure 3.2a). This so-called chicken-wire network is sustained by double hydrogen bonds between the carboxylic acid groups. The shallow cavities formed within the supramolecular network of TMA are routinely used for guest encapsulation. Coronene, a planar polyaromatic molecule, serves as an ideal guest for the TMA host network. A single molecule of coronene is a perfect fit for the hexagonal voids of the TMA network due to size and shape complementarity. Long-range ordered host–guest networks of TMA–COR can be readily obtained either via deposition of premixed solutions of TMA and COR or via a sequential deposition protocol where a solution of COR is added to the pre-formed monolayer of TMA (Figure 3.2b) [22]. C_{60}, on the other hand, can be manipulated using the STM tip within TMA pores due to its spherical nature, where C_{60} guests were found to 'jump' from one TMA pore to the other [25]. It must be noted here that neither COR nor C_{60} forms a stable self-assembled network on their own at the solution/HOPG interface at room temperature. This illustrates the importance of the host network in the immobilisation of the guest species. Larger aromatic tricarboxylic acids have also been explored for host–guest chemistry.

1,3,5-Tris(4-carboxyphenyl)benzene (BTB), a larger analogue of TMA forms a porous network with a larger periodicity (1.7 nm for TMA versus 3.2 nm for BTB). Consequently, the BTB network accommodates three COR molecules per cavity, instead of one as observed for TMA. Furthermore, while COR is immobilised in the TMA pore and is typically imaged as a circular feature, the three COR molecules in a BTB pore remain mobile and the contrast of the guests within the cavities is ill-defined [26]. The higher stability of COR in TMA pores is explained by weak ($\approx$2.2 kJ/mol) but numerous hydrogen bonds formed between immobilised COR and the host network [27]. TMA networks have also been used for immobilisation of exotic guests such as the heterocyclic circulenes namely the 'sulflower' and the 'selenosulflower' (Figure 3.2a,c,d) [23]. Similarly, the immobilisation of C_{60} inside the host cavities formed by terthienobenzenetricarboxylic acid has been investigated. One, two, or three C_{60} molecules per cavity were found to be immobilised thanks to the donor–acceptor interactions between C_{60} (electron acceptor) and thiophene units (electron donor) of the host molecule [28].

The guest molecules often play a more active role than simply occupying the cavities of the host network. At the solution/solid interface, a guest can induce structural changes in the host network upon adsorption. It is thus the guest that templates the host, and not the other way around as discussed in the previous

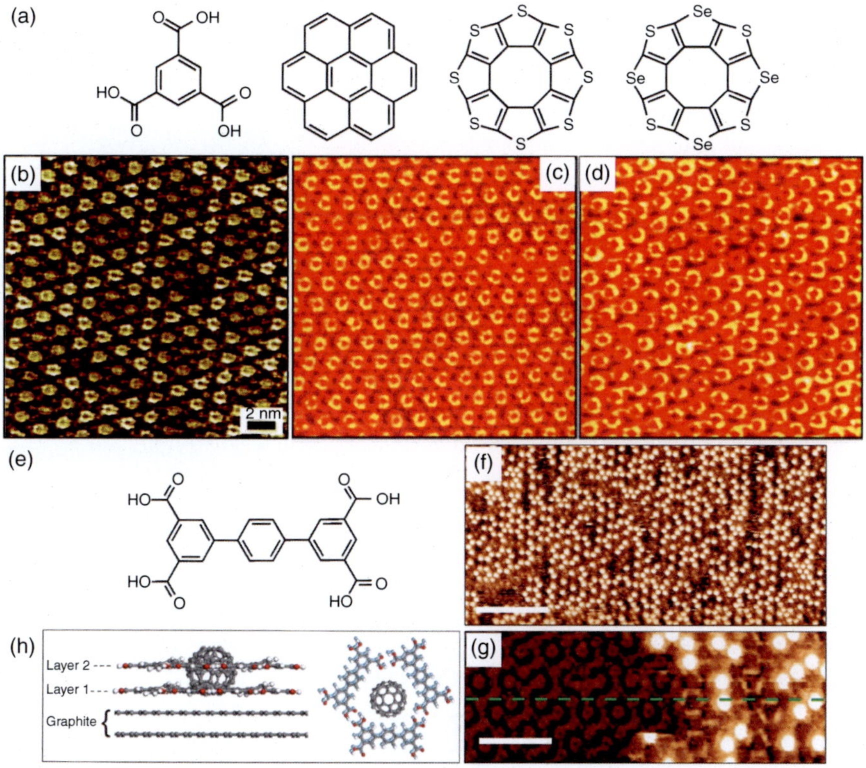

Figure 3.2 Two-component self-assembly using hydrogen-bonded host-networks. (a) Molecular structures of trimesic acid (the host) and coronene, sulflower and selenoflower (the guests). (b–d) STM images showing the two-component host–guest networks obtained using (b) TMA/coronene, (c) TMA/sulflower and (d) TMA/selenoflower. (e) Molecular structure of TPTC. (f) STM image showing TPTC-C_{60} host–guest architecture. (g) STM image of TPTC network immediately after C_{60} deposition. The initial layer of TPTC network is visible with an altered contrast and the TPTC molecules in the second layer appear as bright, rod-like features surrounding the C_{60} molecules which appear as bright blobs. (h) Molecular models showing the side and the top view of QPTC/C_{60} bilayer network. Source: (a, e, h) Refs. [22–24]. (b) Reproduced from Griessl et al. [22] with permission from the American Chemical Society. (c) Reproduced from Ivasenko et al. [23] with permission from the Royal Society of Chemistry. (f) Reproduced from Blunt et al. [24] with permission from Nature Publishing Group.

examples. For instance, the self-assembly of a quaterphenyl tertracarboxylic acid derivative (QPTC) yields a hydrogen-bonded, close-packed network where the molecules adsorb parallel to each other. The addition of COR to such pre-formed monolayer leads to a structural transition where the host network changes from the compact network to a kagome network forming hexa-isophthalate nodes reminiscent of the TMA hexagons. The COR molecules occupy the hexagonal pores of the kagome network [29]. The driving force for the observed guest-induced structural transition comes from the gain in energy caused by the host–guest interactions between the hexa-isophthalate node and COR. Similar structural transitions, where an otherwise compact network transforms into a porous network to accommodate COR, are reported for alkoxy isophthalic acids [30]. Such guest-induced structural transitions are not limited to self-assembly in 2D alone and can initiate the growth of the assembled network in the third dimension orthogonal to the substrate. A shorter analogue of QPTC namely, *p*-terphenyl-3,3″,5,5″-tetracarboxylic acid (TPTC) (Figure 3.2e), forms a monolayer with hexagonal pores. The addition of C_{60} to the TPTC host network not only led to the occupation of the host cavities of C_{60}, but the adsorbed C_{60} molecules promoted the growth of a second TPTC layer (Figure 3.2f–h). The π-stacked bilayer is supported by host–guest interactions in a highly cooperative process, since neither the TPTC bilayer nor the C_{60} could be observed independently. The dynamic nature of the interface allows conversion of the TPTC bilayer back to the monolayer upon the addition of COR. Due to better size complementarity with the voids and higher adsorption energy on the HOPG surface, COR replaces C_{60} as a guest. The bilayer, which was templated by C_{60}, then transforms back to a monolayer wherein the cavities of the host network are occupied by COR [24].

Besides the molecular guests such as COR and C_{60}, guest-induced transitions can also be triggered by ions. An elegant example is the transition of the self-assembled network of an alkyl guanine derivative (Figure 3.3a) from a ribbon-like structure (Figure 3.3b) into another one where the guanine units assemble as guanine quartets (G4) (Figure 3.3c) in response to the addition of K^+ ions. The ribbon-like structure is stabilised by a combination of hydrogen bonds and van der Waals interactions between interdigitated alkyl chains whereas the G4 quartet structure is sustained by hydrogen bonds since the alkyl chains are back-folded into the supernatant solution. The opposite transition was achieved by the addition of a cryptand that can complex K^+ and thus remove it from the quartet. G4 quartets were formed again after the addition of trifluoromethanesulfonic acid, which releases the K^+ ions from the cryptand. The cycle could be continued by the addition of cryptand thus showing the dynamics and reversibility of the transitions (Figure 3.3d) [31]. The design of this switchable bicomponent system is inspired by molecular biology where biopolymers rich in guanine sequences form G4 quartet structure in response to the presence of cations, specifically K^+ ions. Guanine is also involved in Watson–Crick base pairing with cytosine and thus forms an essential component of the double helix structure of DNA. The complementary hydrogen bonding between guanine and cytosine has been employed for the construction of stable macrocycles in solution which can also be transferred to the solid surface. The self-assembled macrocycles can further be used for immobilisation of COR in their cavities [32].

Host networks formed by aromatic tricarboxylic acids also exhibit structural transitions in response to external stimuli such as temperature and electric field. Most often, the networks change from an open porous to a densely packed structure and such structural changes have been employed for realising controlled guest release and capture of guest molecules. Self-assembled networks of BTB undergo temperature-dependent structural transitions both under UHV conditions [33] and at the solution–solid interface [34]. For example, BTB forms a honeycomb network on Ag(111) surface which can be used for capturing a macrocyclic guest under

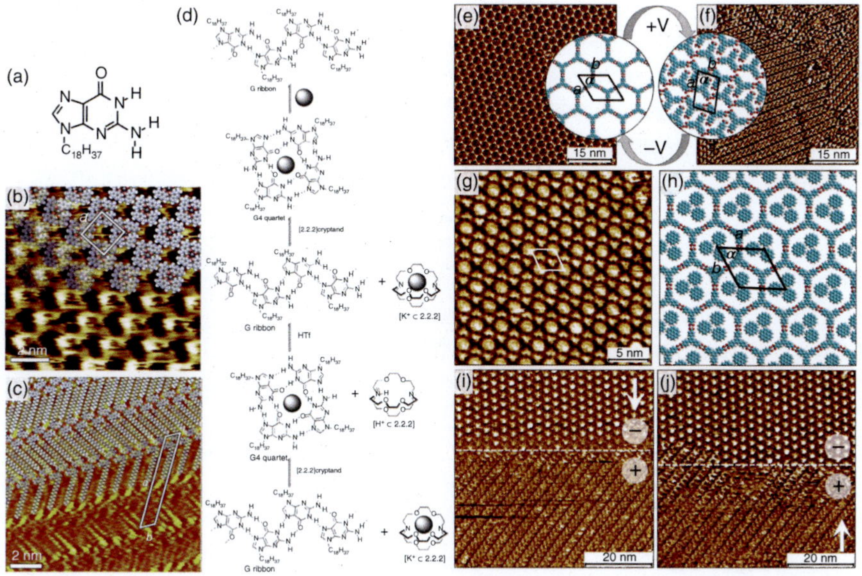

Figure 3.3 Stimulus responsive two-component host–guest systems. (a) Molecular structure of the alkyl guanine derivative. (b) STM image of the ribbon-like self-assembled structure. (c) STM image of the G4 quartet structure obtained after the addition of potassium ions. (d) Schematic showing the reversible transitions in the self-assembled networks via successive addition of reagents. (e, f) Electric field induced phase transitions in the self-assembled networks of BTB. (g) STM image showing a two-component host–guest architecture obtained using BTB and coronene. (h) Molecular model depicting the BTB/coronene self-assembly. (i, j) Bias-induced reversible transitions in the BTB/coronene self-assembled network. Source: (a, d) Lee et al. [26]. (b, c) Reproduced from Ciesielski et al. [31] with permission from the American Chemical Society. (e, f, g, h, i) Reproduced from Lee et al. [26] with permission from Wiley-VCH.

UHV conditions. At lower temperature, the host–guest system remains stable. However, upon increasing the temperature to 320 K, BTB porous network evolves into a compact network thus releasing the guests which could be observed on the Ag(111) surface. The driving force behind the structural transition was attributed to the deprotonation of the acid groups which disrupts the regular hydrogen bonding between them thus leading to the collapse of the network into a densely packed structure. BTB networks also show similar transitions in response to change in the polarity of the substrate bias. The porous to dense transition can be readily achieved by simply changing the sign of the applied substrate bias (Figure 3.3e,f). In contrast to the temperature-induced transitions, the electric field induced structural changes occur rapidly and can be monitored with STM in real-time. Polyaromatic molecules such as coronene (Figure 3.3g–j) could be captured and released in a controlled fashion at the solution/solid interface using such electric field mediated phase transitions in BTB networks [26].

Host Networks Sustained by van der Waals Interactions Between Alkyl Chains The propensity of alkyl chains, or alkanes in general, to adsorb favourably on the graphite surface is well-documented. Although normal alkanes typically form compact lamellar structures upon adsorption on the HOPG surface, it is possible to fabricate low-density porous networks based on the assembly of carefully designed alkylated aromatic building blocks. A common strategy consists in attaching alkyl chains to a rigid conjugated core. Typically, six alkyl chains are attached to a triangular core and separated by an appropriate distance (~1 nm) allowing the insertion of one alkyl chain of an adjacent molecule. Such arrangement of alkyl chains is referred to as *interdigitation* of the chains. Given the favourable interactions between the alkyl chains and the HOPG surface, regular honeycomb host networks with a pore size governed by the length of the alkyl chains are obtained. Self-assembly governed by flexible alky chains is highly versatile and such systems are often used as model systems to study fundamental concepts of self-assembly at solution–solid interface such as concentration dependence, phase transitions, and chirality [35].

One of the early examples of host–guest chemistry in host networks held together via van der Waals interactions between alkyl chains was reported for a threefold symmetric molecule, 1,3,5-tris[(*E*)-2-(3,5-didecyloxyphenyl)-ethenyl]benzene (TSB35). TSB35 consists of a triangular tris-ethenylbenzene core substituted with six decyloxy chains. It forms a honeycomb porous network where the molecules are held together by van der Waals forces between interdigitated decyloxy chains. The porous network could be used for immobilisation of several different polycyclic aromatic guest molecules. Hopping of the guest molecules between neighbouring cavities was observed, indicating 2D diffusion confined within the host matrix. The activation energy for diffusion was found to be intermediate between the values for diffusion on atomically flat surfaces and for the desorption energy. This study illustrated the size and shape dependence of guest adsorption as well as their surface diffusion inside a self-assembled host network [36].

An important consequence of the flexible nature of the host networks built using alkyl chains can be seen in guest-induced structural transitions. Alkoxy substituted dehydrobenzo[12]annulenes (DBA, Figure 3.4a) are planar triangular phenylene–ethynylene macrocycles that have been intensively studied in the context of host–guest chemistry. These building blocks consist of a rigid annulene core substituted with alkoxy or alkyl chains. DBA derivatives show concentration-dependent self-assembly behaviour. Porous honeycomb networks which are stabilised by van der Waals interactions between interdigitated pairs of alkyl chains are formed at lower concentrations. On the other hand, higher concentrations favour the formation of densely packed networks, where one or more chains are back-folded in the supernatant solution phase (Figure 3.4c) [38]. However, such dense networks are readily transformed into the porous phase upon the addition of guests like COR, which stabilise the honeycomb network by occupying the cavities (Figure 3.4b,d). Such structural transition is also guest-selective since it only occurs for planar guest molecules with large π-conjugated cores, indicating that such guests stabilise the open network thanks to their substantial adsorption energies [37]. Relatively simpler building blocks such as isophthalic acids (ISA) with alkoxy groups also exhibit guest-induced phase transitions in their self-assembled monolayers. Isophthalic acid derivatives substituted with long alkoxy chains (decyloxy, tetradecyloxy, and octadecyloxy) show concentration-dependent structural phase transitions (Figure 3.4e, f, h, i). The compact lamellar networks obtained at higher concentrations however respond to the presence of COR in the solution and transition into porous networks where the COR molecules occupy the hexagonal cavities formed by the ISA headgroups and also the triangular ones formed in between the alkoxy chains (Figure 3.4g, j) [30].

Host networks are not only selective to guests based on their size and shape, but the cavities can also be selective to the number of guest molecules. A planar polyaromatic molecule, so-called nanographene (Figure 3.5a) was used as guest in the porous networks formed by DBAs with different chain lengths (from 10 to 20 carbon atoms). The chain length defines the size of the hexagonal cavity and thus determines the number of nanographene guests that can be accommodated. Depending on the chain length, between one and six guest molecules could be immobilised within DBA cavities (Figure 3.5b,c). In some cases, where the cluster of nanographene molecules does not fit perfectly within the cavity, the DBA honeycomb underwent distortion to better accommodate the guests illustrating the flexibility of the host networks built from alkyl chains [39].

Specific guest binding through non-covalent interactions can also induce large structural changes in the host network, provided it is designed to offer enough flexibility. This is, for instance, the case of a 5-(benzyloxy)isophthalic acid derivative (BIC, Figure 3.5d), that offers a lamellar host template sustained by van der Waals interactions between its alkyl chains. The BIC network proved its versatility by accommodating various guest molecules containing pyridylethylnyl moieties that form hydrogen bonds with the isophthalic acid groups. The guests of varying sizes and shapes are forced to stack into linear patterns and act as spacers between the BIC ribbons. Indeed, if the intermolecular distances within the BIC ribbon remain constant for all guests, unit cell parameters along the transverse direction are altered from 3.4 to 7.3 nm (Figure 3.5e–i) [40].

Most of the host–guest systems are based on the size and shape complementarity between the two components. In order to increase the selectivity between the host network and the guest, one strategy consists in adding specific functional groups to the host molecules to enhance the guest binding. A DBA derivative substituted with perfluoroalkyl chains at the end of the three alternating alkoxy chains (Figure 3.6a) was designed and formed a honeycomb network where the rims of all the nanowells are fluorinated (Figure 3.6b). For comparison, a structurally similar DBA derivative with alkyl chains lining the rim was also synthesised (Figure 3.6c). A fluorinated guest molecule, namely an octadecafluoro derivative of hexakis(phenylethynyl)benzene (HPEB) (Figure 3.6a), could be stabilised via fluorophilic interactions with the rims of the host network. A comparison between the combination of fluorophilic and hydrophobic host cavities with fluorophilic and hydrophobic guest molecules revealed that guest binding is influenced by the specific interactions mentioned above (Figure 3.6d–g). When the fluorinated guest was added to a host network formed by a non-fluorinated DBA with the same pore

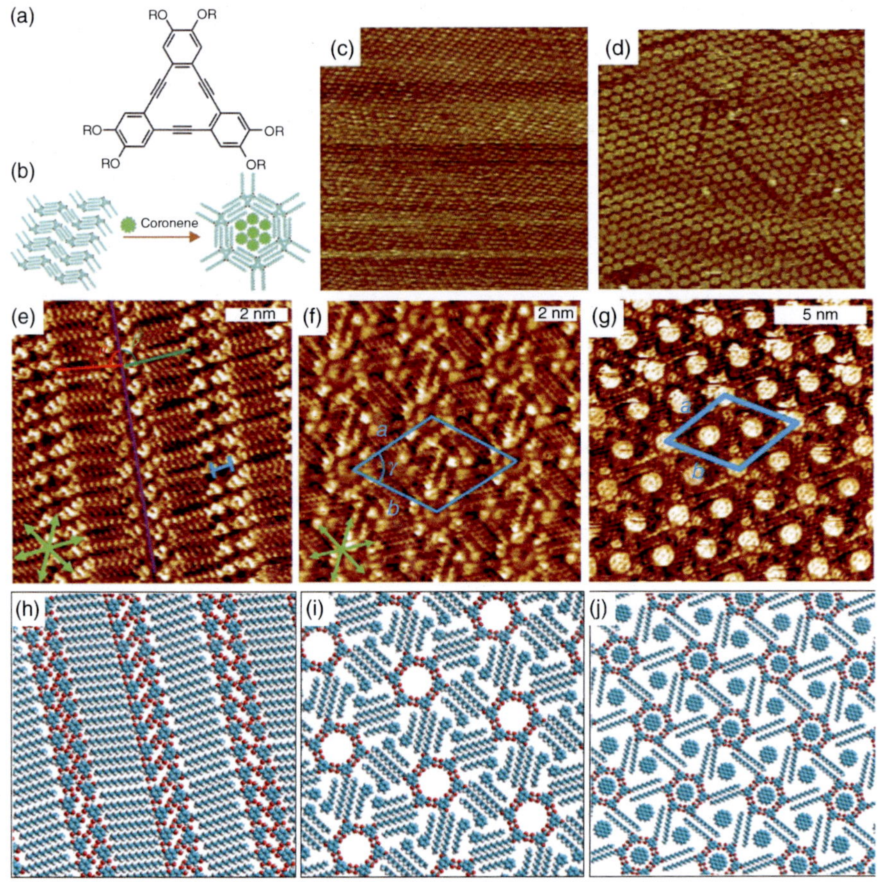

Figure 3.4 Guest-induced phase transitions in self-assembled monolayers. (a) Molecular structure of the DBA derivatives (R = $-C_{14}H_{29}$). (b) A schematic showing the structural transition between the dense and the porous phase of DBA in response to the addition of coronene. (c) STM image of the densely packed phase of DBA-OC14. (d) STM image showing the honeycomb porous network formed after the addition of coronene. (e, f) STM images showing the compact (e), and the open porous (f) network formed by tetradecyloxy isophthalic acid at the 1-phenyloctane/HOPG interface at high and low concentrations, respectively. (g) A two-component host–guest network obtained upon addition of coronene to the preformed compact network of tetradecyloxy isophthalic acid. (h–j) Molecular models corresponding to the self-assembled networks displayed in (e–g). Source: (a, b) Furukawa et al. [37]. (c) Reproduced from Furukawa et al. [37] with permission from Wiley-VCH. (e) Reproduced from Park et al. [30] with permission from the American Chemical Society. (h–j) Park et al. [30].

dimensions, the number of host cavities occupied was reduced drastically (from 56% to 16%, Figure 3.6d,e). Molecular mechanics simulations indicated stronger host–guest interactions between the fluorinated host and fluorinated guest in comparison to those between non-fluorinated host and fluorinated guest (−23.1 versus −12.9 kcal/mol) [41]. A similar strategy was used to show the selectivity in the immobilisation of a phenylene–ethynylene macrocycle with triethylene glycol

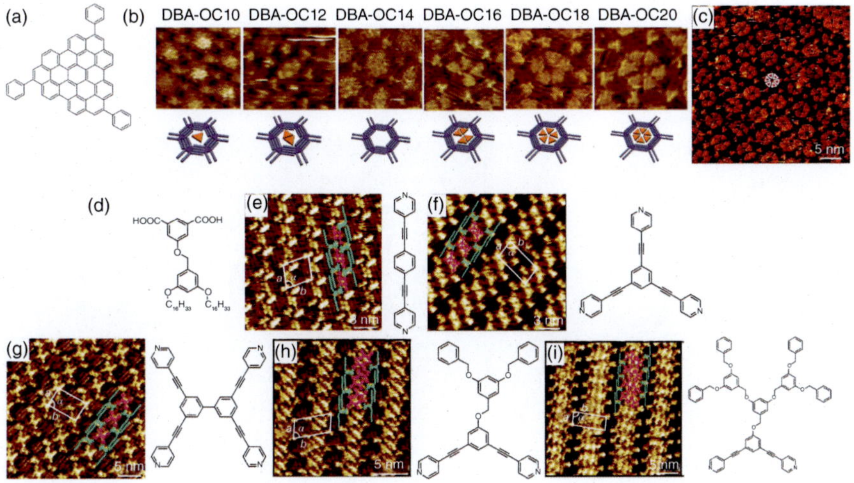

Figure 3.5 Two-component self-assembled systems exhibiting a high degree of flexibility. (a) Molecular structure of nanographene which is used as a guest molecule for supramolecular networks formed by DBA derivatives. (b) STM images showing single hexagonal voids of DBA derivatives occupied by nanographene clusters. (c) STM image showing clusters of six nanographene molecules occupying the hexagonal cavities of DBA-OC20. (d) Molecular structure of the BIC derivative. (e–i) STM images showing the hydrogen-bonded modular co-assembly of the BIC derivative with a number of guests containing pyridylethylnyl groups. The molecular structures of the guests are provided to the right of each STM image. Source: (a, c, d, f–i) Refs. [39, 40]. (b, c) Reproduced from Lei et al. [39] with permission from the American Chemical Society. (e, f, g, h, i) Reproduced from Zhang et al. [40] with permission from the American Chemical Society.

side chains within the pores formed by a DBA with three tetraethylene glycol groups based this time on electrostatic interactions between the host and the guest [42].

The versatility of van der Waals-stabilised networks is also illustrated via stimuli-responsive host–guest systems. Using the same strategy as for the fluorinated DBA, a DBA derivative with three dicarboxyazobenzene units was synthesised. Thus, the six isophthalic acid groups present inside each pore assemble via hydrogen bonding into a hexagon reminiscent of the TMA network that can bind a single COR guest molecule. A light-responsive system was thus built based on the ability of the azobenzene groups present in the molecule to undergo photoisomerisation. When the solution–solid interface was irradiated with 320 nm light, some azobenzene units desorbed following their trans-to-cis isomerisation, allowing the adsorption of additional COR molecules. It resulted in an increase of the number of cavities containing two or more COR molecules. This transition was shown to be reversible when the desorption of the additional COR molecules was induced by the cis-to-trans isomerisation triggered at >400 nm irradiation [43].

3.2.1.3 Two-Component Systems: Host–Guest Architectures Based on Surface-Confined Two-Dimensional Covalent Organic Frameworks (2D-sCOFs)

Covalent organic frameworks (COFs) constitutes a special class of crystalline, all-organic porous materials made up of light elements. Most often, the principles

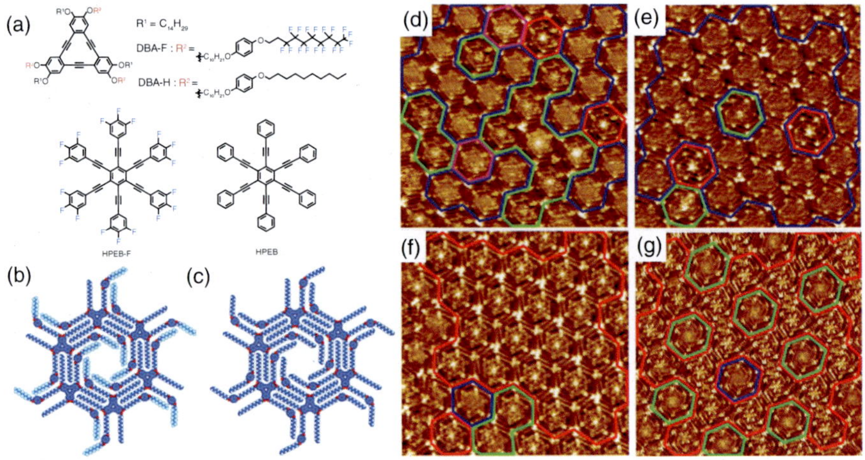

Figure 3.6 Two-component self-assembly using fluorophilic interactions between the host network and the guest molecules. (a) Molecular structures of the partially fluorinated DBA derivative (DBA-F) and a corresponding alkylated DBA derivative (DBA-H) which was used as a reference molecule. The scheme also shows the molecular structures of the fluorinated (HPEB-F) and non-fluorinated (HPEB) guest molecules. (b, c) Molecular models depicting single hexagons which constitute the self-assembled networks of DBA-F and DBA-H. (d) STM image of the two-component monolayer formed by mixing DBA-F and HPEB-F. (e) STM image of the monolayer formed by DBA-H and HPEB-F. (f) STM image of the monolayer formed by mixing DBA-F and HPEB. (g) STM image of the monolayer formed by mixing DBA-H and HPEB. The red, green, blue, and pink hexagons highlight the cavities containing an immobilised guest, fuzzy pores containing a mobile guest, dark pores which do not contain any guest, and undefined pores categorised between the fuzzy and dark pores, respectively. Source: (a–c) Tahara et al. [41]. (d, e, f, g) Reproduced from Tahara et al. [41] with permission from the American Chemical Society.

of dynamic covalent chemistry are employed wherein reversible reactions such as boronic acid self-condensation and Schiff base formation are used for obtaining COFs. Single or few-layer COFs can also be obtained via on-surface synthesis where small molecular building blocks can be linked covalently to obtain defect-free, long-range ordered sheets of 2D-sCOFs. Surface-synthesised, 2D-sCOFs represent a more robust alternative to porous supramolecular networks. The variety of shapes and the development of isoreticular COFs allows precise tuning of the dimensions of the pores in a way that parallels the nanoporous networks discussed earlier. COFs can therefore be classified as single-component hosts able to give rise to host–guest complexes after the addition of a second molecule.

Two-component host–guest systems, where a monolayer of surface-confined 2D-COF serves as a host network for the immobilisation of guest species, have been studied recently. C_{60} is a typical molecule immobilised in single-layer COFs [44]. The 2D-sCOF obtained from benzene-1,4-diboronic acid presents uniform cavities (Figure 3.7a,b) that can trap single C_{60} molecules from 1-phenyloctane solutions with high surface coverage (Figure 3.7c,d) [44b]. The second layer of C_{60} adsorbed on top of the boroxine rings of the COF was also observed once the first layer was (almost) complete (Figure 3.7e,f) [45]. In addition to fullerenes,

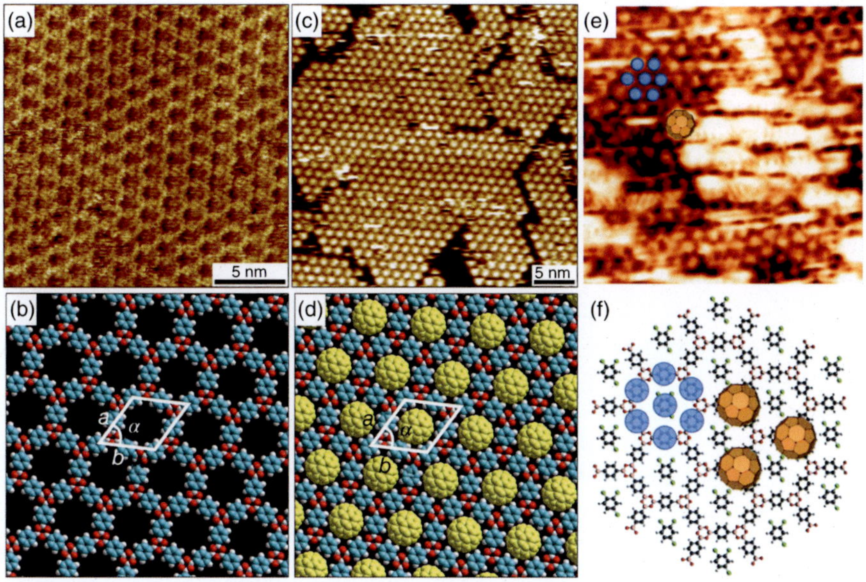

Figure 3.7 Host–guest architectures based on surface-confined 2D-COFs. (a) STM image showing the 2D-sCOF obtained using benzene-1,4-diboronic acid COF imaged at the 1-phenyloctane–HOPG interface. (b) Molecular model for the COF. (c) STM image of the COF–C_{60} host–guest architecture. (d) Molecular model for the COF–C_{60} host–guest system. (e) Formation of the second layer of C_{60} on top of the COF–C_{60} host–guest system. (f) Molecular model depicting the formation of the second layer of C_{60}. Source: (a, c) Reproduced from Plas et al. [44b] with permission from the Royal Society of Chemistry. (b, d) Plas et al. [44b]. (e) Reproduced from Cui et al. [45b] with permission from the Royal Society of Chemistry. (f) Cui et al. [45b].

larger, flat molecular guests can also be adsorbed in the pores of COFs with larger periodicities [46]. Immobilisation of copper phthalocyanine (CuPc) inside the cavities of a boroxine-based COF containing azobenzene groups has been reported [47]. Irradiation of this 2D-COF by light at 365 nm resulted in trans-to-cis isomerisation within the backbone. This process triggered the breaking of some B—O bonds which further caused the release of the phthalocyanine guests. This system closely resembles the light-induced guest release described earlier for the DBA derivatives. This example shows that some of the concepts developed previously for self-assembled nanoporous networks could be potentially implemented in the emerging field of surface-adsorbed 2D-COFs.

3.2.2 Two-Component Systems: Non-Host–Guest Architectures

Some multicomponent networks cannot be necessarily characterised as host–guest networks because none of the two assembling components form an open porous network on their own under any experimental condition. In such networks, the two components co-crystallise on the surface of the substrate due to specific intermolecular interactions. The on-surface co-crystallisation may lead to a periodic arrangement of the two components defined by a precise unit cell or it may result in random mixing. Alternatively, they may co-crystallise into a sandwich-type arrangement wherein one component adsorbs on top of the monolayer formed by the second component.

Non-host–guest bicomponent networks on surfaces are widespread both in UHV and under ambient conditions. As for the single component host systems described above, they can be classified based on the dominant intermolecular interactions responsible for their stabilisation, such as hydrogen bonding [48], halogen bonding [49], or metal–ligand coordination [48c, 50]. Since such networks are already described in numerous reviews [48c, 51], and those porous bicomponent architectures can be used to produce three- or four-component host–guest systems (vide infra), only a few representative examples of two-component systems are mentioned here.

An interesting class of non-host–guest type two-component architectures is the one where the self-assembled networks contain immobilised solvent molecules. This class is a 2D equivalent of a solvate obtained in bulk crystallisation experiments. Immobilisation of solvent molecules has been reported in a number of cases where the solvent molecules co-adsorb on the surface either due to specific supramolecular interactions with the solute molecules or simply occupy the voids left in the 2D networks [52]. Alternatively, non-host–guest type two-component networks have been obtained by mixing chain-like molecules that either co-crystallise or undergo random mixing. A few interesting examples can be found in the co-assembly experiments carried out using dialkyl(iso)phthalates (Figure 3.8a) at the solution–solid interface. Depending on the substitution pattern (ortho/meta) and the chain length, the mixing of the dialkyl(iso)phthalates in appropriate ratios yielded either a two-component network with extraordinarily large periodicity where the molecular columns of the two components alternate (18-ortho + 18-meta, Figure 3.8b,c) or another one in which the two molecules are periodically arranged within the columns (18-ortho and 20-meta, Figure 3.8d,e). Detailed analysis of the packing motifs revealed that no single intermolecular interaction is responsible for co-crystallisation of the two components [53]. Similarly, the mixing of diheptadecyl isophthalate with 4-octadecyloxybenzamide (17-meta + 18-amide, Figure 3.8a) yielded a unique type of co-assembled network that displays random mixing along one axis and periodic ordering along the other giving rise to a so-called one-dimensional (1D) co-crystal (Figure 3.8f,g) [54]. Two-component networks formed by simple alkylamides also exhibit co-crystallisation via either stoichiometric or random mixing [55].

The second class of the non-host–guest type architectures mentioned above consists of sandwich-type arrangement of molecules. Heteromolecular bilayers are formed when one component forms a continuous monolayer on the surface and the other component adsorbs on top of such monolayer forming either a continuous or a discontinuous layer. A popular class consists of various molecules adsorbed on top of buffer layers formed by long-chain alkanes such as *n*-pentacontane [56]. A two-component system can be obtained when C_{60} adsorbs on top of a monolayer formed by another building block [57]. For example, corannulene which can be

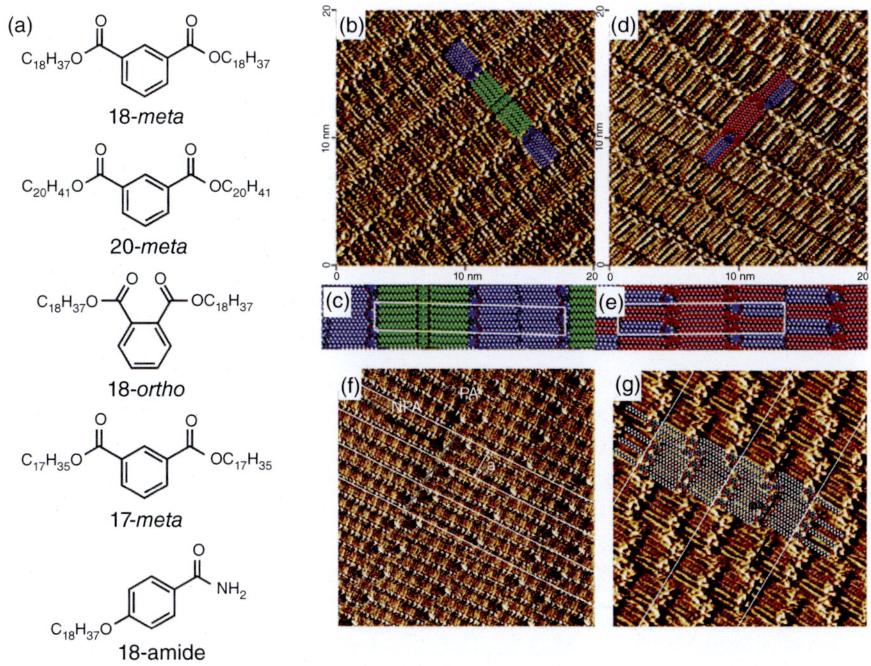

Figure 3.8 Two-component non-host–guest type architectures. (a) Molecular structures of the dialkyl(iso)phthalates and the alkyl amide. (b) Two-component network obtained at the 1-phenyloctane–HOPG interface from a solution mixture of 18-ortho and 18-meta. (c) Molecular model for the 18-ortho (blue) and 18-meta (green) two-component network. (d) STM image of the monolayer obtained from a solution mixture containing 18-ortho and 20-meta. (e) Molecular model for the 18-ortho (blue) and 20-meta (red) co-assembled network. (f) 1D-co-crystal obtained via insertion of dimers of 18-amide into the columns of 17-*meta*-diester at the 1-phenyloctane–HOPG interface ($50 \times 50\,nm^2$). The arrows indicate the periodic axis (PA) and the non-periodic axis (NPA). (g) STM image showing saturation effect on the column length in the case of the 1D co-crystal. Source: (a, c, e) Plass et al. [53]. (b, c, d, e) Reproduced from Plass et al. [53] with permission from the American Chemical Society. (f, g) Reproduced from Ahn et al. [54] with permission from the American Chemical Society.

considered as a bowl-shaped fragment of C_{60}, forms a dense layer on the surface of Cu(110) and C_{60} molecules can be immobilised on top of such layer. The curved structure of corannulene is crucial for the maximisation of the π–π interactions with the C_{60} molecules [57]. A similar type of two-component assembly can be obtained upon mixing perchlorinated hexa-*peri*-hexabenzocoronene (PCHBC) with C_{60} on Au(111) surface. The PCHBC layers are stabilised by halogen–halogen interactions between Cl atoms (believed to be *type II* halogen bonds based on calculations). In contrast to the parent HBC, which is a planar molecule, PCHBC is concave due to the steric hindrance of the peripheral Cl atoms. This curvature allows binding on single C_{60} molecules via π–π stacking. In both cases, the complementarity between the concave aromatic host and the convex fullerene surface turned out to be pivotal to maximise the π–π interactions [58].

A few other systems where the two components undergo random mixing deserve a special mention. These consist of a mixture of an achiral and a structurally similar chiral building block. Such mixtures are employed for induction of a particular handedness in the supramolecular network by merging a small number of chiral molecules in the network predominantly formed by achiral molecules. Both porous [59] and non-porous [60] systems have been reported.

3.3 Three-Component Systems

The difficulty in the fabrication of multicomponent systems increases considerably with the incorporation of each new element. As a result, crystalline networks that consist of more than two assembling components are rarely reported. This is because the fabrication of such heteromeric structures requires efficient recognition between different components and the precise knowledge of intermolecular and molecule–substrate interactions. Such multicomponent networks are extremely rare under UHV and observed in most cases at the liquid–solid interface [48c]. The solution stoichiometry needed for the successful formation of three- or four-component systems is often arrived at based on empirical considerations. Thus, the ratio of molecules in solution that yields a crystalline three- or four-component network is often different from what is anticipated from the structure of the targeted network. This disparity, in part, is a result of dissimilar adsorption energies of the assembling components, and thus, arriving at a correct stoichiometric molar ratio in solution is often challenging. Sub-optimal solution ratios of the components either result in preferential adsorption of one or more components or their phase separation. In contrast to the earlier section where two-component systems were initially classified based on the type of interactions involved, here we avoid such classification because most systems are stabilised by a combination of different interactions. Instead, here they are classified on the basis of the number of components forming the host network and/or the number (type) of guests occupying the host cavities.

3.3.1 Three-Component Systems: Two-Component Host Network + Guest

In this class, we discuss three-component self-assembled systems where the host network is obtained via self-assembly of two different components and the host cavities are occupied by a third component. One of the early examples of a three-component system involved co-assembly of perylene-3,4,9,10-tetracarboxylic di-imide (PTCDI) with melamine by hydrogen bonding to form a hexagonal host network (Figure 3.9a), the cavities of which were used for immobilisation of heptameric clusters of C_{60} (Figure 3.9b,c). This fairly complex system was formed on a Ag/Si(111) surface under UHV conditions. The surface coverage of C_{60} guests was tuned by increasing its dosage which led to the formation of a C_{60} terminated bilayer which was supported by the underlying PTCDI–melamine host network [61]. The PTCDI–melamine bimolecular host networks can also be formed on

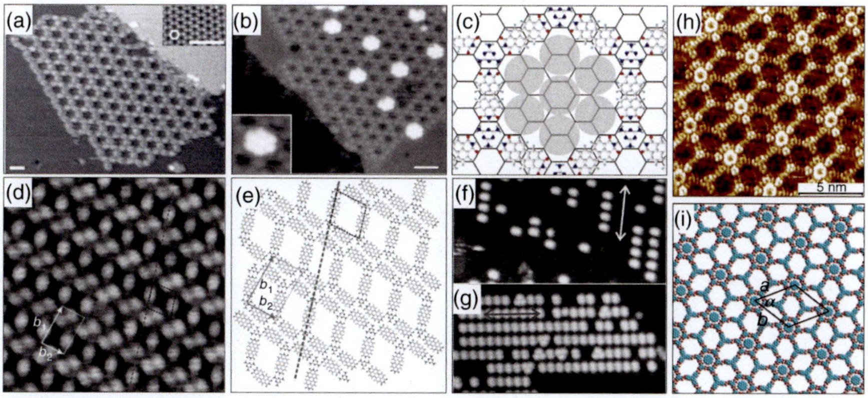

Figure 3.9 Three-component architectures made from two-component host and a guest. (a) STM image of PTCDI–melamine host network on a Ag/Si(111) surface. The inset shows STM image of the substrate lattice. (b) STM image showing the entrapment of heptameric C_{60} clusters in the PTCDI–melamine network. (c) Schematic of the PTCDI–melamine/C_{60} host–guest system. (d) PTCDI–melamine host network on Au(111) surface after annealing at 90 °C. (e) Molecular model showing the parallelogram arrangement of PTCDI–melamine. (f, g) STM images illustrating the correlation between the adsorption of C_{60} dimers at (f) 0.1 monolayer (ML) coverage and at (g) 0.2 ML coverage, along the directions indicated by arrows. (h) STM image of the BTB–TMA–COR host–guest system. (i) Molecular model for the BTB–TMA–COR host–guest system. Source: (a, b) Reproduced from Theobald et al. [61] with permission from Nature Publishing Group. (c) Staniec et al. [62]. (d) Reproduced from Staniec et al. [62] with permission from Nature Publishing Group, Wiley-VCH and the American Chemical Society. (e) Theobald et al. [61]. (f, g) Reproduced from Staniec et al. [62] with permission from Wiley-VCH. (h) Reproduced from Velpula et al. [63] with permission from the American Chemical Society. (i) Velpula et al. [63]

Au(111). However, these hexagonal networks are transformed into a parallelogram arrangement (Figure 3.9d,e) upon annealing the Au(111) substrate at 90 °C. The parallelogram-shaped cavities are smaller than those obtained in the hexagonal network and thus host only two molecules of C_{60} per pore. An interesting aspect of this system is that the filling of the cavities by C_{60} molecules was found to be non-random (Figure 3.9f,g) indicating that the self-assembled network directs a hierarchy of organisation [62]. The PTCDI–melamine host network formed at the Au(111)–vacuum interface has also been used for immobilisation of C_{70} guests. The PTCDI–melamine/C_{70} host–guest system behaves similar to the one involving C_{60} [64].

A few examples of ternary self-assembled systems formed at the solution–solid interface have also been reported. An interesting example consists of a hydrogen-bonded bicomponent host network obtained by mixing TMA and BTB at the HOPG–heptanoic acid interface which can be used for immobilisation of coronene molecules. When mixed in appropriate stoichiometry in solution, the two molecules form mixed self-assembled networks. Six different open porous phases have been identified [65]. One of such bimolecular host networks offers optimal cavities for immobilisation of coronene. The addition of coronene to such a network led to the formation of a flawless three-component network wherein coronene

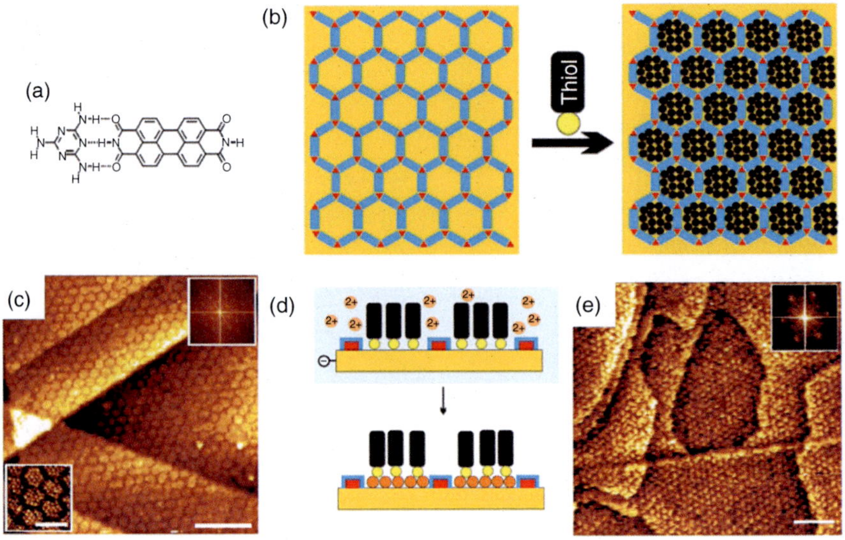

Figure 3.10 Three-component architecture from a two-component host and a chemisorbed guest. (a) The hydrogen bonding motif of melamine and PTCDI. (b) Schematic illustrating the use of PTCDI–melamine network for templated chemisorption of thiols on Au surface. (c) STM image showing adamantane thiol chemisorbed inside the cavities of PTCDI–melamine network. (d) Schematic illustration of electrochemical deposition of Cu in the pores of the network. (e) STM image of the surface after the UDP of Cu. Source: (a, b, d) Madueno et al. [67]. (c, e) Reproduced from Madueno et al. [67] with permission from the Nature Publishing Group.

molecules selectively occupy the hexagonal cavities leaving the other open spaces within the network empty (Figure 3.9h,i) [63]. Similar hydrogen bond-based bimolecular host-systems have been used for immobilisation of coronene molecules [66].

Another ingenious way to construct multicomponent systems involves using a combination of physisorbed host network and chemisorbed guests. The PTCDI–melamine network described earlier serves as a template for chemisorption of alkythiols on Au(111) surface (Figure 3.10a). The physisorbed bicomponent host network survives the chemisorption process of thiol molecules. Templated chemisorption of three different thiol derivatives namely adamantane thiol (Figure 3.10b), dodecane thiol, and ω-(4′-methylbiphenyl-4-yl)propane thiol was successfully achieved. The three-component system was further used for electrochemical deposition of Cu in the underpotential deposition (UDP) region. This caused insertion of a Cu monolayer between the Au substrate and the thiol molecules making the thiol-substrate bond even stronger. Cu was only inserted between the thiol groups and the substrate and not between the melamine–PTCDI network and the substrate (Figure 3.10c–e) [67]. The PTCDI–melamine host network physisorbed on Au surface has since been used as a template for immobilisation of both physisorbed and chemisorbed guest species [68].

3.3.2 Three-Component Systems: One-Component Host Network + Two Different Guests

The three-component systems discussed here are further classified depending on the type of host-cavities they offer: (i) the host network consists of only one type (shape/size) of the cavity and captures two different guest species and (ii) the host network offers two types of cavities that differ in size and shape and thus are capable of capturing two different types of guest species.

The triangular DBA derivatives introduced earlier (Figure 3.4a) exemplify the first category as the honeycomb networks offer uniform hexagonal cavities. The expression of chirality in self-assembled networks of DBA derivatives (Figure 3.11a) is briefly discussed here since it is essential to understand the differences observed in the assembly behaviour observed on HOPG and Au(111) substrates. The rim of each hexagonal nanowell consists of a pair of alkyl chains from one DBA molecule, interdigitated with a pair from an adjacent DBA molecule. The interdigitation pattern becomes chiral upon adsorption at an interface giving rise to two distinct interdigitation motifs, labelled arbitrarily (−) and (+). Note that the combination of interdigitation motifs within a nanowell can produce either a chiral or an achiral nanowell. Chiral nanowells are obtained with a combination of six identical interdigitation motifs. Depending on whether interdigitation motifs are (−) or (+) type, the handedness of the chiral nanowells can be counterclockwise (CCW) or clockwise (CW) [71]. Achiral pores are formed when three (−) and three (+) type interdigitation motifs are arranged in an alternating pattern (Figure 3.11a).

A concentric three-component system was obtained using the hexagonal cavities formed by a DBA derivative with dodecyloxy chains that trap a heteromolecular cluster formed by one molecule of COR and six molecules of ISA. A highly ordered three-component network is obtained at the octanoic acid–HOPG interface. The network consists of a central COR molecule surrounded by a first shell of ISA (reminiscent of the TMA hexagons mentioned earlier), which in turn is trapped by the second shell of DBA-OC10 that also forms the extended honeycomb network (Figure 3.11e,f). The guest-induced transitions in host networks described earlier were crucial in the successful formation of this system. They occur at two different levels in this system. First, at the level of the guest heterocluster: the cyclic hexamers of ISA are formed only in the presence of COR (Figure 3.11b,d), which otherwise forms a densely packed zigzag network (Figure 3.11c) in the absence of COR. Second, at the level of the host network: in the concentration range needed for the successful formation of this system, DBA-OC10 forms a densely packed network at the 1-octanoic acid–HOPG interface. The transition into a porous honeycomb network is induced by the addition of a solution containing COR–ISA. All the guest clusters were found to have the same composition and symmetry, which indicates a highly specific recognition with the host cavity. Importantly, all the nanowells were found to be chiral, which means that the domains consisted of either CW or CCW nanowells [69].

The self-assembly of this multicomponent system at the octanoic acid–Au(111) interface however differs from that observed on HOPG. On Au(111), the

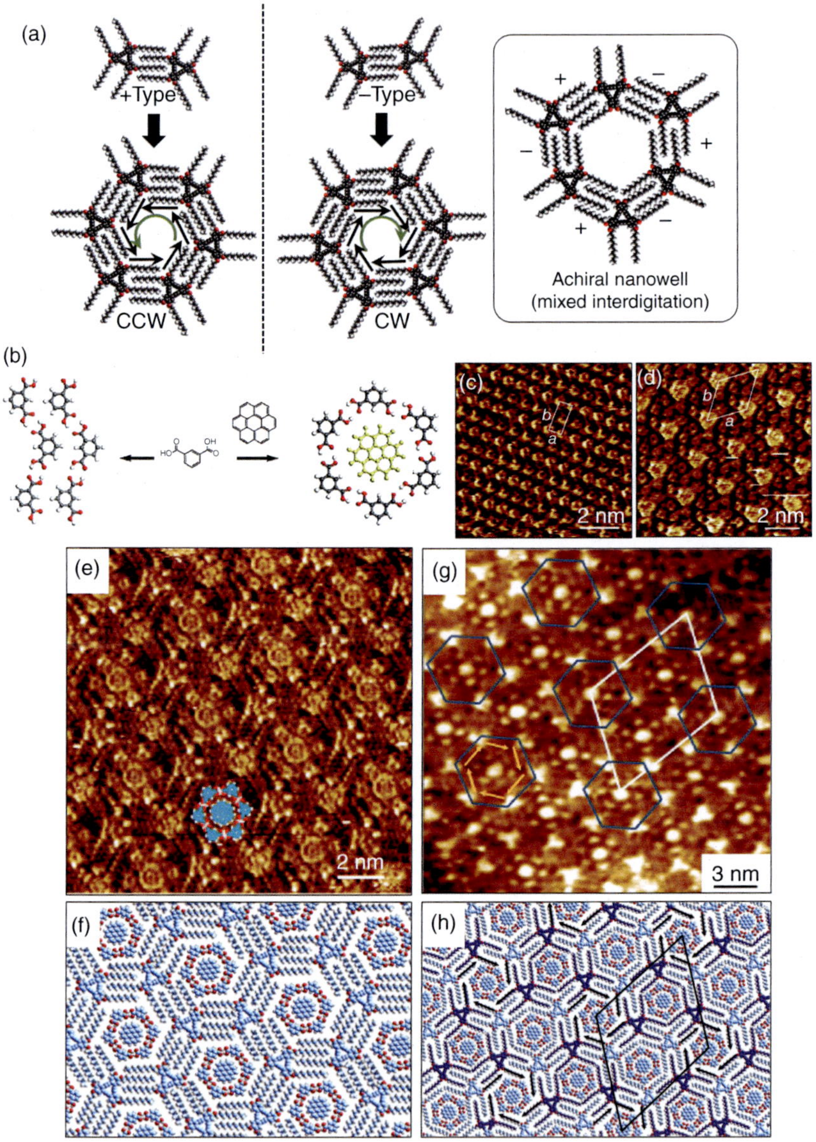

Figure 3.11 Three-component architectures made from a single component host and two different guests. (a) Molecular models depicting the expression of chirality in DBA networks. A combination of three (+)-type and three (−)-type interdigitation patterns yields an achiral pore. (b) Molecular model showing the COR templated assembly of ISA into cyclic hexamers. (c) Zigzag network of ISA obtained in absence of COR. (d) COR–ISA host–guest architecture. (e) STM image of the DBA/COR/ISA system on the HOPG surface. The corresponding molecular model is presented in (f). (g) DBA/COR/ISA system on Au(111) surface. The corresponding molecular model is presented in (h). Source: (a, b) Lei et al. [69]. (c, d, e) Reproduced from Lei et al. [69] with permission from the American Chemical Society. (g) Reproduced from Balandina et al. [70] with permission from the American Chemical Society.

multicomponent network displays an ordered superlattice arrangement of chiral and achiral pores (Figure 3.11g,h). The superlattice structure observed on Au(111) arises due to a lower energetic preference for chiral pores than on HOPG and increased diffusion barriers for guest molecules. The barriers for diffusion are typically higher in the case of metal substrates compared to HOPG. The increased diffusion barriers for guests allow them to act as nucleation sites for the formation of achiral nanowells. Thus, after the nucleation of an achiral pore, restrictions imposed by the accommodation of guests within the porous network enforce the formation of the superlattice structure [70]. This example highlights the importance of the nature of the substrate, which governs the mobility of molecules on the surface. Although annealing the samples at higher temperatures can induce necessary dynamics (often employed under UHV conditions), the temperature window available for experiments carried out at the solution–solid interface is rather limited due to evaporation of the solvent which causes changes in the solution concentration. As a consequence, only a few examples of crystalline multicomponent systems have been reported at the organic solution–Au(111) interface.

It is also possible to employ a combination of a non-covalently assembled host and a covalent host to realise a concentric three-component system. The nanowells within the DBA-OC12 network were used for immobilisation of a shape persistent macrocycle which in turn acts as a covalent porous host for trapping C_{60} molecules inside its cavities [72]. Although the three-component system was fabricated at the 1-phenyloctane–Au(111) interface, no superlattice structures were observed in this case. This could be related to the relatively higher in-plane diffusion of the macrocycle caused by methyl groups present on its backbone which reduce the effective contact with the Au(111) substrate. Another example where a combination of covalent and non-covalent hosts has been used consists of self-assembled networks of TMA or 1,3,5-tris(10-carboxydecyloxy)benzene (TCDB) which were used for immobilisation of macrocyclic oligothiophenes which in turn were used for immobilisation of fullerenes molecules [72, 73].

The second category, where the self-assembly of a single-component generates a host network with two different types of cavities allows site-selective adsorption of guest molecules. A typical example of this type of host network is the kagome network which offers periodically placed hexagonal and triangular cavities. An aromatic tetracarboxylic acid carrying an azobenzene linker (Figure 3.12a) forms a porous kagome network via hydrogen bonding between the carboxyl groups. Coronene molecules specifically occupied the hexagonal voids whereas C_{60} did not show preference when the experiment was carried out by adding the guests separately. However, the addition of all three components to the HOPG surface resulted in a three-component architecture where coronene was entrapped in the hexagonal voids and C_{60} in the smaller triangular voids (Figure 3.12b–d) [74].

However, host networks with only hexagonal cavities can also be designed, which offer different cavities in view of the specific design of the host component. A sophisticated DBA host network has been reported which allows self-sorting of molecular guests based on their size. This network is built using a tailored DBA derivative and consists of periodically functionalised cavities of different size. The building

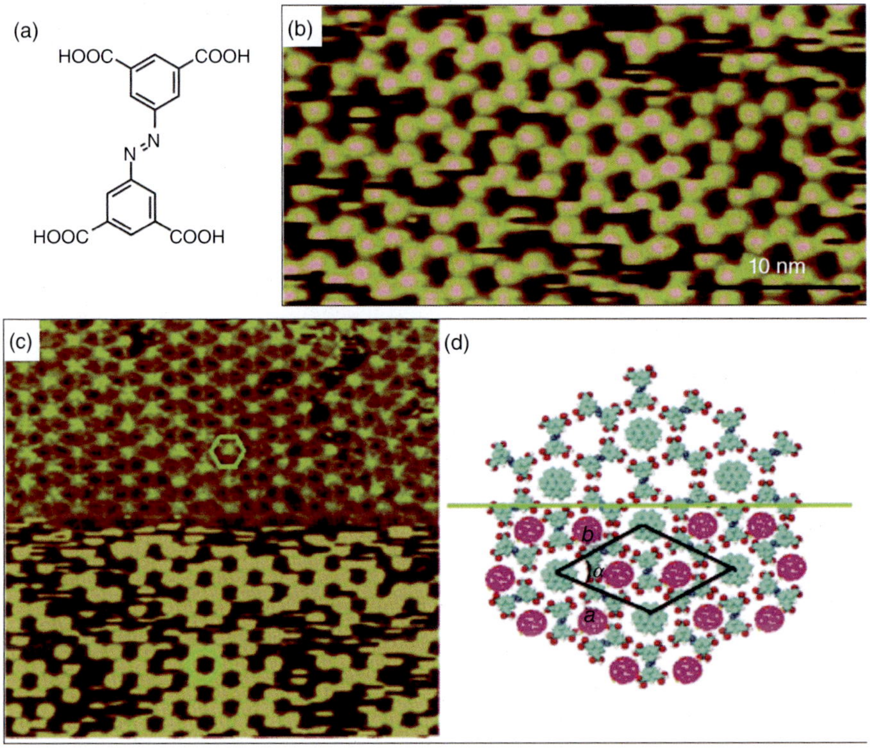

Figure 3.12 Three-component architectures: site-selective guest adsorption. (a) Molecular structure of the azobenzene containing aromatic carboxylic acid derivative. (b) STM image of the ternary architecture. (c) STM image of the architecture under different tunnelling conditions where the upper part reveals the coronene molecules and the lower part reveals the C_{60} molecules. (d) Molecular model of the three-component network. Source: (a) Shen et al. [74]. (b, c) Reproduced from Shen et al. [74] with permission from Wiley-VCH.

block consists of the triangular annulene core substituted with one chain carrying an isophthalic acid with an azobenzene linker and five simple alkoxy chains (Figure 3.13a). The self-assembly of this building block yields a host network with hexagonal cavities in which each nanowell containing the cyclic hexamer of isophthalic acid units is surrounded by six non-functionalised nanowells (Figure 3.13c,d). Although entropically disfavoured, the nanoscale separation of isophthalic acid unit containing cavities is favoured due to the enthalpic gain associated with the formation of hydrogen bonds between six isophthalic acid units. The two types of nanowells differ in size and could be used for site-selective immobilisation of COR and a large aromatic guest HPEPEB (Figure 3.13b,e,f) [75].

A particularly complex and rather unusual example of three-component architecture is the hierarchical self-assembly process involving an alkylated truxenone derivative (TrO23), CuPc), and TCDB (Figure 3.14a). Co-assembly of these three components at appropriate solution mole ratios on the HOPG surface led to the formation of a series of flower-like chiral hierarchical superstructures with the tuneable

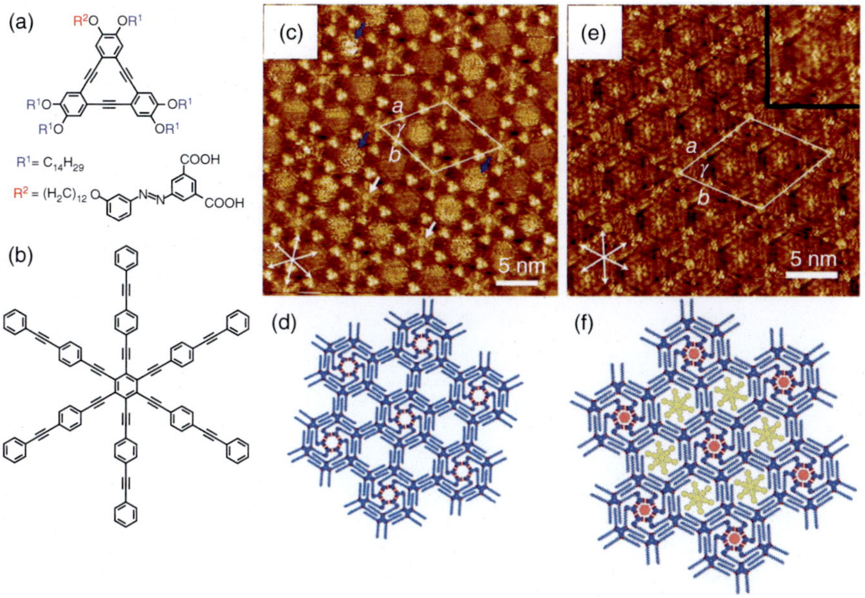

Figure 3.13 Three-component architectures: site-selective guest adsorption. (a) Molecular structure of the modified DBA derivative. (b) Molecular structure of the guest, HPEPEB. (c, d) STM image and molecular model of the self-assembled host network offering two different types of cavities, respectively. (e, f) STM image and molecular model of the DBA/COR/HPEPEB three-component architecture, respectively. Source: (a, b, d, e, f) Tahara et al. [75]. (c, e) Reproduced from Tahara et al. [75] with permission from the American Chemical Society.

periodicity ranging from 7 nm to more than 14 nm (Figure 3.14b–e). The hierarchical architectures were described by a unified configuration in which the lobe of each flower-shaped architecture is composed of a different number of triangular assembling units. Both left- and right-handed flower architectures were observed. Furthermore, the tiling of the triangular units in a vertex sharing configuration led to the systematic expansion of the chiral unit cells (Figure 3.14f–j) [76]. Although TCDB is known to form porous networks, it does not appear to form an open porous structure in this case.

3.3.3 Three-Component Systems: Non-host–Guest Systems

A noteworthy example of a non-host–guest type three-component system is the crystalline co-assembly of C_{60} molecules on a 2D bicomponent monolayer comprising porphyrin and phthalocyanine molecules. This system was characterised at an electrified Au(111)–aqueous electrolyte interface using EC-STM. Immersion of the Au(111) substrate into benzene solutions of the two components, namely zinc phthalocyanine (ZnPc) and zinc octaethyl porphyrin (ZnOEP) (Figure 3.15a) lead to the formation of a 'chessboard'-like supramolecular surface pattern (Figure 3.15b,c) [77]. This pattern templates adsorption of C_{60} molecules (Figure 3.15d–f) into the

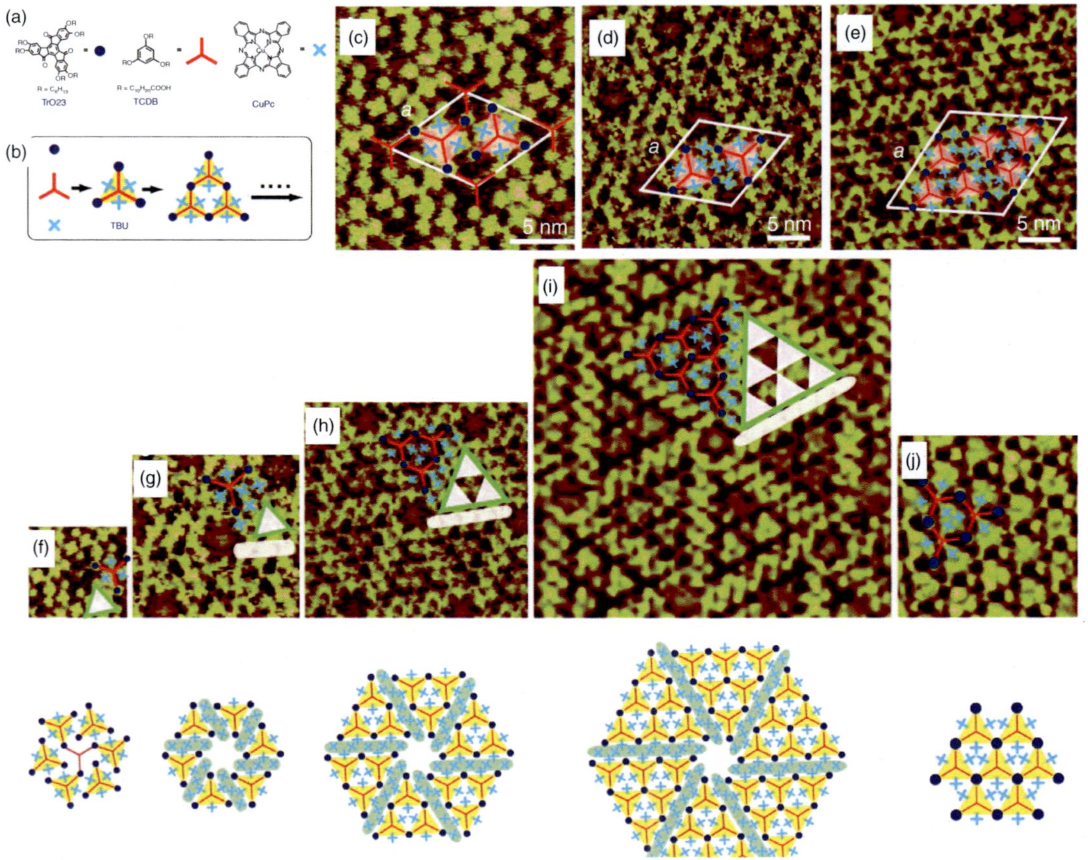

Figure 3.14 Three-component architectures: site-selective guest adsorption. (a) Molecular structures of the assembling components. (b) A schematic showing the hierarchical assembly of the different components. (c–e) STM images showing the assembly of the three components into flower-like structures with increasing periodicity. (f–j) STM images showing the structural evolution into higher-order structures also shown schematically below the STM images. Image sizes are (f) $11 \times 11\ nm^2$; (g) $19 \times 19\ nm^2$; (h) $25 \times 25\ nm^2$; (i) $40 \times 40\ nm^2$; (j) $20 \times 20\ nm^2$. Source: (a, b, d, e, g–j) Liu et al. [76]. (c, f) Reproduced from Liu et al. [76] with permission from the American Chemical Society.

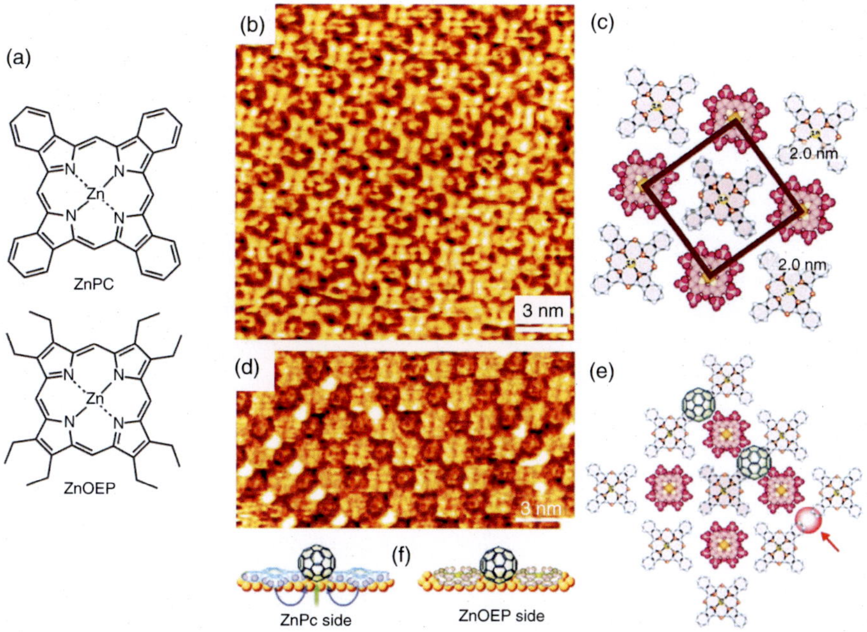

Figure 3.15 Non-host–guest type three-component architectures. (a) Molecular structures of ZnPc and ZnOEP. (b) STM image showing the bimolecular architecture formed by co-adsorption of ZnPc and ZnOEP on Au(111) surface. (c) Schematic representation of the bimolecular lattice. (d) Three-component system formed by co-adsorption of ZnPc/ZnOEP/C_{60} at the electrolyte/Au(111) interface. (e, f) Schematic representation of the three-component system. Source: (a, c–f) Yoshimoto et al. [77]. (b) Reproduced from Yoshimoto et al. [77] with permission from the American Chemical Society.

supramolecular network leading to the formation of a three-component network. Comparative studies carried out on Au(100) surface revealed a strong influence of the crystallographic orientation of the Au lattice on the self-assembly of C_{60} molecules. While the chessboard-like pattern was formed by ZnPc and ZnOEP, no long-range ordered assembly of C_{60} was observed on Au(100) [77].

Another strategy to produce multicomponent assemblies on surfaces is to deposit a 2D supramolecular array on top of another one to create so-called supramolecular heterostructures [78]. Assemblies of TMA or terephthalic acid were formed on top of a bicomponent layer made of cyanuric acid and melamine. The systems showed epitaxy and were stabilised in plane by hydrogen bonding and vertically by mainly van der Waals interactions.

3.4 Four-Component Systems

The complexity of the assembly process increases significantly for three or higher number of components. The highest number of components assembled in a crystalline fashion using on-surface assembly is four. Apart from one case reported under UHV [79], all four component systems have been formed at the solution–solid interface. While most of these systems are based on host–guest architectures, there are exceptions.

3.4.1 Four-Component Systems: Host–Guest Architectures

Changing the shape of the annulene core from triangular to rhombic in DBA derivatives provides access to a kagome network which was used to fabricate a sophisticated four-component system at the octanoic acid–HOPG interface. Similar to the three-component systems based on DBA, the successive guest-induced structural transitions were crucial in the successful realisation of this system. The modified DBA derivative, bisDBA-C12 (Figure 3.16a) does not form a kagome network on its own but transitions into one in the presence of COR–ISA which acts as a guest cluster and occupies the hexagonal cavities. The addition of triphenylene, which is immobilised in the smaller triangular cavities, completes the four-component system (Figure 3.16b,c). It is noteworthy that, despite its complexity, the four-component system emerges spontaneously upon drop-casting of the solution containing the appropriate ratio of the assembling components. Precise control of the concentration of the four components in solution is important and control experiments revealed that the assembly is a highly cooperative process involving the action of all the units at the same time [80].

A second strategy consists in extending the concentric circles approach to the fourth order, namely having the COR–ISA–DBA system nested into an addition host cavity. A DBA with four -OC10 chains (DBA-(4)-OC10) instead of six was synthesised, preventing interdigitation between the DBA hexamers surrounding the ISA shell (Figure 3.16d,e). In the final step, the four-component system was obtained by adding a DBA with six longer chains (DBA-OC26) to the mixture (Figure 3.16e–g). It is important to note that the three shells form dense patterns on their own, showing that they need the presence of the inner core (that plays the role of the template) to adapt the desired supramolecular structure [81].

A special case of a DBA-based four-component system involves the so-called hexagonal star (h-star) structure and consists of co-adsorbed solvent molecules. Conceptually, such a structure was made possible by blocking the interdigitation between adjacent DBAs thanks to solvent coadsorption. Typically, when a DBA-OC15 was deposited from a tetradecane solution, solvent molecules could adsorb between the alkoxy chains, which prevented the formation of the classic honeycomb DBA network. The addition of HPEB and the parent (unsubstituted) dehydrobenzo[12]annulenes (PDBA) to the solution yielded a four-component h-star structure on HOPG. Two types of guest-templated pores were formed by the DBA–tetradecane host network: hexagonal and rhombic voids were filled by single HPEB molecules and two PDBA, respectively. The formation of this cooperative network illustrates how solvent molecules can take an active part in the elaboration of multicomponent assemblies [82].

The PTCDI–melamine host system formed on Au(111) has also been employed for the construction of a four-component system (Figure 3.17a). The bi-component host network is stable under dry conditions and thus allows the deposition of guests in

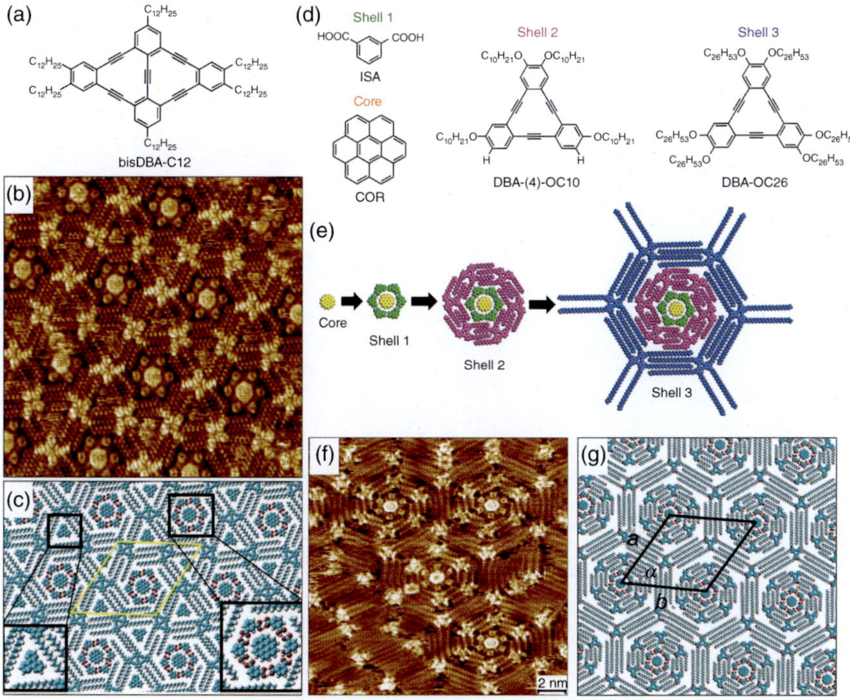

Figure 3.16 Four-component architectures based on DBA derivatives. (a) Molecular structure of the rhombic shaped DBA derivative bisDBA-C12. (b) STM images showing the crystalline four-component architecture obtained at the HOPG–octanoic acid interface using the co-assembly of bisDBA-C12/COR/ISA/triphenylene. (c) A molecular model of the four-component architecture. (d) Molecular structures of the building blocks used for the fabrication of a nested four-component system. (e) Schematic representation of the nested system. (f) STM image of the nested four-component system obtained at the octanoic acid–HOPG interface. (g) Molecular model for the nested four-component system. Source: (a,d,e, c,g) Refs. [80, 81]. (b) Reproduced from Adisoejoso et al. [80] with permission from Wiley-VCH. (f) Reproduced from Velpula et al. [81] with permission from the Royal Society of Chemistry.

a second step using a different solvent. First, a three-component system is obtained by depositing a threefold symmetric star-shaped guest molecule, the hydroxyl substituted tris([biphenyl]-4-ylethynyl)benzene (3BPEB-OH, Figure 3.17b) to the PTCDI–melamine host network. The 3BPEB-OH molecule divides the hexagonal host cavity into three smaller cavities (Figure 3.17a), which are subsequently filled with a second guest to form a four-component system. Such networks were obtained by three cycles of deposition from solution and drying: first for the PTCDI–melamine network, then for 3BPEB-OH and finally for the second guest, either adamantane thiol (Figure 3.17c) or C_{60} (Figure 3.17d). One C_{60} can be physisorbed in each of the three-compartment created by 3BPEB-OH (Figure 3.17d), compared to seven C_{60} per PTCDI–melamine unmodified pore. Adamantane thiol was deposited at low temperature to avoid the complete replacement of the 3BPEB-OH. It chemisorbs on Au forming a S—Au bond. Between one to four adamantane thiol molecules can fit

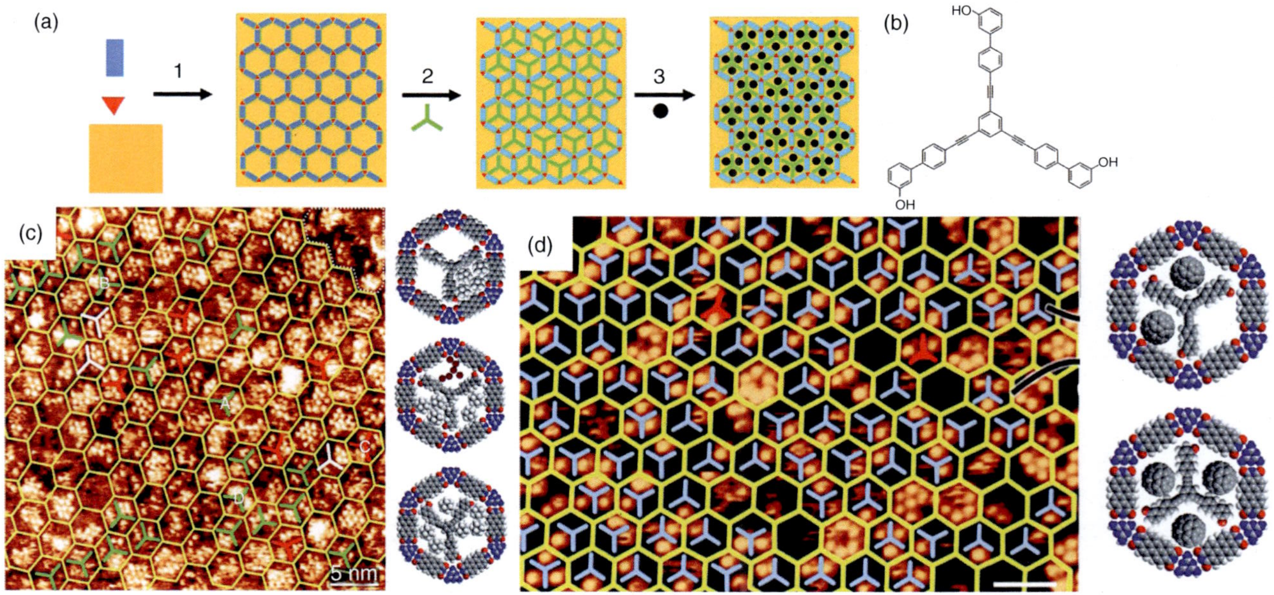

Figure 3.17 Four-component architectures based on PTCDI–melamine. (a) A schematic of the four-component system. (b) Molecular structure of 3BPEB-OH guest. (c) Four-component architecture which combines physisorption (PTCDI/melamine/3BPEB-OH) and chemisorption (adamantane thiol). (d) Four-component architecture ((PTCDI/melamine/3BPEB-OH)/C_{60}) where all components are physisorbed. Source: (a, b, d) Karamzadeh et al. [83]. (c) Reproduced from Karamzadeh et al. [83] with permission from the Royal Society of Chemistry.

in each compartment. However, for both C_{60} and adamantane thiol, the number of guest molecules adsorbed per pore was found to be random [68d, 83].

3.4.2 Four-Component Systems: Non-host–Guest Architectures

Four-component systems based on non-host–guest type interactions are rare and there exists only one example of such system. This system employs a peculiar strategy where the shape complementarity of alkadiyne chains attached to a central anthracene unit (Figure 3.18) is used for efficient recognition between assembling molecules. The key factor is the introduction of a kinked shape in the chains caused by diyne groups. By judiciously choosing the location of this kink within the chains, van der Waals contacts between identical side-chains were reduced, which led to recognition between chains of identical length but complementary shapes. Deposition of four such appropriately designed anthracene derivatives on the surface of HOPG yielded well-ordered monolayers consisting of rows of anthracene units separated from each other by interdigitating alkadiyne chains. The unit cell comprises six molecules arranged in the order S1–D1–D2–S2–D2–D1 and spans 23 nm × 1 nm (Figure 3.18a,b). The domains remain stable even after evaporation of

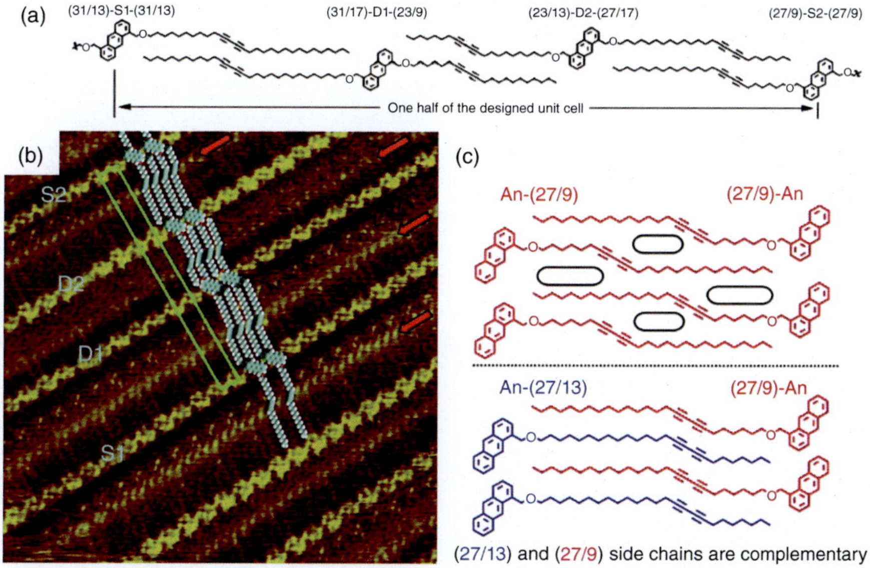

Figure 3.18 Four-component non-host–guest architectures. (a) A schematic showing the relative positioning of the alkadiyne chains exploiting the shape complementarity of the substituted anthracene molecules. (b) STM image of the four-component monolayer obtained at the 1-phenyloctane/HOPG interface showing the relative arrangement of different species S1, D1, D2 and S2 as shown in panel (a). The red arrows indicate the diyne columns. (c) Schematic showing the lack of self-complementarity between two molecules which forms empty areas within the monolayer. On the other hand, two molecules with complementary diacetylene units can lead to a compact arrangement. Source: (a, c) Xue et al. [8]. (b) Reproduced from Xue et al. [8] with permission from the American Chemical Society.

the solvent. Interestingly, no monolayer is formed when S1, D1, or D2 is excluded from the solution, and only disordered domains are formed when S2 is excluded, showing the importance of specific intermolecular recognition to achieve regular arrays [8].

3.5 Summary and Perspectives

The design and fabrication of complex multicomponent supramolecular architectures are at the heart of supramolecular chemistry on surfaces. The field continues to grow at a pace and has offered many opportunities as well as challenges. Diverse multicomponent systems ranging from simple columnar networks to sophisticated multicomponent hexagonal networks confined against a solid surface have been reported. The research on this subject has not only lead to the fabrication of exotic hierarchical networks but has also unearthed a number of fundamental aspects pertinent to supramolecular chemistry on surfaces such as competitive adsorption, guest-induced structural transitions, etc. While remarkable progress has been made in terms of the diversity, complexity and control over multicomponent self-assembly, a few challenges remain in the path ahead. While the inspiration comes from natural systems, the on-surface fabricated architectures are nowhere close to the complexity displayed by natural systems. More importantly, translation of the structure into any sort of function remains to be demonstrated. Since a concerted motion of the building blocks is often involved in functional multicomponent systems found in nature, more emphasis needs to be given to realisation of stimulus-responsive multicomponent architectures. Furthermore, the robustness of such supramolecular systems needs to be evaluated systematically to ensure their utility in real-life applications. These challenges make the research on multicomponent self-assembly an emerging research area with the ultimate goal of rivalling the craftsmanship of Mother Nature.

References

1 Whitesides, G.M. and Grzybowski, B. (2002). *Science* 295: 2418–2421.
2 Okesola, B.O. and Mata, A. (2018). *Chem. Soc. Rev.* 47: 3721–3736.
3 De Greef, T.F.A., Smulders, M.M.J., Wolffs, M. et al. (2009). *Chem. Rev.* 109: 5687–5754.
4 Desiraju, G.R. (2013). *J. Am. Chem. Soc.* 135: 9952–9967.
5 Elemans, J.A.A.W., Lei, S., and De Feyter, S. (2009). *Angew. Chem. Int. Ed.* 48: 7298–7332.
6 Mali, K.S., Pearce, N., De Feyter, S., and Champness, N.R. (2017). *Chem. Soc. Rev.* 46: 2520–2542.
7 MacLeod, J.M., Lipton-Duffin, J., Fu, C. et al. (2017). *ACS Nano* 11: 8901–8909.
8 Xue, Y. and Zimmt, M.B. (2012). *J. Am. Chem. Soc.* 134: 4513–4516.

9 Teyssandier, J., Feyter, S.D., and Mali, K.S. (2016). *Chem. Commun.* 52: 11465–11487.
10 Schmaltz, B., Rouhanipour, A., Räder, H.J. et al. (2009). *Angew. Chem. Int. Ed.* 48: 720–724.
11 Geng, Y., Liu, M., Xue, J. et al. (2015). *Chem. Commun.* 51: 6820–6823.
12 Ohira, A., Sakata, M., Hirayama, C., and Kunitake, M. (2003). *Org. Biomol. Chem.* 1: 251–253.
13 Yoshimoto, S., Suto, K., Tada, A. et al. (2004). *J. Am. Chem. Soc.* 126: 8020–8027.
14 Mena-Osteritz, E. and Bäuerle, P. (2006). *Adv. Mater.* 18: 447–451.
15 Yoshimoto, S., Suto, K., Itaya, K., and Kobayashi, N. (2003). *Chem. Commun.*: 2174–2175.
16 Hirsch, B.E., McDonald, K.P., Qiao, B. et al. (2014). *ACS Nano* 8: 10858–10869.
17 Pan, G.-B., Liu, J.-M., Zhang, H.-M. et al. (2003). *Angew. Chem. Int. Ed.* 42: 2747–2751.
18 Sherman, J.C. (2004). Preorganization and complementarity. In: *Encyclopedia of Supramolecular Chemistry*, vol. 2 (eds. J.L. Atwood and J.W. Steed), 1158–1160. New York: CRC Press.
19 Thontasen, N., Levita, G., Malinowski, N. et al. (2010). *J. Phys. Chem. C* 114: 17768–17772.
20 Wieland, M.B., Perdigao, L.M.A., Kondratuk, D.V. et al. (2014). *Chem. Commun.* 50: 7332–7335.
21 Tahara, K., Lei, S., Mossinger, D. et al. (2008). *Chem. Commun.*: 3897–3899.
22 Griessl, S.J.H., Lackinger, M., Jamitzky, F. et al. (2004). *Langmuir* 20: 9403–9407.
23 Ivasenko, O., MacLeod, J.M., Chernichenko, K.Y. et al. (2009). *Chem. Commun.*: 1192–1194.
24 Blunt, M.O., Russell, J.C., del Carmen Gimenez-Lopez, M. et al. (2011). *Nat. Chem.* 3: 74–78.
25 Griessl, S.J.H., Lackinger, M., Jamitzky, F. et al. (2004). *J. Phys. Chem. B* 108: 11556–11560.
26 Lee, S.-L., Fang, Y., Velpula, G. et al. (2015). *ACS Nano* 9: 11608–11617.
27 Eder, G., Kloft, S., Martsinovich, N. et al. (2011). *Langmuir* 27: 13563–13571.
28 MacLeod, J.M., Ivasenko, O., Fu, C. et al. (2009). *J. Am. Chem. Soc.* 131: 16844–16850.
29 Blunt, M., Lin, X., Gimenez-Lopez, M.d.C. et al. (2008). *Chem. Commun.*: 2304–2306.
30 Park, K.-W., Adisoejoso, J., Plas, J. et al. (2014). *Langmuir* 30: 15206–15211.
31 Ciesielski, A., Lena, S., Masiero, S. et al. (2010). *Angew. Chem. Int. Ed.* 49: 1963–1966.
32 Bilbao, N., Destoop, I., De Feyter, S., and González-Rodríguez, D. (2016). *Angew. Chem. Int. Ed.* 55: 659–663.
33 Ruben, M., Payer, D., Landa, A. et al. (2006). *J. Am. Chem. Soc.* 128: 15644–15651.
34 Gutzler, R., Sirtl, T., Dienstmaier, J.R.F. et al. (2010). *J. Am. Chem. Soc.* 132: 5084–5090.
35 Tahara, K., Lei, S., Adisoejoso, J. et al. (2010). *Chem. Commun.* 46: 8507–8525.

36 Schull, G., Douillard, L., Fiorini-Debuisschert, C. et al. (2006). *Adv. Mater.* 18: 2954–2957.

37 Furukawa, S., Tahara, K., De Schryver, F.C. et al. (2007). *Angew. Chem. Int. Ed.* 46: 2831–2834.

38 Lei, S., Tahara, K., De Schryver, F.C. et al. (2008). *Angew. Chem. Int. Ed.* 47: 2964–2968.

39 Lei, S., Tahara, K., Feng, X. et al. (2008). *J. Am. Chem. Soc.* 130: 7119–7129.

40 Zhang, X., Chen, T., Yan, H.-J. et al. (2010). *ACS Nano* 4: 5685–5692.

41 Tahara, K., Katayama, K., Blunt, M.O. et al. (2014). *ACS Nano* 8: 8683–8694.

42 Iritani, K., Ikeda, M., Yang, A. et al. (2018). *Langmuir* 34: 6036–6045.

43 Tahara, K., Inukai, K., Adisoejoso, J. et al. (2013). *Angew. Chem. Int. Ed.* 52: 8373–8376.

44 a) Blunt, M.O., Russell, J.C., Champness, N.R., and Beton, P.H. (2010). *Chem. Commun.* 46: 7157–7159. b) Plas, J., Ivasenko, O., Martsinovich, N. et al. (2016). *Chem. Commun.* 52: 68–71.

45 a) Cui, D., MacLeod, J.M., Ebrahimi, M. et al. (2015). *Chem. Commun.* 51: 16510–16513. b) Cui, D., MacLeod, J.M., Ebrahimi, M., and Rosei, F. (2017). *CrystEngComm* 19: 4927–4932.

46 Liu, X.-H., Yue, J.-Y., Mo, Y.-P. et al. (2016). *J. Phys. Chem. C* 120: 15753–15757.

47 Liu, C., Zhang, W., Zeng, Q., and Lei, S. (2016). *Chem. Eur. J.* 22: 6768–6773.

48 a) Liu, L., Besenbacher, F., and Dong, M. (2014). Self-assembly of DNA Bases via hydrogen bonding studied by scanning tunneling microscopy. In: *Nucleic Acid Nanotechnology* (eds. J. Kjems, E. Ferapontova and K.V. Gothelf), 3–21. Berlin, Heidelberg: Springer. b) Ding, Y., Wang, X., Xie, L. et al. (2018). *Chem. Commun.* 54: 9259–9269. c) Bouju, X., Mattioli, C., Franc, G. et al. (2017). *Chem. Rev.* 117: 1407–1444; d) Slater, A.G., Perdigão, L.M.A., Beton, P.H., and Champness, N.R. (2014). *Acc. Chem. Res.* 47: 3417–3427.

49 a) Zheng, Q.-N., Liu, X.-H., Chen, T. et al. (2015). *J. Am. Chem. Soc.* 137: 6128–6131. b) Mukherjee, A., Teyssandier, J., Hennrich, G. et al. (2017). *Chem. Sci.* 8: 3759–3769.

50 a) Schlickum, U., Decker, R., Klappenberger, F. et al. (2007). *Nano Lett.* 7: 3813–3817. b) Écija, D., Urgel, J.I., Papageorgiou, A.C. et al. (2013). *Proc. Natl. Acad. Sci. U.S.A.* 110: 6678–6681.

51 Ciesielski, A., Palma, C.A., Bonini, M., and Samorì, P. (2010). *Adv. Mater.* 22: 3506–3520.

52 a) Zhang, X., Chen, T., Chen, Q. et al. (2009). *Chem. Eur. J.* 15: 9669–9673. b) Zhang, X., Chen, Q., Deng, G.-J. et al. (2009). *J. Phys. Chem. C* 113: 16193–16198. c) Mamdouh, W., Uji-i, H., Ladislaw, J.S. et al. (2005). *J. Am. Chem. Soc.* 128: 317–325.

53 Plass, K.E., Engle, K.M., Cychosz, K.A., and Matzger, A.J. (2006). *Nano Lett.* 6: 1178–1183.

54 Ahn, S. and Matzger, A.J. (2009). *J. Am. Chem. Soc.* 131: 13826–13832.

55 Mali, K.S., Van Averbeke, B., Bhinde, T. et al. (2011). *ACS Nano* 5: 9122–9137.

56 Piot, L., Marchenko, A., Wu, J. et al. (2005). *J. Am. Chem. Soc.* 127: 16245–16250.

57 Xiao, W., Passerone, D., Ruffieux, P. et al. (2008). *J. Am. Chem. Soc.* 130: 4767–4771.

58 Zhang, Y., Zhang, Y., Li, G. et al. (2015). *J. Chem. Phys.* 142: 101911.

59 Fang, Y., Ghijsens, E., Ivasenko, O. et al. (2016). *Nat. Chem.* 8: 711–717.

60 Raval, R. (2009). *Chem. Soc. Rev.* 38: 707–721.

61 Theobald, J.A., Oxtoby, N.S., Phillips, M.A. et al. (2003). *Nature* 424: 1029–1031.

62 Staniec, P.A., Perdigão, L.M.A., Saywell, A. et al. (2007). *ChemPhysChem* 8: 2177–2181.

63 Velpula, G., Teyssandier, J., De Feyter, S., and Mali, K.S. (2017). *ACS Nano* 11: 10903–10913.

64 Silly, F., Shaw, A.Q., Porfyrakis, K. et al. (2007). *Appl. Phys. Lett.* 91: 253109.

65 Kampschulte, L., Werblowsky, T.L., Kishore, R.S.K. et al. (2008). *J. Am. Chem. Soc.* 130: 8502–8507.

66 Li, Y., Ma, Z., Deng, K. et al. (2009). *Chem. Eur. J.* 15: 5418–5423.

67 Madueno, R., Räisänen, M.T., Silien, C., and Buck, M. (2008). *Nature* 454: 618–621.

68 a) Silien, C., Räisänen, M.T., and Buck, M. (2009). *Angew. Chem. Int. Ed.* 48: 3349–3352. b) Räisänen, M.T., Slater, A.G., Champness, N.R., and Buck, M. (2012). *Chem. Sci.* 3: 84–92. c) Perdigão, L.M.A., Staniec, P.A., Champness, N.R., and Beton, P.H. (2009). *Langmuir* 25: 2278–2281. d) Karamzadeh, B., Eaton, T., Cebula, I. et al. (2014). *Chem. Commun.* 50: 14175–14178. e) Lombana, A., Battaglini, N., Tsague-Kenfac, G. et al. (2016). *Chem. Commun.* 52: 5742–5745.

69 Lei, S., Surin, M., Tahara, K. et al. (2008). *Nano Lett.* 8: 2541–2546.

70 Balandina, T., Tahara, K., Sändig, N. et al. (2012). *ACS Nano* 6: 8381–8389.

71 Tahara, K., Yamaga, H., Ghijsens, E. et al. (2011). *Nat. Chem.* 3: 714–719.

72 Cui, K., Schlütter, F., Ivasenko, O. et al. (2015). *Chem. Eur. J.* 21: 1652–1659.

73 a)Shen, Y.T., Deng, K., Zeng, Q.D., and Wang, C. (2010). *Small* 6: 76–80; b) Cojal González, J.D., Iyoda, M., and Rabe, J.P. (2017). *Nat. Commun.* 8: 14717.

74 Shen, Y.-T., Li, M., Guo, Y.-Y. et al. (2010). *Chem. Asian J.* 5: 787–790.

75 Tahara, K., Nakatani, K., Iritani, K. et al. (2016). *ACS Nano* 10: 2113–2120.

76 Liu, J., Chen, T., Deng, X. et al. (2011). *J. Am. Chem. Soc.* 133: 21010–21015.

77 Yoshimoto, S., Honda, Y., Ito, O., and Itaya, K. (2008). *J. Am. Chem. Soc.* 130: 1085–1092.

78 Korolkov, V.V., Baldoni, M., Watanabe, K. et al. (2017). *Nat. Chem.* 9: 1191.

79 Gardener, J.A., Shvarova, O.Y., Briggs, G.A.D., and Castell, M.R. (2010). *J. Phys. Chem. C* 114: 5859–5866.

80 Adisoejoso, J., Tahara, K., Okuhata, S. et al. (2009). *Angew. Chem. Int. Ed.* 48: 7353–7357.

81 Velpula, G., Takeda, T., Adisoejoso, J. et al. (2017). *Chem. Commun.* 53: 1108–1111.

82 Tahara, K., Kaneko, K., Katayama, K. et al. (2015). *Langmuir* 31: 7032–7040.

83 Karamzadeh, B., Eaton, T., Torres, D.M. et al. (2017). *Faraday Discuss.* 204: 173–190.

4

Complexity in Two-Dimensional Assembly: Using Coordination Bonds

Nian Lin and Jing Liu

The Hong Kong University of Science and Technology, Department of Physics, Clear Water Bay, Kowloon, Hong Kong

4.1 Introduction

Complexity is generally in describing the behaviour of systems whose components interact with each other in multiple ways, rendering a higher order of emergence greater than the sum of the components. Complexity is expressed abundantly in supramolecular self-assembly, in which molecular components are held together via non-covalent bonds to form highly complex materials [1–3]. Reversible forming and breaking of the non-covalent bonds enable self-organisation, self-recognition, and self-selection, driven by thermodynamics as well as kinetics in the supramolecular self-assembly processes [4]. As a result, supramolecular self-assembly exhibits highly selective and spontaneous build-up of specific outputs. A particular type of non-covalent interaction is metal–ligand coordination. Comparing with other types of non-covalent interactions such as hydrogen bonding, van der Waals interaction, electrostatic forces, or π–π interaction, metal–ligand coordination has the advantages of high directionality, good selectivity, and it is reversible yet strong. These features make metal–ligand coordination a superb choice for constructing supramolecular architectures [5–26].

The concepts of supramolecular coordination self-assembly have been applied on the surface [27–29] when molecules and atoms are adsorbed yet free to move on a surface due to attractive yet moderate interaction between these species and the surface. This on-surface strategy provides efficient protocols for constructing low dimensional coordination architectures including coordination chains [12, 30–39], coordination polygons [37, 40–42], and coordination networks [4, 6, 43–59]. In many cases, knowing the structural and chemical features of the molecules and the coordination configuration enables one to predicate the outcome structures of the supramolecular self-assembly. We name the systems following this principle as simple coordination structures. Such predictability fails in certain situations when the molecular and atomic components interact with each other in multiple ways, which often results in supramolecular structures that express high organisational orders. We term these systems as complex coordination structures. Here we list

Supramolecular Chemistry on Surfaces: 2D Networks and 2D Structures, First Edition.
Edited by Neil R. Champness.

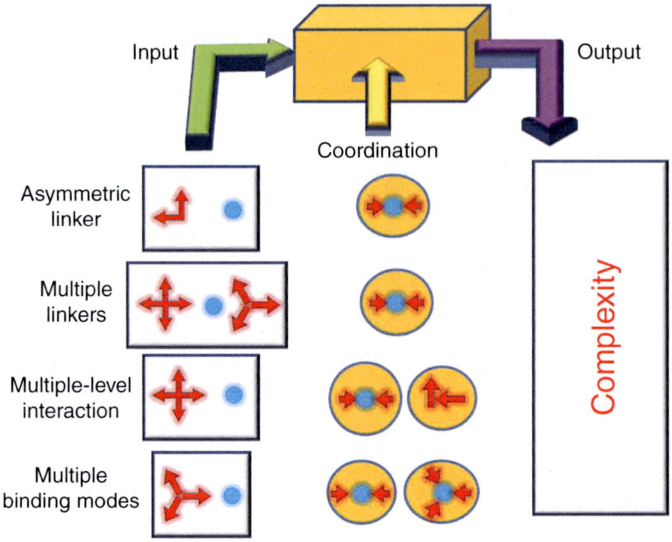

Scheme 4.1 Four categories of complexity discussed in this work.

four types of complex coordination structures, as illustrated in Scheme 4.1: (i) The molecular component (linker) has an asymmetric shape. (ii) Multiple types of molecular components (linkers) participate in the self-assembly; (iii) Besides metal–ligand coordination, other types of non-covalent interactions are at work in the self-assembly; (iv) The self-assembly involves multiple types of metal–ligand coordination modes. Instead of giving a comprehensive review, we will select few representative examples of each of these four categories to highlight the complexity that emerged in two-dimensional (2D) supramolecular coordination self-assembly. Scheme 4.2 lists the molecular linkers in these examples. Interested readers are referred to Refs. [13, 28, 29].

4.2 Asymmetric Linkers

Fractal structures exhibit self-similarity at different scales and can be found in nature such as snowflakes, leaves, and trees. Such phenomena can be expressed on a molecular scale also [60–70]. Xu and coworkers used molecular linker **1a**, which features a 120°-bent backbone functionalized with carbonitrile ligands at its two termini, to form fractal structures of Sierpiński triangles on Au(111) surface [64]. In these structures, the carbonitrile ligand coordinates with Ni in a threefold coordination mode. The asymmetrical backbone prohibits forming long-range periodic structures but results in fractal Sierpiński triangles. Figure 4.1a shows that different generations of fractal Sierpiński triangles are formed. The high-resolution images (Figure 4.1b–d) show the zero, the first, and the second-generation triangles. The three generations consist of 6 molecules of **1a** and 3 Ni atoms, 15 molecules of **1a** and 9 Ni atoms, and 42 molecules of **1a** and 27 Ni atoms, respectively. Wang's group

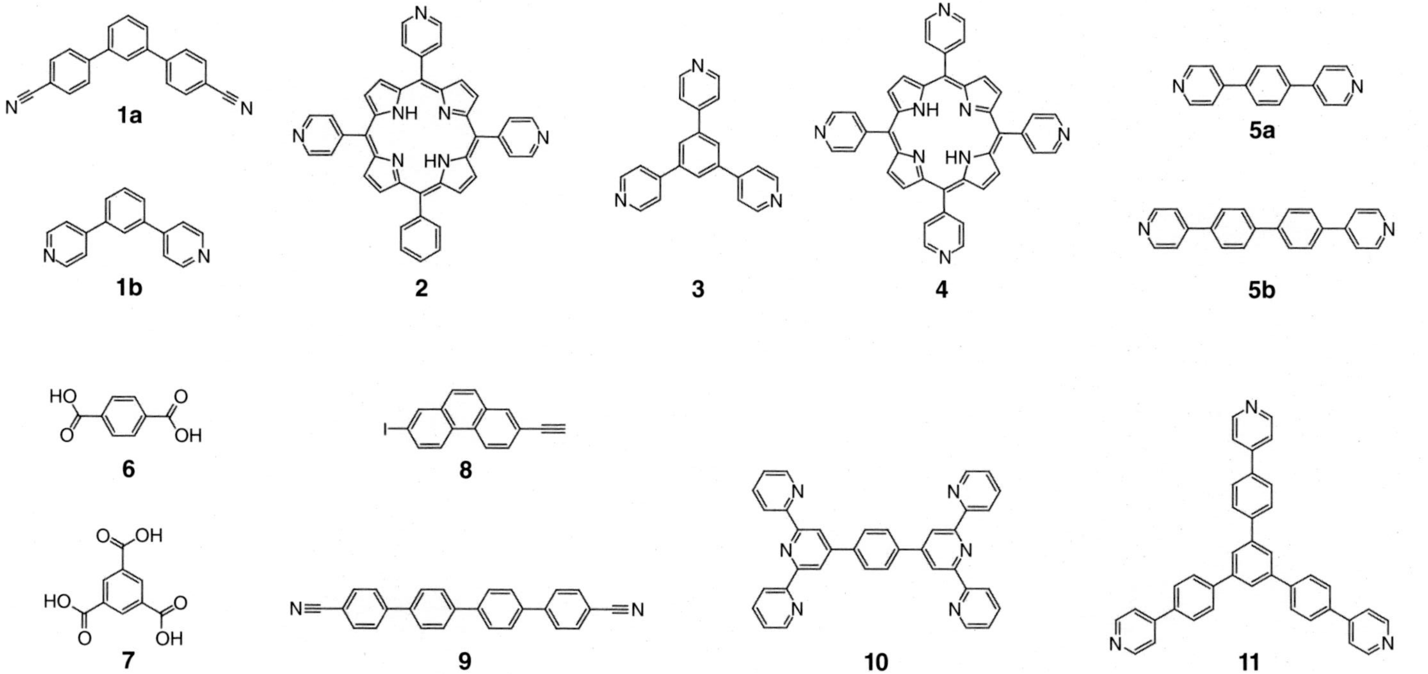

Scheme 4.2 Molecular linkers in the selected examples.

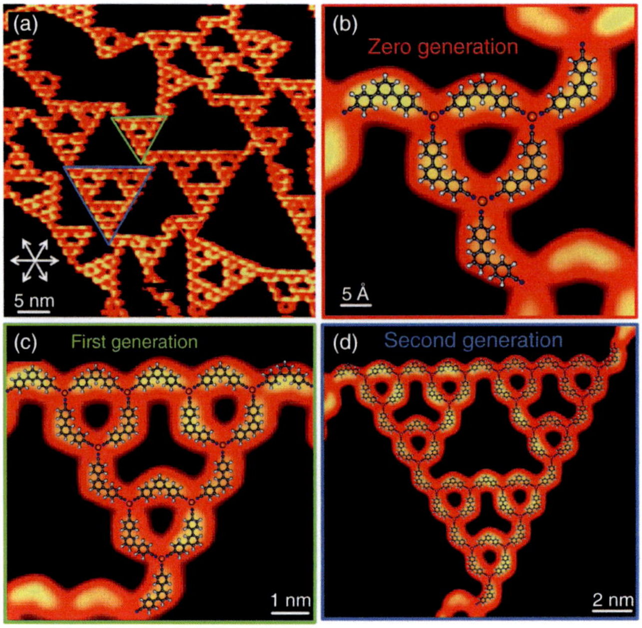

Figure 4.1 (a) STM images showing the formation of the metal–organic Sierpiński triangles after co-deposition of **1a** and Ni atoms on Au(111). (b–d) High-resolution STM images of zero, first, and second generation Sierpiński triangles with the theoretically proposed models. Source: Adapted with permission from Sun et al. [64] with permission of The Royal Society of Chemistry.

used pyridyl functionalized 120°-bent molecular linker **1b** and Co to assemble Sierpiński triangles on Au(111) [70]. Figure 4.2a shows the fourth-generation Sierpiński triangle. This structure has 123 molecules of **1b** and 81 Co atoms. Figure 4.2b shows the structures formed with a high coverage of **1b**, displaying ordered packing of pairs of second-generation Sierpiński triangles (marked by white dashed triangles) as one-dimensional molecular ribbons. The high-resolution scanning tunnelling microscopy (STM) image and chemical structure are presented in Figure 4.2c,d. When three molecular linkers coordinate to a central metal, their 120°-bent backbone can orientate either clockwise or anticlockwise. The threefold coordination does not favour any specific backbone orientation. However, the formation of the Sierpiński triangles requires two clockwise orientated linkers and one anticlockwise orientated linker (or vice versa). Consequently, the fractal structures must be traced to co-operativity at the supramolecular level.

5,10,15-Tri(4-pyridyl)-20-phenylporphyrin (**2**) features a fourfold symmetric backbone while three out of the four end groups are functionalized with pyridyl ligands, representing an asymmetric molecular linker. Self-assembly of low-coverage **2** with Fe on Au(111) results in discrete rosette-like structures, as shown in Figure 4.3a [71]. The rosette structure consists of six interconnected triangular units as highlighted by the triangle in Figure 4.3b. Each rosette is composed of six molecules forming an inner hexagon core and twelve molecules in the outer circle. Each molecule in

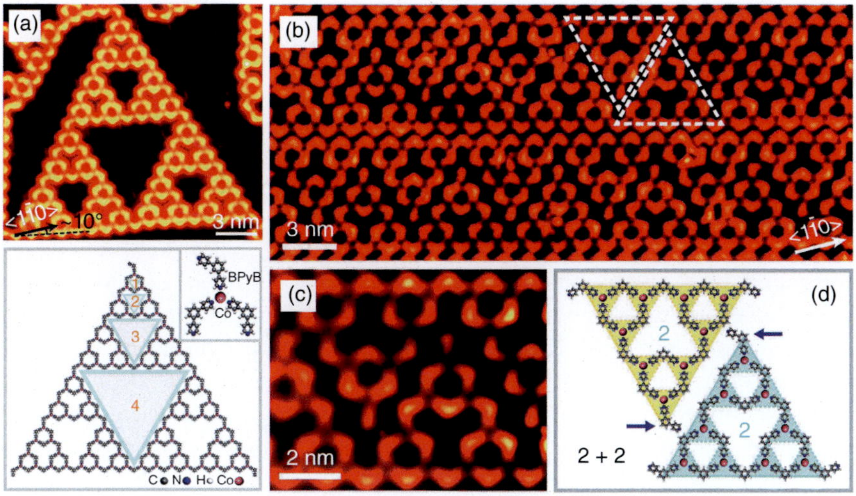

Figure 4.2 (a) STM image (up) and structural model (bottom) of a fourth generation Sierpiński triangle. (b) 1D molecular ribbons formed at high molecular coverage. (c) STM image and (d) structural model of two interlocking triangles. Source: (a–d) Adapted with permission from Zhang et al. [70] with permission of The Royal Society of Chemistry.

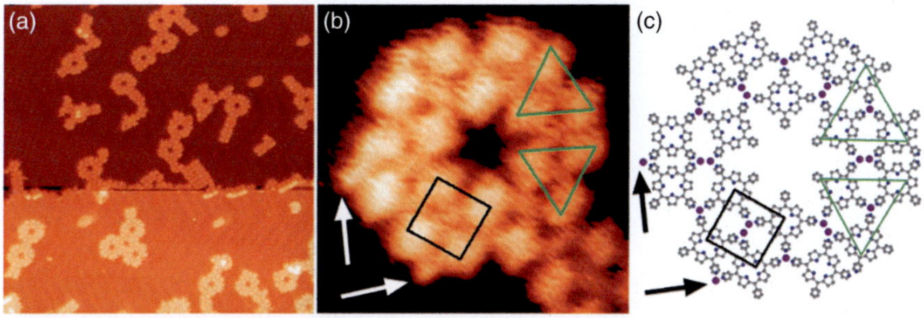

Figure 4.3 (a) STM topograph (100 nm × 100 nm) of the rosette structures formed by the coordination of **2** with Fe at low coverage. (b) High-resolution image (size: 10 nm × 10 nm) and (c) structural model of the rosette structure (Fe in purple and N in blue). Source: (a–c) Adapted with permission from Mao et al. [71] with permission of The Royal Society of Chemistry.

the inner hexagon is connected with six neighbouring molecules – two in the inner hexagon and four in the outer circle. We can also view the rosette-like structure as constituting trimeric units (highlighted by the green triangles). A structural model is shown in Figure 4.3c, which shows that a trimeric unit consists of three molecules of **2** coordinating to a central Fe through threefold pyridyl–Fe coordination, and the neighbouring trimeric units are linked by pyridyl–Fe coordination afforded by a pair of Fe atoms, as marked with the square in Figure 4.3c. The model is based on geometric consideration: four pyridyl groups point inwardly, where the neighbouring N atoms can bind to a pair of Fe atoms. The remaining pyridyl groups might be coordinated by Fe too, as marked by the arrows in Figure 4.3b,c. The formation of the rosette-like structures indicates that the self-assembly is a self-limited process. This behaviour is ubiquitous in nature. Here we propose the asymmetric linker molecules rendering a closed-shell effect that results in self-limited growth.

4.3 Multiple Types of Linkers

Self-assembly involving multiple types of linker molecules often leads to multiple phases whose structures are less predictable. As an example, our group studied self-assembly of a threefold symmetric linker 1,3,5-trispyridylbenzene (**3**) and a fourfold symmetric linker 5,10,15,20-tetra(4-pyridyl) porphyrin (**4**) with Cu [72]. Pyridyl functions coordinate with Cu in a twofold pyridyl–Cu–pyridyl coordination motif. When the linker molecules coordinate with Cu on Au(111) separately, **3** forms a honeycomb network and **4** forms a nearly square network. Mixing the threefold and the fourfold linkers is unpredictable since many different phases might be formed. When a mixture of **3** and **4** was deposited onto Au(111) together with Cu, however, we observed only one network structure which constitutes both linker molecules alongside the two single-component networks. Figure 4.4a is a representative STM image showing the two-component structure. As shown in the high-resolution STM image in Figure 4.4b, columns of **4** (larger ones) are linked by dimers of **3** (smaller ones), forming a rectangular lattice. The rectangular frame in Figure 4.4b marks the unit cell. The structural model in Figure 4.4c illustrates that the unit cell of this phase contains two molecules of **3**, one molecule of **4**, and five Cu atoms.

Multi-component self-assembly often undergoes self-selection. Langner et al. demonstrated self-selection in three-component self-assembly [4]. The three components are two ditopic bi-pyridyl linkers (**5a**, **5b**) and a ditopic bi-carboxylic linker (**6**). The mixture of **5a**, **5b**, and **6** with Fe on Cu(100) (Figure 4.5a) self-assembles into well-ordered rectangular compartments (Figure 4.5b). The coordination motif

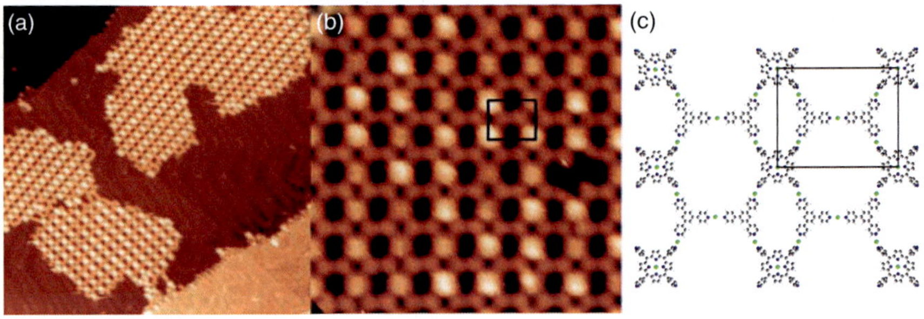

Figure 4.4 (a) An overview STM image (100 nm × 100 nm) showing the two-component network. (b) A high-resolution image (22.8 nm × 22.8 nm) and (c) structural model of the two-component network. The black frame marks the unit cell (C: grey, N: blue, H: white, Cu: green). Source: (a–c) Adapted with permission from Shi et al. [72] with permission of The Royal Society of Chemistry.

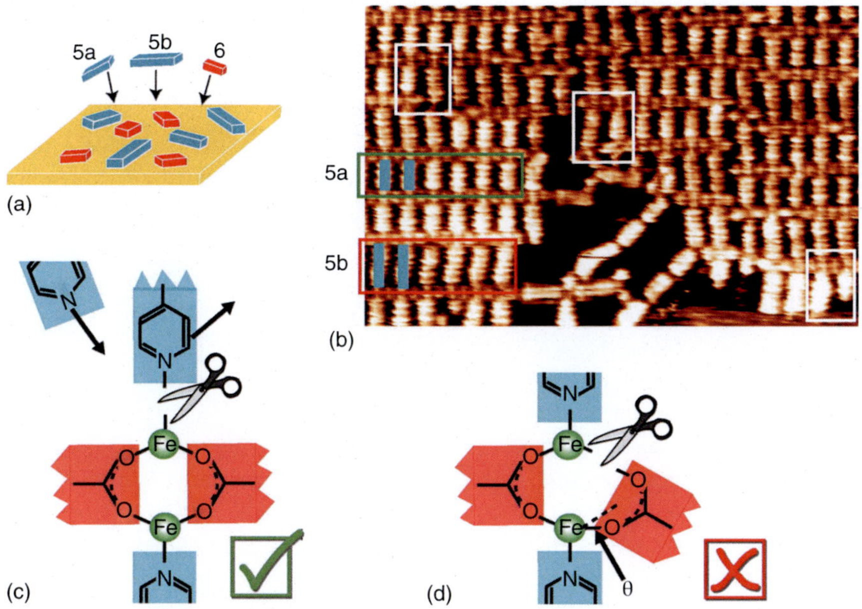

Figure 4.5 (a) The mixture of three molecular linkers **5a**, **5b**, and **6**. (b) STM image (22 nm × 14 nm) showing the local segregation of the mixture into highly ordered subdomains containing linker **5a** (green box) or **5b** (red box). Two defects of neighbouring bi-pyridyl liners of different lengths are highlighted (white boxes). (c) Schematic diagram of the reversible pyridyl–Fe bonding, the basis for active error correction by self-selection of ligands **5a** and **5b** into highly ordered subdomains. (d) Random packing would require breaking of one of the COO–Fe bonds and distortion of the other bond. Source: (a, c, d) Langner et al. [4]. (b) Adapted with permission from Langner et al. [4]. Copyright (2007) National Academy of Sciences.

consists of a di-iron centre with two bridging carboxylates and two axial pyridyls. A closer view reveals local spatial segregation into subdomains of **5a** × **6** and **5b** × **6**, as highlighted by the green and red rectangular frames, respectively. These subdomains within the 2D network can be considered as quasi-1D 'ladder' rows, within which the two 'side posts' consist of rows of **6** – Fe chains (horizontal in Figure 4.5b) bridged with bi-pyridyl linker 'rungs' of uniform size (either **5a** or **5b** in a given domain) at the di-iron nodes.

This structural outcome demonstrates that the self-selection is sufficiently strong to segregate linkers **5a** and **5b**, rather than form domains of higher entropy configurations with randomly distorted structures consisting of all three linkers. Therefore, the self-assembly of the mixture must involve both self-selection and self-recognition. With this mixture composition, the critical error correction process relies on the reversible coordination of the more labile pyridyl ligand (Figure 4.5c). The release of the pyridyl–Fe coordination in the case of negative self-recognition is favoured compared with a distortion of the carboxylic–Fe dimer interaction (Figure 4.5d). Presumably, if **5a** is next to **5b** within a network domain, the shorter linker **5a** is singly bonded and is easy to detach and be replaced by a long linker **5b**.

In contrast, a bi-pyridyl linker is stabilised when bonded at its two ends. The growth of the architecture will be through the gain of cooperativity, i.e. by the attachment of a larger number of ligands of the same length. The segregation is finally driven by the cooperativity of a coupled self-selection/self-assembly process featuring efficient error correction mechanisms. As a result, the read-out in the self-assembly process occurs with a higher degree of self-selection, and drives the self-assembly into locally segregated subdomains.

4.4 Multiple-Level (Hierarchical) Interaction

When different types of bonding modes with different strengths are at work, hierarchical structures can be formed, which often introduce complexity in the self-assembly. Here we show an example of a system involving metal–ligand coordination and hydrogen bonding. A threefold symmetric molecule, trimesic acids (TMA, **7**), forms clover-leave $Fe(TMA)_4$ complexes with Fe on Cu(100) (Figure 4.6a) [73]. As shown in the models illustrated in Figure 4.6a, the correlated attachment of the carboxylate ligands adapts a clockwise or anticlockwise orientation, defining the handedness of the complexes. In Figure 4.6a, two mirror-symmetric complexes are denoted as ***R*** and ***S***, respectively. The resulting symmetry break accounts for the chirality of the complexes. The corresponding rotation of the carbon backbone is strictly correlated for all TMA molecules in a given complex and accounts for intermolecular hydrogen bonds. This example shows that 2D chirality can be expressed in metal-coordination assemblies of archiral molecular linkers when secondary interactions involve.

The mononuclear chiral complexes shown in Figure 4.6a are antecedents for the higher level polynuclear nanogrids after annealing treatment at 350 K (see Figure 4.6b) [74]. Each of these nanogrids consists of 16 molecules of **7** and 9 Fe. These nanogrids inherit the chiral nature of the central $Fe(TMA)_4$ complexes. Furthermore, after annealing treatment at 400 K, the nanogrids are interconnected by hydrogen bonds forming mesoscale networks comprising a regular arrangement of homochiral nanocavities. Figure 4.6c shows two homochiral domains assembled by enantiomeric nanogrids (labelled ***R*** and ***S***), marked by coloured rectangles. The only control parameter in the assembly is temperature. TMA molecules and Fe atoms represent the primary units that are employed for the formation of secondary chiral complexes. The complexes are antecedents for tertiary polynuclear nanogrids which are in turn the supramolecular motifs for the assembly of homochiral nanocavity arrays. The control of self-assembly schemes involving hierarchical structures represents an appealing possibility for the bottom-up fabrication of complex functional materials.

Klappenberger and coworkers have studied the self-assembly of ethynyl-iodophenanthrene (**8**) on Ag(111) [75]. This molecule features dissymmetry in the geometry of its backbone (phenanthrene) and the reactivity of its functional groups (an H-terminated alkynyl group in the 2nd position and an iodine substitution in the 7th position). At a low annealing temperature (≤ 130 K), **8** forms a honeycomb

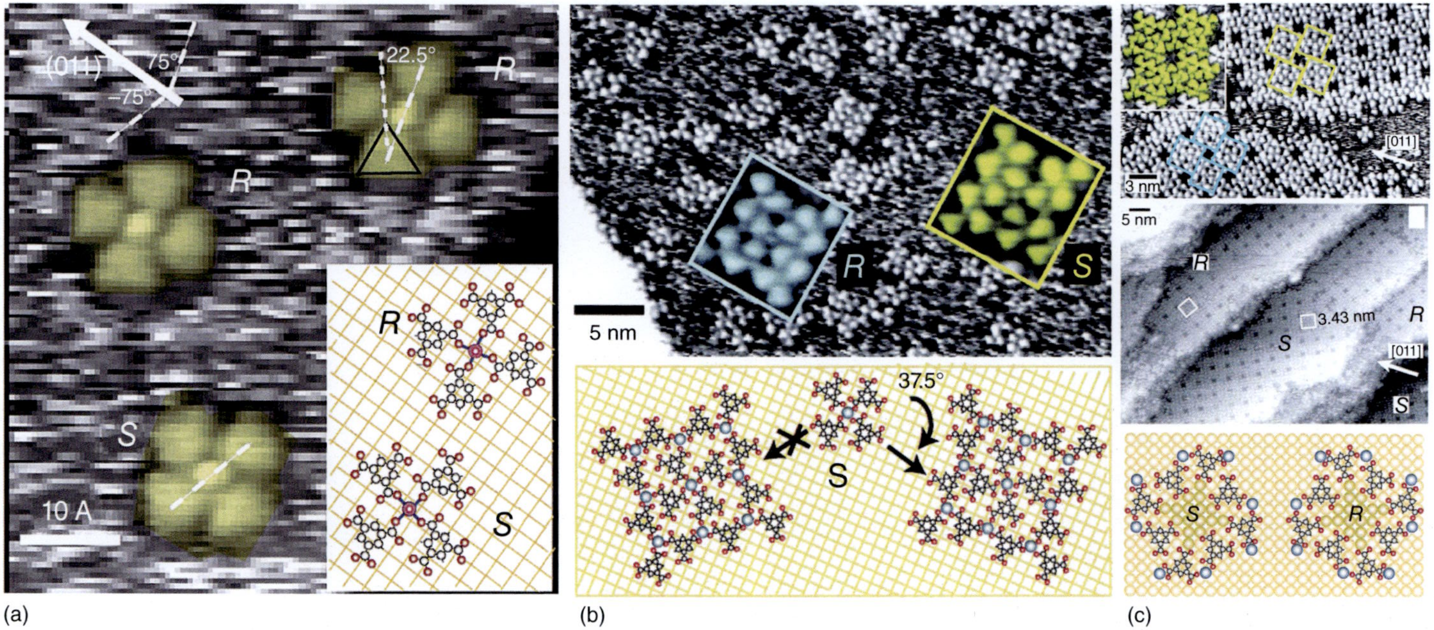

Figure 4.6 (a) STM image showing the two $Fe(TMA)_4$ stereoisomers on the Cu(100) surface, labelled ***R*** and ***S***, representing mirror-symmetric species. The corresponding model depicts a unidentate coordination of the carboxylate ligands to the central Fe atom. (b) Assembly of polynuclear nanogrids evolves upon annealing at 350 K. The insets and model below reveal the respective core units of the chiral nanogrids. The respective mirror-symmetric configurations (labelled ***S*** and ***R***) are indicated with a yellow and turquoise background. (c) Formation of extended nanocavity arrays triggered by 400 K annealing. Two homochiral domains are assembled consisting of pure enantiomers (labelled ***R*** and ***S***), marked by coloured rectangles. Source: (a) Adapted with permission from Messina et al. [73]. Copyright (2002) American Chemical Society. (b) and (c) are adapted with permission from Spillmann et al. [74]. Copyright (2003) American Chemical Society.

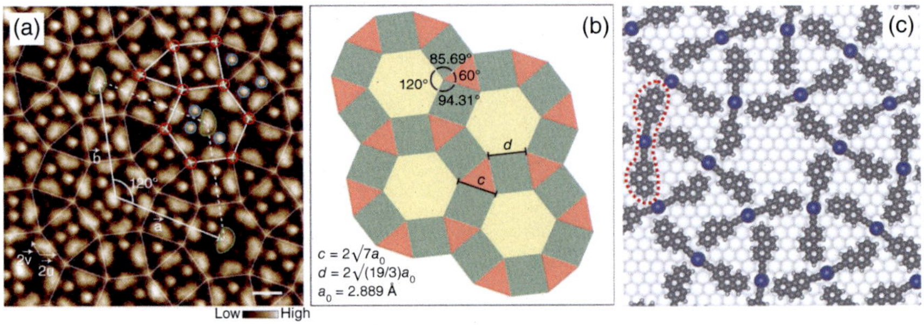

Figure 4.7 (a) High-resolution STM image of (3.4.6.4) semi-regular titling (Scale bar: 1 nm) with a commensurate model as well as proposed registry superimposed. Red circles highlight Ag complex nodes, blue circles highlight trapped adatoms, and green outlines highlight trapped single molecules of **8**. (b) Tile representation with geometric parameters. (c) DFT-calculated structure on Ag(111). The building block is highlighted with a red dashed outline. Ag, blue; C, grey; H, white. Scale bars: 10 Å. Source: (a) Adapted by permission from Springer Customer Service Centre GmbH: Springer Nature, Nature Chemistry Zhang et al. [75], copyright 2018. (b) Zhang et al. [75].

network. Following thermal annealing up to 350 K, a well-ordered semi-regular (3.4.6.4) tiling emerges, as shown in Figure 4.7a,b. They used complementary STM, X-ray spectroscopy (XS), and density functional theory (DFT) to reveal the chemical nature of the phase transition and elucidate a convergent reaction pathway, including an H transfer from the terminal alkyne to the dehalogenated site on the phenanthrene backbone mediated by in situ generated catalytic Ag complexes at low temperature. The semi-regular (3.4.6.4) tiling is constructed from interwoven 12-segmented rings (Figure 4.7c). The crossings of the interwoven rings (red circles in Figure 4.7a) represent the fourfold vertices of the architecture consisting of three types of a polygon: triangles, tetragons, and hexagons. This structure is stabilised by non-covalent interactions between the C–H moieties and the electron-rich transition metal [76–78] as well as the alkynyl π-systems [79].

4.5 Multiple Binding Modes

When multiple binding modes are involved in self-assembly, the output structures are intricate to predict. Our group studied the self-assembly of Eu atoms coordinated with two heterotypic ligands of quarterphenyl-4,4″-dicarbonitrile (**9**) and 4′,4″″-(1,4-phenylene)bis(2,2′:6′,2″-terpyridine) (**10**) on Au(111) surface [80]. At a molecular density ratio of [**9**]/[**10**] = 0.9, the self-assembly generates 1D sinuous supramolecular chains and three-way junctions (Figure 4.8a,b). As illustrated in the model (Figure 4.8c), the 1D chains are based on fourfold coordination nodes, whereby an Eu node is coordinated complementarily by a terpyridyl and a carbonitrile moiety, respectively. A closer inspection of Figure 4.8b reveals that a three-way junction comprises a terpyridyl attached to two carbonitrile groups. As detailed in the model shown in Figure 4.8c, in a three-way junction an Eu atom is surrounded by five nitrogen atoms, three from the tridentate terpyridyl group

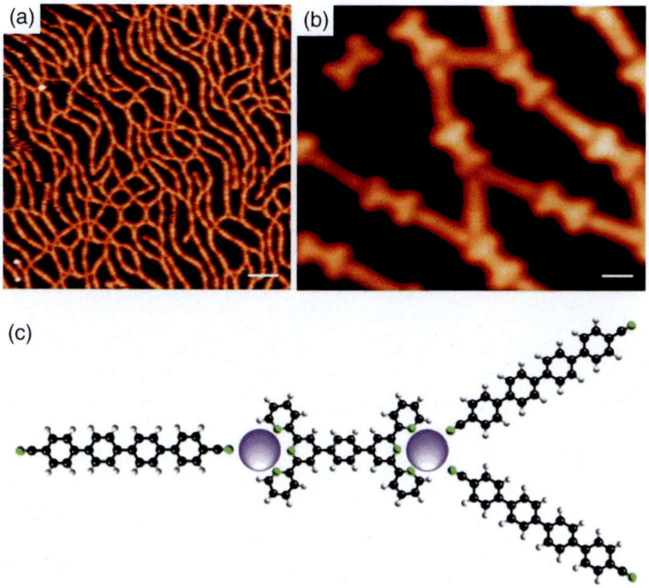

Figure 4.8 Mixture of 1D supramolecular chains and three-way junctions inter-connected assemblies. (a) A STM topograph showing mixed 1D supramolecular chains and inter-connected networks afforded by the three-way junctions (Scale bar: 10 nm). (b) A magnified STM view highlighting the two plausible Eu coordination schemes: fourfold corresponding to 1D supramolecular chains and fivefold assigned to the three-way junctions (Scale bar: 1 nm). (c) An atomistic model of the Eu coordination schemes of the fourfold and fivefold motifs. Colour code: C, black; N, green; H, white; Eu, purple. Scanning temperature: 77 K. Source: (a) Adapted with permission from Lyu et al. [80] with permission of The Royal Society of Chemistry. (c) Lyu et al. [80].

and two from two carbonitrile groups, thus giving rise to a fivefold metal–organic coordination centre.

We attribute the fourfold and fivefold coordination schemes to the large ionic radii of lanthanide elements which allows remarkable adaptability of their coordination sphere to the variation of the coordination number. We have varied the molecular density ratio of the two linkers and found that the population of the two types of coordination varies accordingly. At [**9**]/[**10**] = 0.7, Eu atoms are primarily ligated in a fourfold coordination scheme. With increasing [**9**]/[**10**] ratio, the fourfold coordination decreases while the fivefold coordination increases. Eventually fivefold coordination reaches 95% at [**9**]/[**10**] = 1.8. Hence, regulating the molecular density ratio of the two components allows us to choose specific coordination with very high selectivity.

With a stoichiometric ratio of [**9**]/[**10**] = 1.8, the fivefold coordination prevails. Two types of coordination frameworks are formed. Figure 4.9a depicts the 2D-MOF exhibiting hexagonal geometry with a periodicity of 8.10 ± 0.05 nm. A high-resolution STM image (Figure 4.9b) reveals pores of dodecagonal shape whereas each dodecagon has six molecular linkers **9** and six molecular linkers **10**, arranged alternatively, as its sides. Three Eu nodes are joined by an equilateral

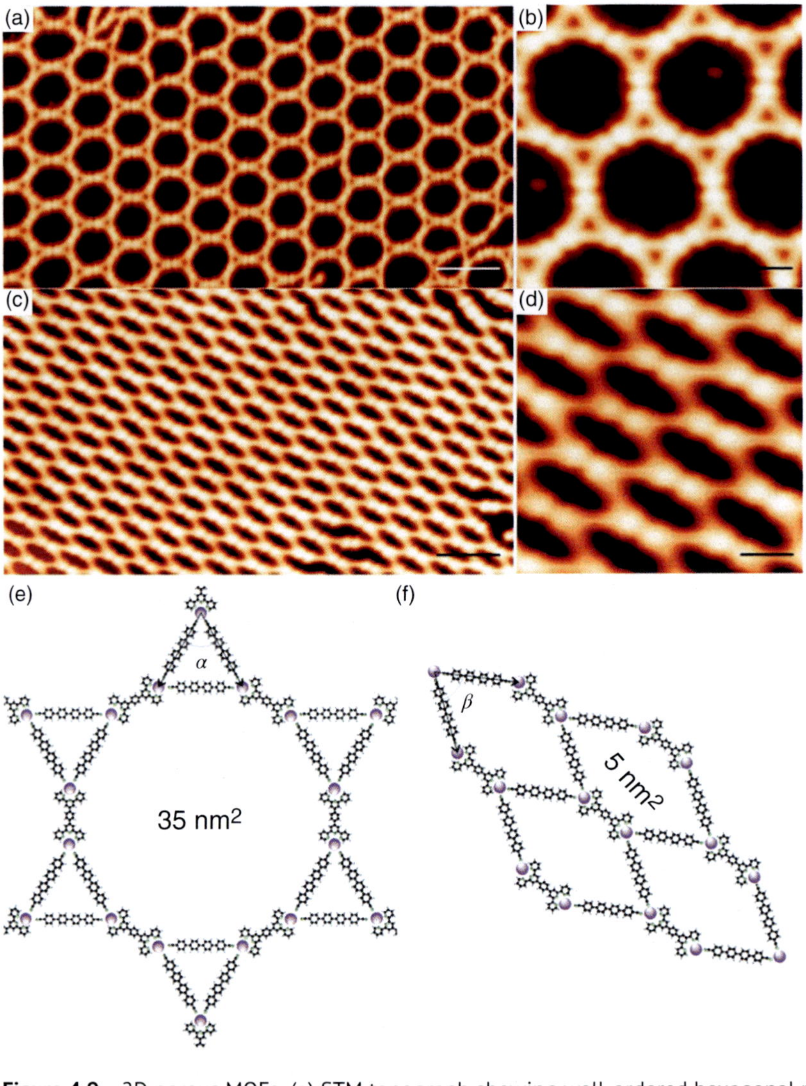

Figure 4.9 2D porous MOFs. (a) STM topograph showing well-ordered hexagonal porous 2D MOFs formed at [**9**]/[**10**] = 1.8 (Scale bar: 10 nm). (b) High resolution STM image of (a) (Scale bar: 3 nm). (c) STM topograph displaying a fishing-net 2D MOFs formed at the same molecular ratio as (a) (Scale bar: 10 nm). (d) High resolution STM image of (c) (Scale bar: 2 nm). (e, f) Atomistic models of the hexagonal and the fishing-net structures respectively. Source: (a–f) Adapted with permission from Lyu et al. [80] with permission of The Royal Society of Chemistry.

triangle comprising three molecular linkers **9**. An atomistic model is shown in Figure 4.9e, featuring the dodecagonal pore and the triangles. The inner area of a dodecagonal nanopore is ~35 nm^2. The surface molecular density is 10 molecular linkers (**9**) and 5 molecular linkers (**10**) per 100 nm^2. Figure 4.9c reveals another 2D-MOF structure, a fishing-net 2D-MOF. A high-resolution STM topograph and an atomistic model of this phase are shown in Figure 4.9d,f. The fishing-net structure contains elongated hexagonal pores, which are surrounded by two molecular linkers **10** and four molecular linkers **9**.

We tested the thermal stability of the two network structures after 120 °C annealing. The hexagonal structure was partially distorted and disrupted, while the fishing-net phase preserved its structural integrity. After 170 °C annealing, no complete hexagonal units were detected, indicating at this temperature, the hexagonal phase became unstable. In contrast, large areas of the fishing-net phase were observed. Apparently, the fishing-net phase is energetically more stable, whereas the hexagonal phase is metastable. A plausible mechanism is a porous structure containing open voids and less stable as compared to denser structures. The hexagonal phase contains 2D pores of ~35 nm^2 void, whilst the fishing-net phase contains 2D pores of ~5 nm^2 void. As a result, the former structure is less stable than the latter one.

In another work, our group studied the self-assembly of 1,3,5-tris[4-(pyridin-4-yl)phenyl)]benzene (**11**)), shown in Figure 4.10a, on Cu(111) [81]. Figure 4.10b shows the molecules are connected with one another at their three tips, forming a complex yet periodic 2D framework articulated with twofold and threefold

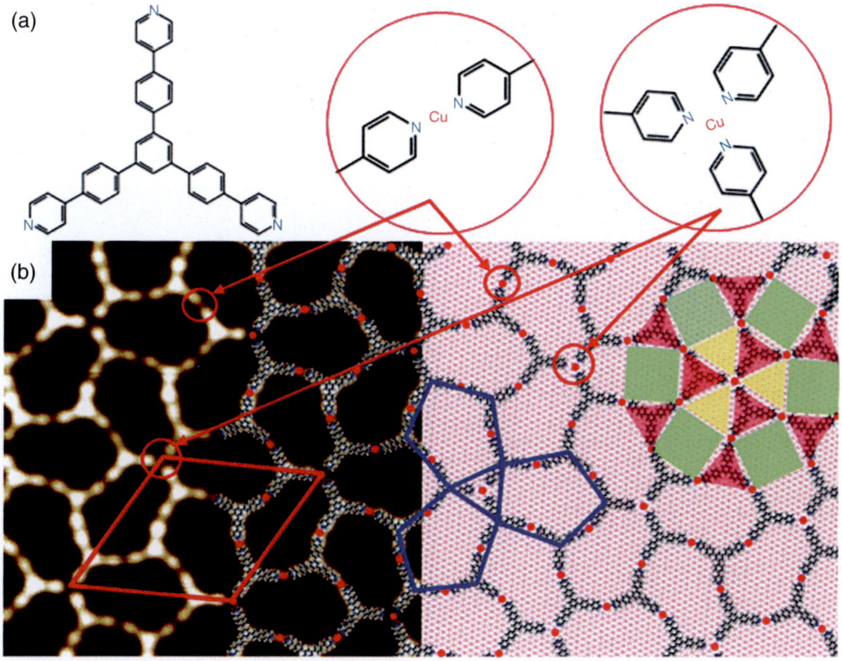

Figure 4.10 (a) Chemical structure of **11** and twofold and threefold Cu–pyridyl coordination motifs. (b) High-resolution STM image (15 nm × 15 nm, imaged at 5 K) of the framework self-assembled on Cu(111) and structural model. The red diamond-shaped frame marks a unit cell. The edges of the coloured squares and triangles link the neighbouring Cu adatoms to highlight the (3^6; 4, 3^2, 4, 3) non-Archimedean demi-regular tessellation. Source: (a) Yan et al. [81]. (b) Adapted with permission from Yan et al. [81] with permission of The Royal Society of Chemistry.

intermolecular connections (Figure 4.10b, circles) with a molecular model laid on top of the STM topograph. The diamond-shaped frame in Figure 4.10b depicts a unit cell, which renders a supercell of $\left(13\sqrt{3} \times 13\sqrt{3}\right)$ R30° with respect to the Cu(111) lattice. The unit cell contains five molecules **11** and seven Cu adatoms, six of which are twofold coordinated and one is threefold coordinated [81]. The network comprises pentagonal 2D pores, as indicated by the blue lines linking the adjacent molecules. To highlight the 2D Cu adatom lattice, the neighbouring Cu adatoms are linked by the edges of the coloured squares and triangles in the right part of Figure 4.10b. The lattice consists of equilateral triangles and squares fused together in an edge-to-edge manner. In the lattice, the central Cu atom is a vertex of six triangles, and the surrounding six Cu atoms are vertices of two squares and three triangles each. The lattice can be defined as (3^6; 4, 3^2, 4, 3), corresponding to one of the 20 2-uniform demi-regular tessellations [82]. Hereafter we refer to this structure as the D-phase.

The single domains of the D-phase homogeneously cover the entire surface, such that the domain boundaries exist only at atomic steps. Within the single domains, the D-phase is nearly defect-free (Figure 4.11a). A typical defect is shaded in green, which comprises three molecules that are rotated by 60° around the central Cu atom from their normal position. Such defects amount to less than 4% of the total area. It is worthwhile to point out that such defects do not disrupt the Cu lattice structure. Globally, the Cu lattice bears p6m symmetry; locally, one can recognise 12-fold rotational symmetry (cf. the blue frame in Figure 4.11a). The inset in Figure 4.11a is a 2D Fourier transformation (FT) of the Cu adatom lattice. The 12 spots are circled

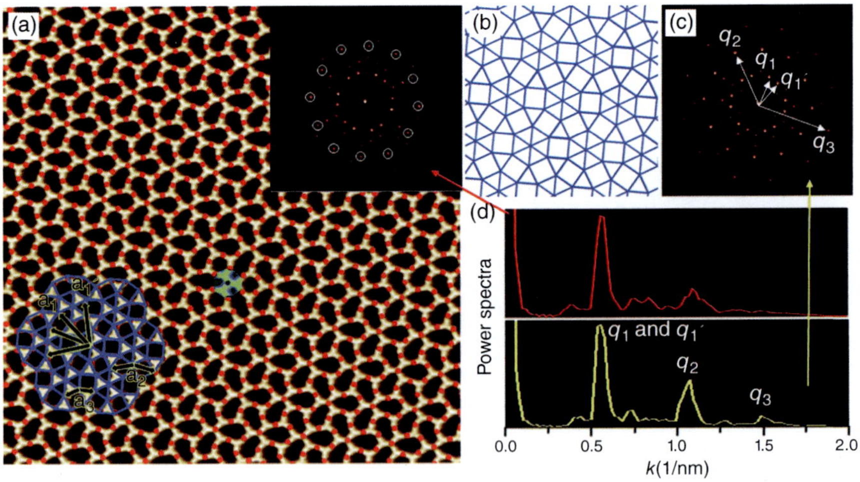

Figure 4.11 Comparison of the real-space arrangement and reciprocal pattern of the Cu adatom lattice and a perfect (3^6; 4, 3^2, 4, 3) tessellation. (a) STM topograph of the D-phase (60 nm × 60 nm, imaged at 5 K). Cu adatoms are marked by red dots. A defect is shaded in green. Inset: 2D FT of the Cu adatom lattice. (b) A perfect (3^6; 4, 3^2, 4, 3) tessellation. (c) 2D FT of the point lattice in (b). (d) FT power spectra of the inset in (a) and (c). Source: (a, c) Adapted with permission from Yan et al. [81] with permission of The Royal Society of Chemistry.

in the FT pattern, manifesting dodecagonal symmetry. Figure 4.11c is a 2D FT of a perfect lattice of the (3^6; 4, 3^2, 4, 3) tessellation shown in Figure 4.11b. The two FT patterns are nearly identical. The FT power spectra of the experimental structure and the ideal (3^6; 4, 3^2, 4, 3) lattice are plotted in Figure 4.11d for comparison. The good agreements confirm that the Cu adatoms in the D-phase represent a factual manifestation of the (3^6; 4, 3^2, 4, 3) lattice.

As the last example, we demonstrate that up to four types of coordination modes can be achieved by tuning the ratio of linker molecule **9** and Eu atoms [83]. This arises from the large size of the Eu centre and the flexible coordination chemistry of lanthanide elements. When the ratio of [**9**]/[Eu] is around 1.5, we observe the formation of irregular pores, as shown in Figure 4.12a. The porous structure is based on threefold coordination, as illustrated by the model. When the ratio of [**9**]/[Eu] is increased to 1.6, mixture structures consist of threefold rectangular pores and fourfold square pores are formed, hinting that uniform MOFs comprising of fourfold coordination could be achieved by increasing the ratio of [**9**]/[Eu]. When the ratio of [**9**]/[Eu] is 2.0, a square MOF structure is formed, as shown in Figure 4.12b. Each Eu vertex is linked by four rod-like molecules in fourfold coordination (as marked by the blue bars). The square MOFs are nearly defect-free and extend to 100 nm on the surface. Upon increasing the ratio of [**9**]/[Eu] to 2.5, most Eu centres are linked by five molecules in fivefold coordination, as shown in Figure 4.12c. The five-vertex stars interconnect in such a way that a random tilling of two squares and three equilateral triangles is defined. Further increasing the ratio of [**9**]/[Eu] to 3.0, irregular MOFs, including distorted squares and hexagonal lattices are observed, as shown in Figure 4.12d. It can be seen that six molecules are joined together in sixfold coordination. The network is thus a triangular network constructed by the molecules, as highlighted by the blue bars in Figure 4.12d.

When the ratio of [**9**]/[Eu] is about 2.6, we identify distinct coordination modes interconnected by the molecules and spanning complex, fully reticulated coordination networks (Figure 4.13a). The individual Eu centres are surrounded by four, five,

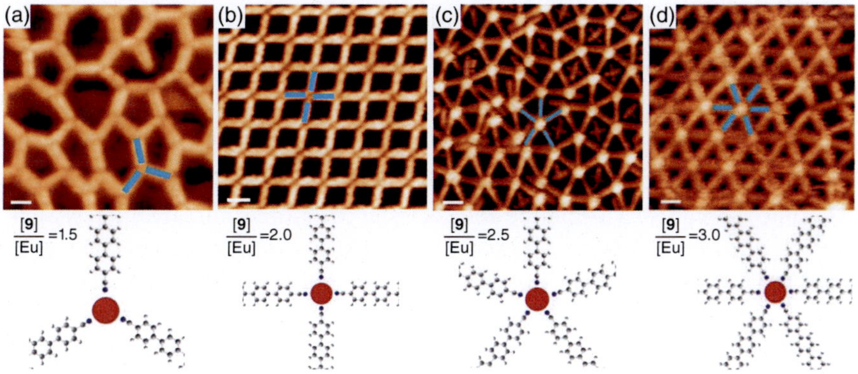

Figure 4.12 STM images of the 2D-MOFs assembled at different ratios of [**9**]/[Eu]: (a) 1.53. (b) 2.0. (c) 2.5. (d) 3.0. Scale bars: 2 nm. Source: (a–d) Adapted by permission from Springer Customer Service Centre GmbH: Springer Nature, Nature Chemistry Urgel et al. [83], copyright 2016.

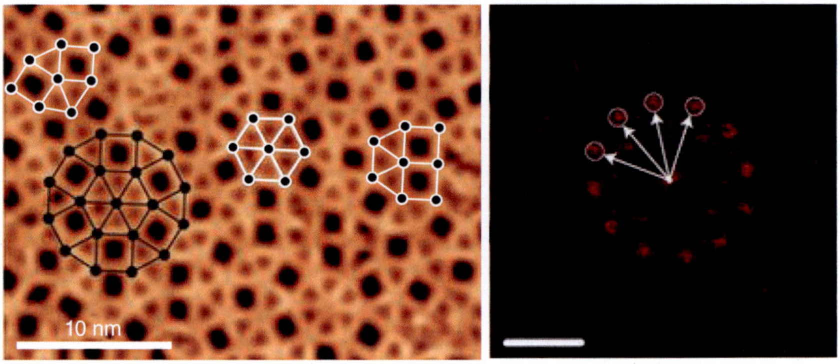

(a) (b)

Figure 4.13 (a) STM image of the dodecagonal quasicrystal MOF assembled on Au(111) by depositing molecule **9** and Eu with the ratio of [**9**]/[Eu] at about 2.6. (b) FFT of the quasicrystal MOF (Scale bar: 0.84 nm^{-1}). Source: Adapted by permission from Springer Customer Service Centre GmbH: Springer Nature, Nature Chemistry Urgel et al. [83], copyright 2016.

or six molecules, and the sixfold node is the centre of a dodecagonal distribution of Eu centres. Moreover, the Eu vertexes are distributed in such a fashion that the design can be interpreted as a tessellation based on a random tiling of squares and triangles. The overall arrangement is closely reminiscent of the square–triangle random tiling dodecagonal quasicrystals described in previous works [84–87], in which it was also recognised by the 12-fold rotational symmetry fast Fourier transform (FFT) image (Figure 4.13b). The statistical analysis gives a triangle to square number ratio of 2.36 ± 0.01 and comes close to the expectation for a perfect quasicrystal ($\approx 4/\sqrt{3} = 2.31$) [85]. The expressed dodecagonal units are not only constituents of random tiling quasicrystals but also central building blocks of perfect Stämpfli–Gähler quasicrystalline lattices [88, 89]. The FFT of STM image (Figure 4.13b) clearly reveals a set of spots reminiscent of modulated ring patterns with dodecagonal symmetry. This structure is rationalised as a random square–triangle tiling that comprises a repetition of dodecagonal structural units with adequate distribution of triangles and squares. We conclude that the Eu-directed assembly of polyphenyl-dicarbonitrile linkers on Au(111) drives the expression of dodecagonal random tiling quasicrystals.

It is known that two major issues affect supramolecular architectures: (i) intermolecular interactions and (ii) interactions between molecules and substrate. On Au(111), the spontaneous formation and coexistence of three, four, five, and six-fold planar coordination modes are an exceptional situation. This phenomenon can be ascribed to (i) the large and flexible coordination sphere of the lanthanide centre and (ii) weak substrate-molecule effects, which are reflected by the conservation of the substrate's herringbone reconstruction. This result demonstrates that the flexible lanthanide coordination chemistry facilitates the rational design of versatile metal–organic architectures.

4.6 Summary and Outlook

To date, the majority of the on-surface coordination self-assembly reported in the literature is simple systems [29]. The examples presented in this article, however, demonstrate that emergence of complexity in on-surface coordination of self-assembly is a norm rather than the exception when multiple components or multiple interactions are involved. On one hand, complexity limits our capability of designing a structure per se because the self-assembly is less or even not predictable. On the other hand, complexity offers rich opportunities for forming unexpected structures, and often, surprisingly beautiful. In this regard, it provides a playground for satisfying curiosity. As many complex structures do not have long-range ordering or are very complex, as exemplified in this article, it's a big challenge to determine their structures using diffraction techniques. The on-surface coordination structures can be resolved by the advanced microscopic technique at a sub-molecular scale in real space. This technique is thus of great advantage to study the complex processes and such studies may help us to understand the underlying mechanisms of complex phenomena. To this end, there are plenty of rooms to explore the complexity in on-surface coordination self-assembly.

References

1 Lehn, J.M. (1993). Supramolecular chemistry. *Science* 260: 1762–1763.
2 Lehn, J.M. (2002). Toward complex matter: supramolecular chemistry and self-organization. *Proc. Natl. Acad. Sci. U.S.A.* 99: 4763–4768.
3 Whitesides, G.M. and Grzybowski, B. (2002). Self-assembly at all scales. *Science* 295: 2418–2421.
4 Langner, A., Tait, S.L., Lin, N. et al. (2007). Self-recognition and self-selection in multicomponent supramolecular coordination networks on surfaces. *Proc. Natl. Acad. Sci. U.S.A.* 104: 17927–17930.
5 Gambardella, P., Stepanow, S., Dmitriev, A. et al. (2009). Supramolecular control of the magnetic anisotropy in two-dimensional high-spin Fe arrays at a metal interface. *Nat. Mater.* 8: 189–193.
6 Umbach, T.R., Bernien, M., Hermanns, C.F. et al. (2012). Ferromagnetic coupling of mononuclear Fe centers in a self-assembled metal-organic network on Au(111). *Phys. Rev. Lett.* 109: 267207.
7 Abdurakhmanova, N., Tseng, T.C., Langner, A. et al. (2013). Superexchange-mediated ferromagnetic coupling in two-dimensional Ni-TCNQ networks on metal surfaces. *Phys. Rev. Lett.* 110: 027202.
8 Giovanelli, L., Savoyant, A., Abel, M. et al. (2014). Magnetic coupling and single-ion anisotropy in surface-supported Mn-based metal–organic networks. *J. Phys. Chem. C* 118: 11738–11744.

9 Lin, T., Kuang, G., Wang, W., and Lin, N. (2014). Two-dimensional lattice of out-of-plane dinuclear iron centers exhibiting Kondo resonance. *ACS Nano* 8: 8310–8316.

10 Fabris, S., Stepanow, S., Lin, N. et al. (2011). Oxygen dissociation by concerted action of di-iron centers in metal-organic coordination networks at surfaces: modeling non-heme iron enzymes. *Nano Lett.* 11: 5414–5420.

11 Grumelli, D., Wurster, B., Stepanow, S., and Kern, K. (2013). Bio-inspired nanocatalysts for the oxygen reduction reaction. *Nat. Commun.* 4: 2904.

12 Feng, M., Sun, H., Zhao, J., and Petek, H. (2014). Self-catalyzed carbon dioxide adsorption by metal–organic chains on gold surfaces. *ACS Nano* 8: 8644–8652.

13 Stepanow, S., Lingenfelder, M., Dmitriev, A. et al. (2004). Steering molecular organization and host–guest interactions using two-dimensional nanoporous coordination systems. *Nat. Mater.* 3: 229–233.

14 Stepanow, S., Lin, N., Barth, J.V., and Kern, K. (2006). Non-covalent binding of fullerenes and biomolecules at surface-supported metallosupramolecular receptors. *Chem. Commun.*: 2153–2155.

15 Kühnea, D., Klappenberger, F., Krennera, W. et al. (2010). Rotational and constitutional dynamics of caged supramolecules. *Proc. Natl. Acad. Sci. U.S.A.* 107: 21332–21336.

16 Pivetta, M., Pacchioni, G.E., Schlickum, U. et al. (2013). Formation of Fe cluster superlattice in a metal-organic quantum-box network. *Phys. Rev. Lett.* 110: 086102.

17 Nowakowska, S., Wackerlin, A., Kawai, S. et al. (2015). Interplay of weak interactions in the atom-by-atom condensation of xenon within quantum boxes. *Nat. Commun.* 6: 6071.

18 Zhang, Q., Kuang, G., Pang, R. et al. (2015). Switching molecular Kondo effect via supramolecular interaction. *ACS Nano* 9: 12521–12528.

19 Zhang, R., Lyu, G., Chen, C. et al. (2015). Two-dimensional superlattices of Bi nanoclusters formed on a Au(111) surface using porous supramolecular templates. *ACS Nano* 9: 8547–8553.

20 Klappenberger, F., Kuhne, D., Krenner, W. et al. (2011). Tunable quantum dot arrays formed from self-assembled metal-organic networks. *Phys. Rev. Lett.* 106: 026802.

21 Wang, S., Wang, W., Tan, L.Z. et al. (2013). Tuning two-dimensional band structure of Cu(111) surface-state electrons that interplay with artificial supramolecular architectures. *Phys. Rev. B* 88.

22 Zhang, J., Shchyrba, A., Nowakowska, S. et al. (2014). Probing the spatial and momentum distribution of confined surface states in a metal coordination network. *Chem. Commun.* 50: 12289–12292.

23 Wang, W., Hong, Y., Shi, X. et al. (2010). Inspecting metal-coordination-induced perturbation of molecular ligand orbitals at a submolecular resolution. *J. Phys. Chem. Lett.* 1: 2295–2298.

24 Wang, W., Shi, X., Wang, S. et al. (2013). Cooperative modulation of electronic structures of aromatic molecules coupled to multiple metal contacts. *Phys. Rev. Lett.* 110: 046802.

25 Wang, Z.F., Liu, Z., and Liu, F. (2013). Organic topological insulators in organometallic lattices. *Nat. Commun.* 4: 1471.

26 Zhang, X. and Zhao, M. (2015). Robust half-metallicity and topological aspects in two-dimensional Cu-TPyB. *Sci. Rep.* 5: 14098.

27 Stepanow, S., Lin, N., and Barth, J.V. (2008). Modular assembly of low-dimensional coordination architectures on metal surfaces. *J. Phys. Condens. Matter* 20: 184002.

28 Lin, N., Stepanow, S., Ruben, M., and Barth, J.V. (2009). Surface-confined supramolecular coordination chemistry. *Top. Curr. Chem.* 287: 1–44.

29 Dong, L., Gao, Z.A., and Lin, N. (2016). Self-assembly of metal–organic coordination structures on surfaces. *Prog. Surf. Sci.* 91: 101–135.

30 Tait, S.L., Langner, A., Lin, N. et al. (2007). One-dimensional self-assembled molecular chains on Cu(100): interplay between surface-assisted coordination chemistry and substrate commensurability. *J. Phys. Chem. C* 111: 10982–10987.

31 Boscoboinik, J.A., Calaza, F.C., Habeeb, Z. et al. (2010). One-dimensional supramolecular surface structures: 1,4-diisocyanobenzene on Au(111) surfaces. *Phys. Chem. Chem. Phys.* 12: 11624–11629.

32 Heim, D., Écija, D., Seufert, K. et al. (2010). Self-assembly of flexible one-dimensional coordination polymers on metal surfaces. *J. Am. Chem. Soc.* 132: 6783–6790.

33 Wang, W., Shi, X., Wang, S. et al. (2011). Single-molecule resolution of an organometallic intermediate in a surface-supported Ullmann coupling reaction. *J. Am. Chem. Soc.* 133: 13264–13267.

34 Fan, Q., Wang, C., Han, Y. et al. (2014). Surface-assisted formation, assembly, and dynamics of planar organometallic macrocycles and zigzag shaped polymer chains with C—Cu—C bonds. *ACS Nano* 8: 709–718.

35 Shchyrba, A., Wackerlin, C., Nowakowski, J. et al. (2014). Controlling the dimensionality of on-surface coordination polymers via endo- or exoligation. *J. Am. Chem. Soc.* 136: 9355–9363.

36 Skomski, D., Tempas, C.D., Smith, K.A., and Tait, S.L. (2014). Redox-active on-surface assembly of metal-organic chains with single-site Pt(II). *J. Am. Chem. Soc.* 136: 9862–9865.

37 Knor, M., Gao, H.-Y., Amirjalayer, S. et al. (2015). Stereoselective formation of coordination polymers with 1,4-diaminonaphthalene on various Cu substrates. *Chem. Commun.* 51: 10854–10857.

38 Liu, J., Chen, Q., Xiao, L. et al. (2015). Lattice-directed formation of covalent and organometallic molecular wires by terminal alkynes on Ag surfaces. *ACS Nano* 9: 6305.

39 Skomski, D., Tempas, C.D., Cook, B.J. et al. (2015). Two- and three-electron oxidation of single-site vanadium centers at surfaces by ligand design. *J. Am. Chem. Soc.* 137: 7898–7902.

40 Heim, D., Seufert, K., Auwarter, W. et al. (2010). Surface-assisted assembly of discrete porphyrin-based cyclic supramolecules. *Nano Lett.* 10: 122–128.

41 Wang, W., Wang, S., Hong, Y. et al. (2011). Selective supramolecular assembly of multifunctional ligands on a Cu(111) surface: metallacycles, propeller trimers and linear chains. *Chem. Commun.* 47: 10073–10075.

42 Lin, T., Shang, X.S., Liu, P.N., and Lin, N. (2013). Multicomponent assembly of supramolecular coordination polygons on a Au(111) surface. *J. Phys. Chem. C* 117: 23027–23033.

43 Stepanow, S., Lin, N., Barth, J.V., and Kern, K. (2006). Surface-template assembly of two-dimensional metal-organic coordination networks. *J. Phys. Chem. B* 110: 23472–23477.

44 Stepanow, S., Lin, N., Payer, D. et al. (2007). Surface-assisted assembly of 2D metal–organic networks that exhibit unusual threefold coordination symmetry. *Angew. Chem.* 119: 724–727.

45 Matena, M., Riehm, T., Stöhr, M. et al. (2008). Transforming surface coordination polymers into covalent surface polymers: linked polycondensed aromatics through oligomerization of N-heterocyclic carbene intermediates. *Angew. Chem. Int. Ed. Engl.* 120: 2448–2451.

46 Tait, S.L., Langner, A., Lin, N. et al. (2008). Assembling isostructural metal-organic coordination architectures on Cu(100), Ag(100) and Ag(111) substrates. *ChemPhysChem* 9: 2495–2499.

47 Shi, Z. and Lin, N. (2009). Porphyrin-based two-dimensional coordination Kagome lattice self-assembled on a Au(111) surface. *J. Am. Chem. Soc.* 131: 5376–5377.

48 Shi, Z. and Lin, N. (2010). Structural and chemical control in assembly of multicomponent metal-organic coordination networks on a surface. *J. Am. Chem. Soc.* 132: 10756–10761.

49 Shi, Z. and Lin, N. (2010). Self-assembly of a two-dimensional bimetallic coordination framework and dynamic control of reversible conversions to homo-metallic hydrogen-bond arrays. *ChemPhysChem* 11: 97–100.

50 Stepanow, S., Ohmann, R., Leroy, F. et al. (2010). Rational design of two-dimensional nanoscale networks by electrostatic interactions at surfaces. *ACS Nano* 4: 1813–1820.

51 Shi, Z., Liu, J., Lin, T. et al. (2011). Thermodynamics and selectivity of two-dimensional metallo-supramolecular self-assembly resolved at molecular scale. *J. Am. Chem. Soc.* 133: 6150–6153.

52 Walch, H., Dienstmaier, J., Eder, G. et al. (2011). Extended two-dimensional metal-organic frameworks based on thiolate-copper coordination bonds. *J. Am. Chem. Soc.* 133: 7909–7915.

53 Li, Y., Xiao, J., Shubina, T.E. et al. (2012). Coordination and metalation bifunctionality of Cu with 5,10,15,20-tetra(4-pyridyl)porphyrin: toward a mixed-valence two-dimensional coordination network. *J. Am. Chem. Soc.* 134: 6401–6408.

54 Lin, T., Kuang, G., Shang, X.S. et al. (2014). Self-assembly of metal-organic coordination networks using on-surface synthesized ligands. *Chem. Commun.* 50: 15327–15329.

55 Matena, M., Björk, J., Wahl, M. et al. (2014). On-surface synthesis of a two-dimensional porous coordination network: unraveling adsorbate interactions. *Phys. Rev. B* 90: 125408.

56 Hieulle, J., Peyrot, D., Jiang, Z., and Silly, F. (2015). Engineering two-dimensional hybrid NaCl-organic coordinated nanoarchitectures on metal surfaces. *Chem. Commun.* 51: 13162–13165.

57 Lyu, G., Zhang, R., Zhang, X. et al. (2015). On-surface assembly of low-dimensional Pb-coordinated metal–organic structures. *J. Mater. Chem. C* 3: 3252–3257.

58 Rodriguez-Fernandez, J., Lauwaet, K., Herranz, M.A. et al. (2015). Temperature-controlled metal/ligand stoichiometric ratio in Ag-TCNE coordination networks. *J. Chem. Phys.* 142: 101930.

59 Urgel, J.I., Schwarz, M., Garnica, M. et al. (2015). Controlling coordination reactions and assembly on a Cu(111) supported boron nitride monolayer. *J. Am. Chem. Soc.* 137: 2420–2423.

60 Newkome, G.R., Wang, P., Moorefield, C.N. et al. (2006). Nanoassembly of a fractal polymer: a molecular "Sierpiński hexagonal gasket". *Science* 312: 1782–1785.

61 Sarkar, R., Guo, K., Moorefield, C.N. et al. (2014). One-step multicomponent self-assembly of a first-generation Sierpiński triangle: from fractal design to chemical reality. *Angew. Chem. Int. Ed. Engl.* 53: 12182–12185.

62 Wang, M., Wang, C., Hao, X.Q. et al. (2014). Hexagon wreaths: self-assembly of discrete supramolecular fractal architectures using multitopic terpyridine ligands. *J. Am. Chem. Soc.* 136: 6664–6671.

63 Shang, J., Wang, Y., Chen, M. et al. (2015). Assembling molecular Sierpiński triangle fractals. *Nat. Chem.* 7: 389–393.

64 Sun, Q., Cai, L., Ma, H. et al. (2015). On-surface construction of a metal-organic Sierpiński triangle. *Chem. Commun.* 51: 14164–14166.

65 Zhang, X., Li, N., Gu, G.-C. et al. (2015). Controlling molecular growth between fractals and crystals on surfaces. *ACS Nano* 9: 11909–11915.

66 Gu, G., Li, N., Liu, L. et al. (2016). Growth of covalently bonded Sierpiński triangles up to the second generation. *RSC Adv.* 6: 66548–66552.

67 Rastgoo-Lahrood, A., Martsinovich, N., Lischka, M. et al. (2016). From Au-thiolate chains to thioether Sierpiński triangles: the versatile surface chemistry of 1,3,5-tris(4-mercaptophenyl)benzene on Au(111). *ACS Nano* 10: 10901–10911.

68 Zhang, X., Li, N., Liu, L. et al. (2016). Robust Sierpiński triangle fractals on symmetry-mismatched Ag(100). *Chem. Commun.* 52: 10578–10581.

69 Li, C., Zhang, X., Li, N. et al. (2017). Construction of Sierpiński triangles up to the fifth order. *J. Am. Chem. Soc.* 139: 13749–13753.

70 Zhang, X., Gu, G., Li, N. et al. (2018). One-dimensional molecular chains formed by Sierpiński triangles on Au(111). *RSC Adv.* 8: 1852–1856.

71 Mao, X.F., Lin, T., Adisoejoso, J. et al. (2013). Coordination self-assembly of bromo-phenyl and pyridyl functionalized porphyrins with Fe on an Au(111) surface. *Phys. Chem. Chem. Phys.* 15: 12447–12450.

72 Shi, Z., Lin, T., Liu, J. et al. (2011). Regulating a two-dimensional metallo-supramolecular self-assembly of multiple outputs. *Cryst. Eng. Comm.* 13: 5532–5534.

73 Messina, P., Dmitriev, A., Lin, N. et al. (2002). Direct observation of chiral metal-organic complexes assembled on a Cu(100) surface. *J. Am. Chem. Soc.* 124: 14000–14001.

74 Spillmann, H., Dmitriev, A., Lin, N. et al. (2003). Hierarchical assembly of two-dimensional homochiral nanocavity arrays. *J. Am. Chem. Soc.* 125: 10725–10728.

75 Zhang, Y.-Q., Paszkiewicz, M., Du, P. et al. (2018). Complex supramolecular interfacial tessellation through convergent multi-step reaction of a dissymmetric simple organic precursor. *Nat. Chem.* 10: 296–304.

76 Brammer, L., Zhao, D., Ladipo, F.T., and Braddock-Wilking, J. (1995). Hydrogen bonds involving transition metal centres – a brief review. *Acta Crystallogr. B* 51: 632–640.

77 Braga, D., Grepioni, F., and Desiraju, G.R. (1998). Crystal engineering and organometallic architecture. *Chem. Rev.* 98: 1375–1406.

78 Steiner, T. (2002). The hydrogen bond in the solid state. *Angew. Chem. Int. Ed.* 41: 48–76.

79 Zhang, Y.-Q., Björk, J., Weber, P. et al. (2015). Unusual deprotonated alkynyl hydrogen bonding in metal-supported hydrocarbon assembly. *J. Phys. Chem. C* 119: 9669–9679.

80 Lyu, G., Zhang, Q., Urgel, J.I. et al. (2016). Tunable lanthanide-directed metallosupramolecular networks by exploiting coordinative flexibility through ligand stoichiometry. *Chem. Commun.* 52: 1618–1621.

81 Yan, L., Kuang, G., Zhang, Q. et al. (2017). Self-assembly of a binodal metal-organic framework exhibiting a demi-regular lattice. *Faraday Discuss.* 204: 111–121.

82 Grünbaum, B. and Shephard, G.C. (1987). *Tilings and Patterns.* New York: W. H. Freeman & Co.

83 Urgel, J.I., Ecija, D., Lyu, G. et al. (2016). Quasicrystallinity expressed in two-dimensional coordination networks. *Nat. Chem.* 8: 657–662.

84 Chen, H., Li, D.X., and Kuo, K.H. (1988). New type of two-dimensional quasicrystal with twelvefold rotational symmetry. *Phys. Rev. Lett.* 60: 1645–1648.

85 Hayashida, K., Dotera, T., Takano, A., and Matsushita, Y. (2007). Polymeric quasicrystal: mesoscopic quasicrystalline tiling in ABC star polymers. *Phys. Rev. Lett.* 98: 195502.

86 Talapin, D.V., Shevchenko, E.V., Bodnarchuk, M.I. et al. (2009). Quasicrystalline order in self-assembled binary nanoparticle superlattices. *Nature* 461: 964–967.

87 Xiao, C., Fujita, N., Miyasaka, K. et al. (2012). Dodecagonal tiling in mesoporous silica. *Nature* 487: 349–353.

88 Stampfli, P. (1986). A dodecagonal quasiperiodic lattice in two dimensions. *Helv. Phys. Acta* 59: 1260–1263.

89 Gähler, F. (1988). *Quasicrystalline Materials.* World Scientific.

5

Complexity in Two-Dimensional Assembly: Quasicrystalline Structures

S. Alex Kandel

University of Notre Dame, Department of Chemistry and Biochemistry, Notre Dame, IN 46556, USA

5.1 History

The first quasicrystal was discovered in 1982 by Shechtman [1, 2]. The discovery stemmed from the observation of fivefold rotational symmetry in the electron diffraction pattern of an Al–Mn alloy that had been rapidly cooled to induce crystallisation from a bimetallic melt. The finding was initially a surprising one, as the sharp diffraction pattern indicated a crystalline structure, but the translational symmetry of crystals is only compatible with one, two, three, four, and six-fold rotational axes. The facile explanation, that multiple twinned crystals were present, could be ruled out on the basis of other data, leaving Shechtman and coworkers with a puzzle to reconcile. As a consequence, they did not publish the data for two years, waiting on a theory or model that could explain the observations [3]. This was developed in 1984 by Blech, who proposed a model involving parallel icosahedra attached by edges, lacking translational symmetry but possessing long-range orientational order. The Shechtman data and the Blech model effected a lasting change in the understanding of crystallography and solid-state physics; namely, the realisation that crystallinity – as defined by the presence of translational symmetry – was not a necessary precondition for a sharp diffraction pattern.

The International Union of Crystallography defines a quasicrystal as a material exhibiting an essentially sharp diffraction pattern in the absence of translational periodicity [4], where often, the lack of translational symmetry is inferred from the existence of a 'forbidden' n-fold axis in the diffraction pattern. Since 1984, discoveries of quasicrystalline materials include dozens of bimetallic and trimetallic alloys [5], materials derived from these alloys by adsorption and epitaxy [6, 7], terpolymers [8], micelles and liquid crystals [9, 10], and assemblies of nanoparticles [11, 12].

An alternate definition comes from Levine and Steinhardt, who state that quasicrystals are orientationally ordered and quasiperiodic [13]. The latter term

Supramolecular Chemistry on Surfaces: 2D Networks and 2D Structures, First Edition.
Edited by Neil R. Champness.

describes a mathematical function composed of two or more periodic functions, where the fundamental frequencies of those functions are related by a ratio that is an irrational number. Under this definition, a quasiperiodic function can never repeat; if it does, then p oscillations of one function would necessarily align with q oscillations of another, and the quotient p/q would be rational. For a real crystal of finite size, this quotient cannot be proven irrational, though this is not any more a limitation of the definition than the requirement that the diffraction pattern be *essentially* sharp.

5.2 Random Tilings

This chapter considers quasicrystalline structures that are formed from self-assembly processes on surfaces. Specifically, the discussion will be restricted to molecular systems where the assembly is driven by non-covalent interactions, and where the self-assembled layer is quasicrystalline but the underlying substrate is not. Also beyond the scope of this chapter are nanoparticle-based quasicrystals [11, 12, 14] and quasicrystals assembled from strands of DNA [15].

In all cases the surface-adsorbed layer is planar, and either definition of a quasicrystal generalises well to a two-dimensional system. Of the available surface diffraction techniques, low-energy electron diffraction (LEED) is the most commonly used, and a sharp LEED pattern showing a forbidden n-fold symmetry provides evidence of quasicrystallinity. Alternately, characterisation of a quasicrystal in real space using scanning tunnelling microscopy (STM) can be followed by Fourier analysis to generate the equivalent of a diffraction pattern. Analysis of quasiperiodicity likewise proceeds in a similar fashion to three dimensions.

Non-covalent bonding interactions include van der Waals forces, hydrogen bonding, metal–organic coordination, and π–π stacking. For assemblies of molecules on surfaces, these compete and cooperate with surface–adsorbate interactions to determine the overall degree of order (or disorder) in the molecular overlayer. In many cases, the structure that self-assembles is the thermodynamic minimum free energy for the system, though depending on the conditions of self-assembly one or more kinetically trapped structures can form [16]. Even small molecules with few interacting functional groups can produce surface tessellations with significant complexity; examples are shown in Figure 5.1 [17–20].

The first system that has many of the properties associated with quasicrystallinity is the self-assembly of *p*-terphenyl-3,5,3′,5′-tetracarboxylic acid (TPTC; Figure 5.2a) on graphite (and specifically, at the nonanoic acid/graphite interface), reported by Blunt and coworkers in *Science* in 2008[21].The molecule lies flat on the graphite surface, with the predominant intermolecular interaction being strong reciprocal hydrogen bonding through the four carboxylic acid groups. Each pair of TPTC molecules can then be arranged in one of two geometries, as shown in Figure 5.2 in panels (b) and (c). As the geometries differ only in the relative rotation around the

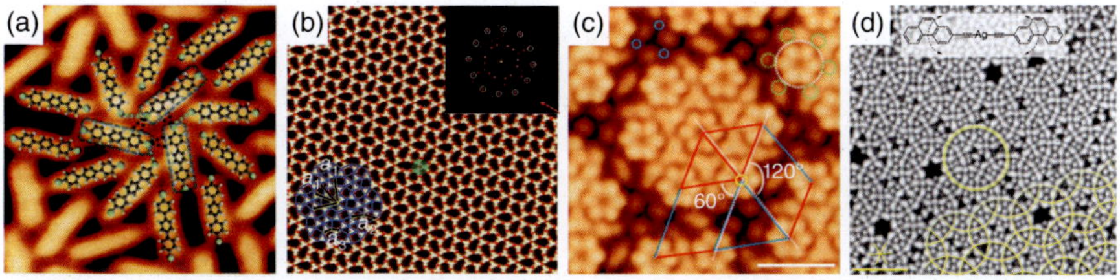

Figure 5.1 Complex but periodic self-assembled structures. (a) 4,4″-dichloro-1,1′:4′,1″-terphenyl on Ag(111); (b) 1,3,5-tris[4-(pyridin-4-yl)phenyl]benzene on Cu(111); (c) hexakis(4-iodophenyl)benzene on Au(111); (d) ethynyl-iodophenanthrene on Ag(111). Source: (a) Shu et al. [17]; (b) Yan et al. [18]; (c) Cheng et al. [19]; (d) Zhang et al. [20].

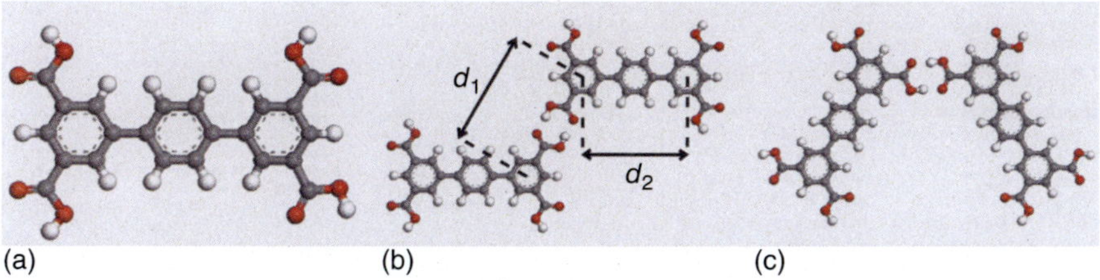

Figure 5.2 Structure of *p*-terphenyl-3,5,3′,5′-tetracarboxylic acid (TPTC, panel (a)). Illustrations of intermolecular interactions show two different binding configurations (b and c) that will be nearly isoenergetic. Source: Blunt et al. [21].

bond to the COOH group, they are guaranteed to be roughly equal in energy, leading to the conformational flexibility that characterises assemblies of these molecules.

This flexibility is demonstrated in Figure 5.3, which shows STM images of self-assembled TPTC molecules on the graphite surface. Mutually hydrogen-bonded molecules can be grouped according to their configuration around an open pore, and there are five distinct configurations possible, illustrated in Figure 5.3b–f with each panel showing an example of the configuration from the data, a molecular model, and a schematic diagram in which each molecule is replaced by a rhombus. Colour-coding the rhombuses according to their orientation encodes the STM data in Figure 5.3a into the tiling shown in Figure 5.3h. Shown this way, the tiling immediately appears to be random in nature, and indeed, more detailed analysis shows this to be the case. A more recent study indicates that the degree of order, as well as the relative sizes of tiles and pores, is a sensitive function of the COOH group positioning [22].

Randomness in tiling comes from a different source in the assembly of 1,3,5-benzenetricarboxylic acid (trimesic acid or TMA) on Ag(111).[23] In this system, TMA molecules can be deprotonated through annealing, and the resulting ratio of COOH to COO^- groups can be controlled using the annealing temperature. COO^- groups have the potential not only to interact via hydrogen bonding but also to participate in strong metal–organic binding with silver adatoms on the surface. With a 2 : 1 ratio of COOH to COO^- (one COO^- per molecule), TMA molecules

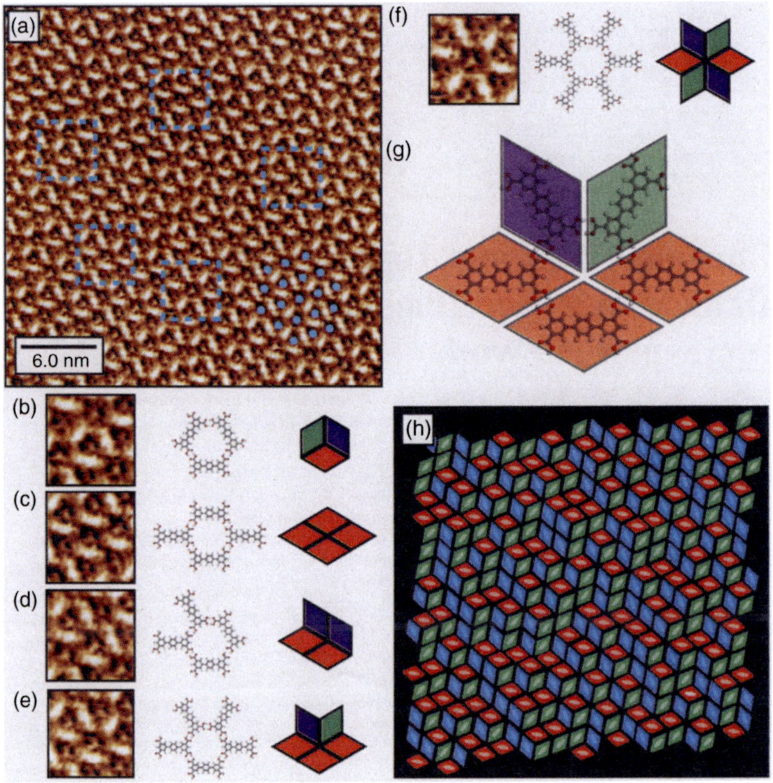

Figure 5.3 STM image of TPTC at the graphite/nonanoic acid interface (a). Panels (b–f) show excerpts from this image, alongside structural diagrams that show the arrangement of molecules, alongside a schematic diagram in which each molecule is replaced by a rhombus and coloured according to its orientation (detailed in panel (g)). The full 2-D tessellation corresponding to the image in (a) is shown in panel (h). Source: From Blunt et al. [21].

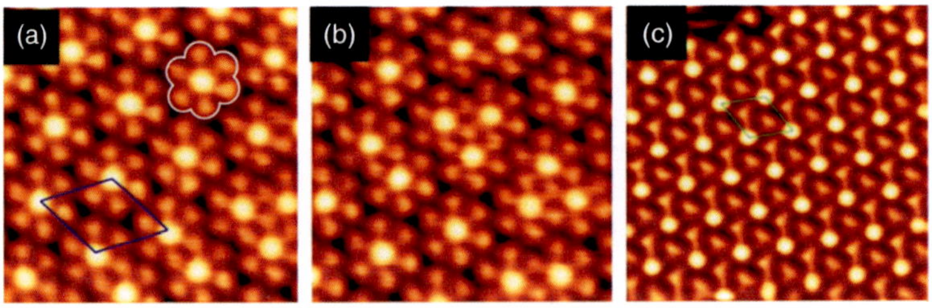

Figure 5.4 STM image of TMA assembled on Ag(111). The ratio of COOH to COO^- groups controls whether the molecules adopt the discrete pinwheel phase (a), the granular alloy (b), or the knitted pinwheel phase (c). Source: From Lipton-Duffin et al. [23].

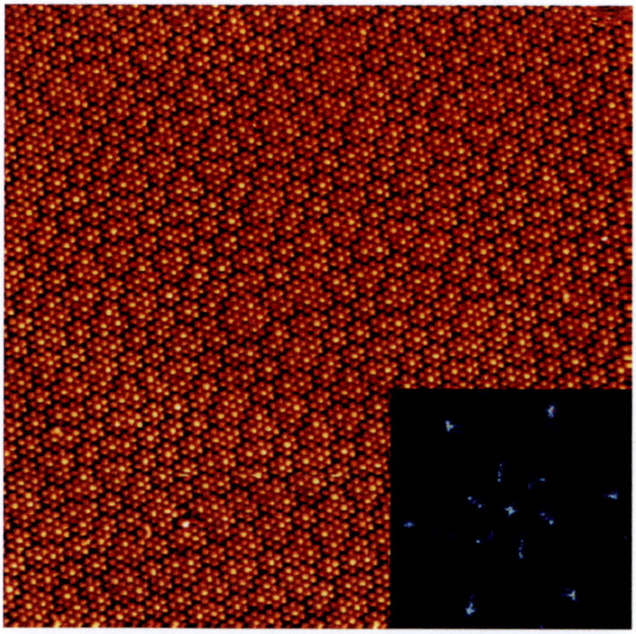

Figure 5.5 Large-scale (600 Å × 600 Å) STM image of TMA granular alloy phase, with 2-D Fourier transform inset. Source: From Lipton-Duffin et al. [23].

surround a silver adatom symmetrically, each molecule with a COO^- coordinating to the central adatom, producing a highly ordered 'discrete pinwheel' phase, as shown in Figure 5.4a. A different highly ordered phase, the 'knitted pinwheel' phase (Figure 5.4c), is created when each TMA molecule is fully deprotonated. Levels of deprotonation intermediate between these two limits produce an admixture of the two phases: the 'granular alloy' phase shown in Figure 5.4b. The granular alloy is characterised by short-range but not long-range translational symmetry, with the transitions between phases creating irregular translational defects throughout (Figure 5.5).

As a random tiling, both the self-assembled TPTC and TMA structures lack translational symmetry while preserving long-range orientational order, two of the essential criteria for quasicrystalline systems. The overall symmetry remains hexagonal, however, which prompts the question of what random tiling based on shapes other than rhombuses might produce by way of surface structure. A simple view is that pentagonal assembly units would have a preference for forming fivefold-symmetric structures, resulting in a quasicrystalline self-assembled system. In most cases, however, self-assembly of pentagonal molecules results in ordered monolayers [24–26]; this is because the densest possible packing of pentagons is in a hexagonal lattice, with each adjacent pentagon offset along their shared side, leaving only small gaps. An additional geometric constraint, along with entropic considerations, is needed to produce a quasicrystalline surface.

5.3 Quasicrystalline Tilings

Self-assembled monolayers of ferrocenecarboxylic acid (FcCOOH) on the Au(111) surface were described in 2014 by Wasio and coworkers [27]. FcCOOH molecules also contain a carboxylic acid group, with the possibility of strong reciprocal hydrogen bonding to form a dimer; however, under the conditions used for sample preparation, a large number of cyclic hydrogen-bonded FcCOOH pentamers are also formed. This can be seen in the STM image in Figure 5.6, where the bright, round features – each corresponding to one FcCOOH molecule – are predominantly in regular pentagonal groups. Dim features interstitial to these groups are also present in the image, and will also be discussed.

FcCOOH pentamers have a chemically interesting structure, in which the strain created in the O—H…O hydrogen bonds by the cyclic geometry is compensated because this geometry also brings adjacent C—H bonds into proximity to the oxygen atoms to form weak C—H…O hydrogen bonds. This is demonstrated by density functional theory (DFT) calculations of FcCOOH dimer energies versus an applied bending angle, shown in Figure 5.7. With no bending constraint added, the ∠COH angle is approximately 120°, corresponding to a deep well for the reciprocally bound dimer. As the ∠COH angle is opened up, energy initially rises as one of the O—H…O hydrogen bonds are broken and the other strained. Then, energy levels off, and a small well at ∠COH = 224.8° is created by the new C–H…O interaction. The minimum of this well is almost exactly at the angle needed to form a cyclic pentamer; the angles for trimer, tetramer, and hexamer formation are marked and all correspond to higher calculated energies.

The relative shallowness of the secondary well in Figure 5.7 is potentially misleading, as this is the computed energy for a dimer structure, and it leaves hydrogen bonds dangling that would be fully coordinated in the cyclic pentamer geometry.

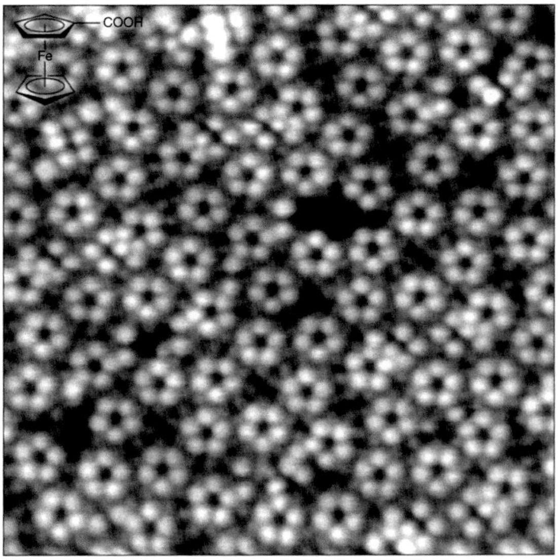

Figure 5.6 STM image of a self-assembled monolayer of ferrocenecarboxylic acid (FcCOOH) on Au(111). Each bright spot corresponds to one molecule, and regular pentamers are the dominant structure observed. Source: From Wasio et al. [27].

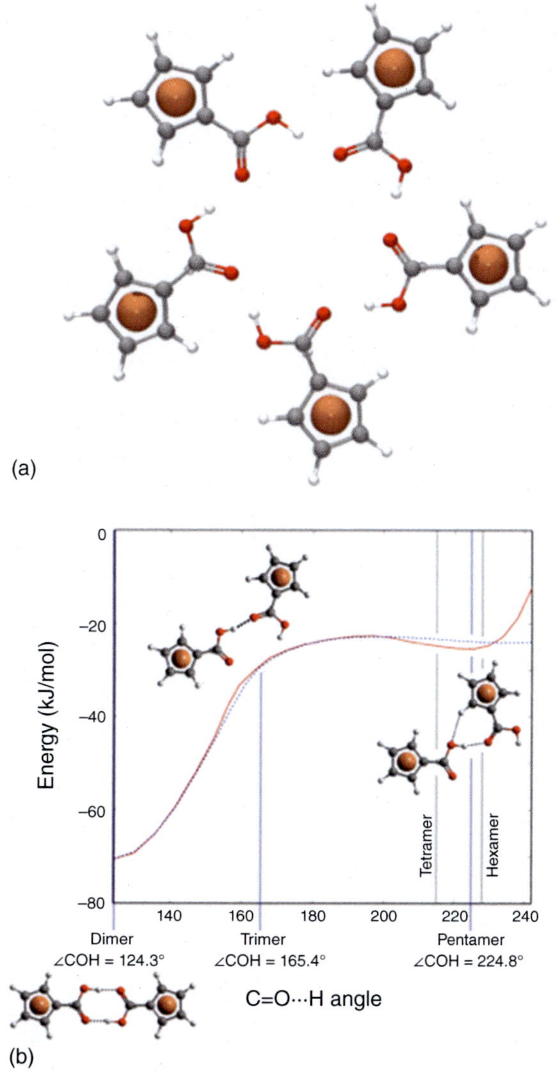

Figure 5.7 Illustration of cyclic hydrogen bonding forming the regular pentamer structure for FcCOOH (a).This structure is supported by C−H· · ·O hydrogen bonds, evidence for which is shown in the calculation (b) by the presence of a second well at a COH angle of 224.8°. Source: Based on Wasio et al. [27].

DFT calculations on the full pentamer structure give per-molecule binding energy that is similar to the dimer: depending on the density functional used, the cyclic pentamer is between 3.86 kJ/mol more stable and 1.62 kJ/mol less stable than the dimer structure. Considering that, samples are prepared at room temperature, where kT is ~2.5 kJ/mol, the expectation is that dimers and pentamers could coexist on the surface side by side.

Studies of contrast in STM images of ferrocene-containing molecules have shown that ferrocene appears bright when the cyclopentadienyl (Cp) ring is parallel to the surface and appears dimmer when the Cp is oriented perpendicular to the surface. With this additional information, a model can be proposed that assigns the bright image features to FcCOOH pentamers and the dim, interstitial features to FcCOOH

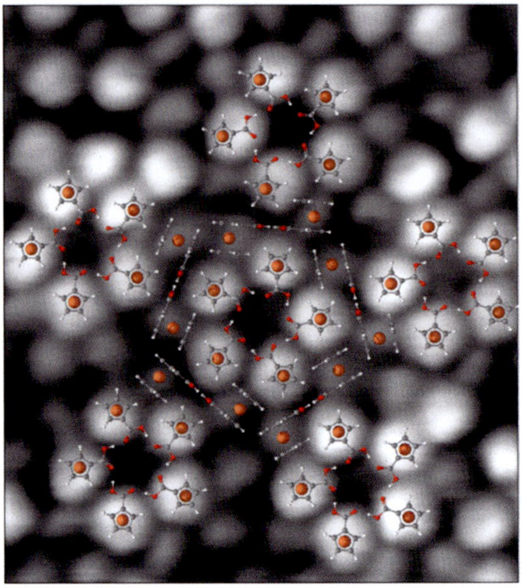

Figure 5.8 High-resolution image of FcCOOH on Au(111), with structural model superimposed. Bright features correspond to the ferrocene's Cp rings adsorbed parallel to the surface, while dim features are from Cp rings perpendicular to the surface. Interlocking pentamers and dimers propagate orientational order and fivefold rotational symmetry in the system. Source: From Wasio et al. [27].

dimers; this is shown in Figure 5.8. Dimers and pentamers are presumably bound together by multiple C–H…π interactions, and we consider the packing unit to be the pentamer plus the surrounding dimers; this forms a larger pentagon that surrounds the pentamer. A key feature of this model is that any given dimer is shared in its interactions with two adjacent pentamers. This applies a geometric constraint to the pentagonal packing units, which are incorporated in a tiling where they not only share an edge but additionally are matched vertex-to-vertex. This constraint also imposes orientational ordering, as adjacent pentagons necessarily face in exactly opposite directions.

Figure 5.9 shows STM data overlaid with the pentagonal tiling. For reference, a portion of a Penrose tiling, known to be a periodic, is shown below, and the two have marked similarities in medium-scale features (highlighted in white). A Fourier transform of the STM data shows sharp peaks and 10-fold rotational symmetry, as does a two-dimensional correlation function of pentamer positions. The self-assembled FcCOOH monolayer thus satisfies the International Union of Crystallography definition for a quasicrystal, and closer analysis of the correlation function shows that it satisfies the Levine and Steinhardt definition, as well. This is demonstrated in Figure 5.10, which shows a one-dimensional section along a high-symmetry direction; the irregularly spaced spikes show fundamentally crystalline (as opposed to liquid) behaviour for a length scale of hundreds of ångstroms. A frequency analysis of these peaks shows a comb of spatial frequencies, each separated by a factor of the irrational number $\varphi = (1 + \sqrt{5})/2$.

The vertex-to-vertex constraint prevents hexagonal packing of pentagons and ensures long-range orientation, but it alone is not sufficient to ensure a quasicrystalline pentagon tiling. Ordered 'chickenwire' arrays of pentagons cover the plane at 85.4% density; however, the Penrose tiling in comparison has an 83.6%

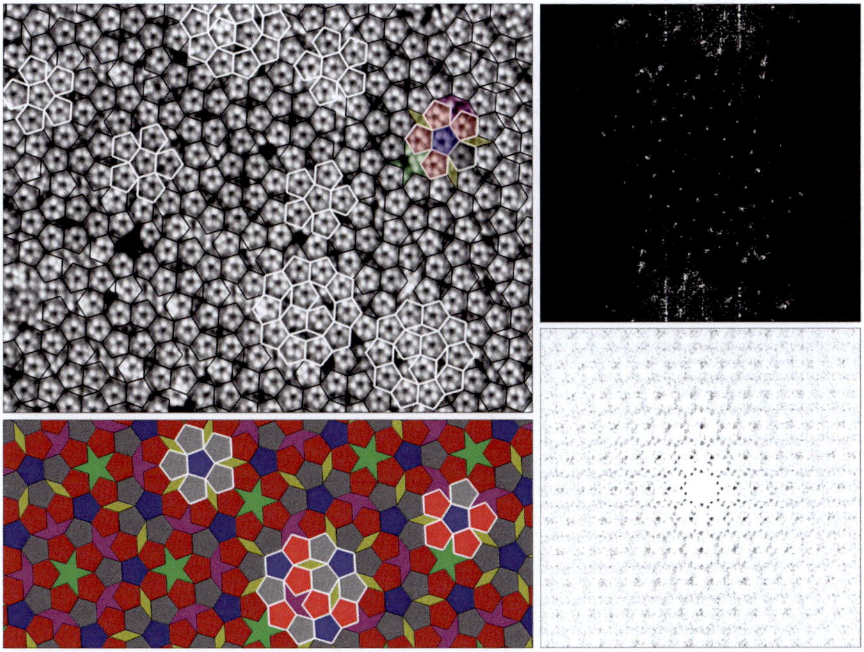

Figure 5.9 STM image of FcCOOH on Au(111) with pentagonal tiles superimposed, qualitatively matching the 2-D Penrose P1 tiling. Both the Fourier transform of the data (upper right) and the 2-D spatial correlation function (lower right) show the 10-fold rotational symmetry of the system. Source: From Wasio et al. [27].

coverage of pentagons, and the closeness of these two numbers means that entropic considerations tip the scale in favour of a random pentagonal tiling that appears quasiperiodic. Note that the data does not and would not be expected to match the Penrose tiling exactly, which contains matching rules that extend beyond nearest-neighbour interactions and would be unlikely to arise from molecular interactions. Furthermore, the Penrose tiling is unique, subject to translation, so it is not entropically favoured compared to generally matching but defected random tilings.

The sample preparation method used to deliver molecules to the surface likely also plays a role in the formation of entropically favoured structures. FcCOOH is adsorbed on the Au(111) surface using pulsed deposition in a vacuum, a method where microliter droplets of solution are injected directly into a vacuum and onto the surface. The rapid evaporation of solvent increases concentration and lowers the temperature, creating an environment that has been shown to promote the formation of metastable structures; indeed, for a related molecule, indole-2-carboxylic acid, pulse deposition produces hydrogen-bonded pentamers that are demonstrably metastable [28], as they are not formed through vacuum deposition of the same molecule [29]. Potentially, pulse deposition is responsible for either the large numbers of FcCOOH pentamers, or for their assembly into a non-ordered tiling, or both.

In 2016, Urgel and coworkers reported the observation of a self-assembled two-dimensional quasicrystal based upon metal–organic coordination [30].

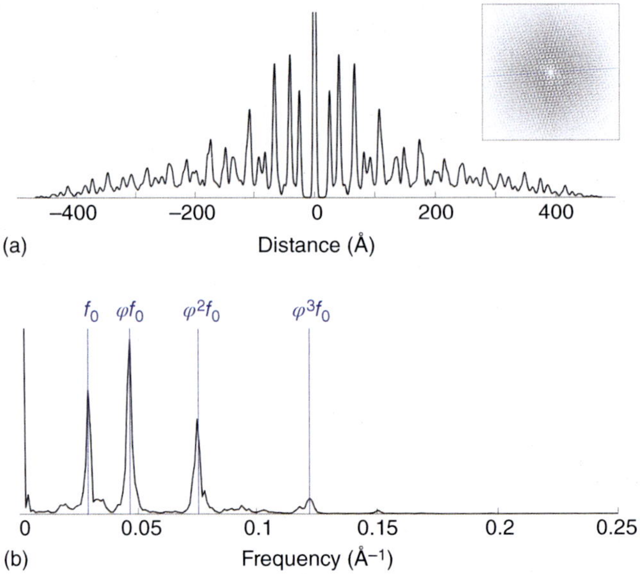

Figure 5.10 (a) A 1-D section of the spatial correlation function shows long-range crystallinity in the sharpness of peaks and their return to baseline. No periodicity is apparent. (b) The Fourier transform of (a) shows it is composed of a comb of spatial frequencies related to each other by multiples of the irrational number $\varphi = (1 + \sqrt{5})/2$. Source: Wasio et al. [27].

In this study, europium atoms were co-deposited with an organic linker, *para*-quaterphenyl-dicarbonitrile (qdc), a linear molecule with terminal C≡N groups. Metal–organic bonding in this system is largely ionic, and only partially covalent. Because of this, neither the number of bonds made nor their geometry are fixed at specific values; instead, metal–organic assemblies are flexible in their structure and dependent on the exact stoichiometry between europium atoms and linkers.

Changing the relative proportion of Eu atoms and qdc linkers produces a large and qualitative change in the types of self-assembled structures produced. This is shown in Figure 5.11. When Eu atoms and qdc linkers are in 2 : 4 or 2 : 6 ratios, STM images show highly ordered assemblies, with near-square and hexagonal lattices, respectively. When the Eu-to-qdc ratio is 2 : 3, a disordered, porous network results. Finally, with a 2 : 5 ratio, highly structured but not obviously ordered surface results, based on a mixture of square and triangular tiling elements. It is at this stoichiometry that the metal–organic coordination networks are quasicrystalline.

A large-scale STM image of the 2 : 5 Eu-to-qdc coordination network is shown in Figure 5.12. The 2-D Fourier transform of the image is inset, and it contains sharp peaks with 12-fold rotational symmetry; a 12-fold axis is, of course, not compatible with the periodic tiling of the plane. This symmetry is also evident in the waggon-wheel dodecagonal elements that appear across the imaged area; two of these are highlighted in the figure.

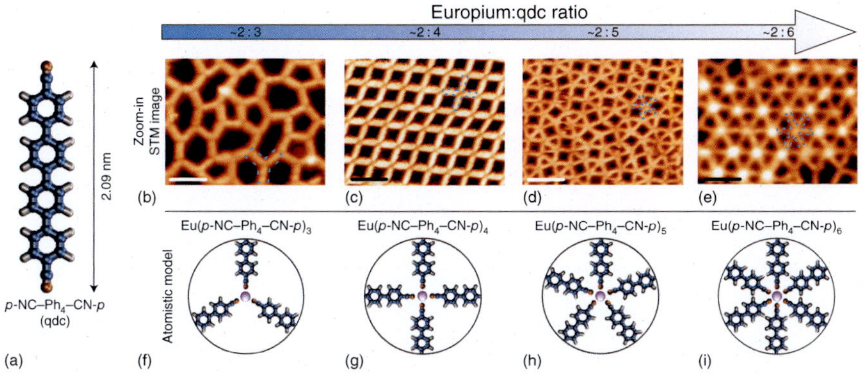

Figure 5.11 (a) Chemical structure of the molecule *para*-quaterphenyl-dicarbonitrile (qdc) used as a metal-organic linker. (b–e) STM images of Eu/qdc metal–organic coordination assemblies recorded at different stoichiometries of Eu to qdc. (f–i) Structural models showing the dominant coordination number of Eu atoms that make up the structures in the associated images. Source: Urgel et al. [30].

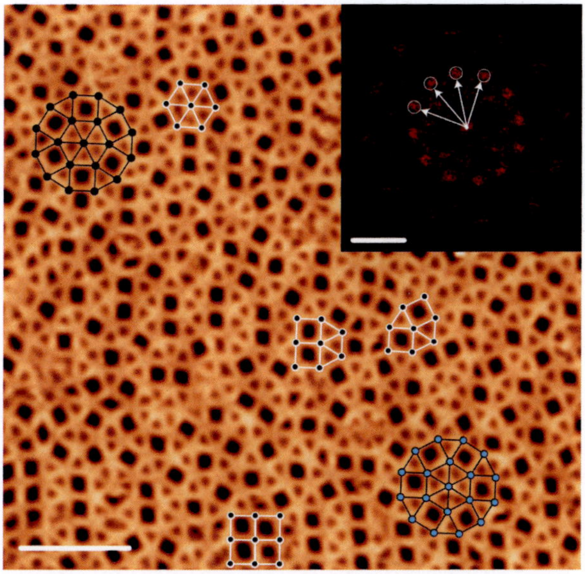

Figure 5.12 A large-scale STM image of a Eu/qdc metal–organic coordination network at 2 : 5 Eu-to-qdc stoichiometry. The Fourier transform (inset) shows essentially sharp peaks as well as 12-fold rotational symmetry. Source: From Urgel et al. [30].

The presence of fivefold coordination sites is on its own insufficient to ensure quasicrystallinity in Eu–qdc metal–organic assemblies. As evidence of this, the authors explore substrate dependence; as described above, assemblies on Au(111) are quasicrystalline at a 2 : 5 stoichiometry, with a mixture of four, five, and six-fold nodes. In contrast, assemblies on Ag(111) are periodic in nature and consist exclusively of fivefold nodes, lacking the sixfold coordination sites necessary for dodecagonal rotational symmetry. The reason for this is the increased interaction strength between the qdc linker and the Ag(111) surface, which forces a more planar geometry of the qdc phenyl rings. This in turn creates steric hindrance in the vicinity of the Eu atom centers and precludes the formation of sixfold-coordinate Eu. The interaction between the linker and the Au(111) surface is weaker, allowing for more rotation of the qdc phenyl rings out of plane, and thus reducing the enthalpic penalty for forming sixfold sites. Quasicrystallinity thus arises out of a balance between entropy and enthalpy: the weakness of the Au/qdc linker interaction allows for a range of coordination numbers, at which point entropy favours the inherent randomness of the quasicrystalline tiling.

Looking to the future, two questions about self-assembled quasicrystalline monolayers arise: what new systems can be created, and what applications can be found for them?

Of the various approaches discussed, metal–organic coordination networks have the most promises to be developed into a whole class of 2-D quasicrystals. There are likely other metals that have the same flexibility of coordination as europium, and the length and functionality of the organic linker are almost infinitely variable. This is in contrast to the TPTC random tiling and the FcCOOH quasicrystal, systems where any change in molecular size, shape, or functionality would almost certainly affect the self-assembly process.

In terms of applications, bulk quasicrystalline materials have a number of practically useful properties, including low thermal conductivity, high hardness, and low friction. The specific quasicrystalline self-assembled monolayers discussed here are unlikely to be robust enough for their hardness or frictional properties to be exploitable. Thermal conductivity will only be unusually low (if, indeed, it is) within the 2D plane where the monolayer is quasicrystalline, so applications in thermally insulating coatings are not practical. Electronic, optical, or magnetic properties remain potential areas to explore. Perhaps the most promising direction for future research will be the application of the ideas of quasiperiodic self-assembly to the construction of layer-by-layer materials, as is done in atomic layer epitaxy.

References

1 Shechtman, D., Blech, I., Gratias, D., and Cahn, J.W. (1984). Metallic phase with long-range orientational order and no translational symmetry. *Phys. Rev. Lett.* 53 (20): 1951–1953.

2 Shechtman, D. and Blech, I.A. (1985). The microstructure of rapidly solidified Al_6Mn. *Metall. Mater. Trans. A* 16 (6): 1005–1012.

3 Blech, I.A., Cahn, J.W., Gratias, D. et al. (2012). Reminiscences about a Chemistry Nobel Prize won with metallurgy: comments on D. Shechtman and I. A. Blech; Metall. Trans. A, 1985, vol. 16A, pp. 1005–12. *Metall. Mater. Trans. A* 43A (10): 3411–3422.

4 International Union of Crystallography (1992). Report of the executive committee for 1991. *Acta Crystallogr., Sect. A: Found. Crystallogr.* 48 (6): 922–946. https://doi.org/10.1107/S0108767392008328.

5 Gierer, M., Van Hove, M.A., Goldman, A.I. et al. (1998). Fivefold surface of quasicrystalline AlPdMn: structure determination using low-energy-electron diffraction. *Phys. Rev. B* 57 (13): 7628–7641.

6 Smerdon, J.A., Parle, J.K., Wearing, L.H. et al. (2008). Nucleation and growth of a quasicrystalline monolayer: Bi adsorption on the fivefold surface of i-Al(70)Pd(21)Mn(9). *Phys. Rev. B* 78 (7): 075407.

7 Kalashnyk, N., Ledieu, J., Gaudry, E. et al. (2018). Building 2D quasicrystals from 5-fold symmetric corannulene molecules. *Nano Res.* 11 (4): 2129–2138.

8 Hayashida, K., Dotera, T., Takano, A., and Matsushita, Y. (2007). Polymeric quasicrystal: mesoscopic quasicrystalline tiling in ABC star polymers. *Phys. Rev. Lett.* 98 (19): 195502.

9 Fischer, S., Exner, A., Zielske, K. et al. (2011). Colloidal quasicrystals with 12-fold and 18-fold diffraction symmetry. *Proc. Natl. Acad. Sci. U.S.A.* 108 (5): 1810–1814.

10 Zeng, X.B., Ungar, G., Liu, Y.S. et al. (2004). Supramolecular dendritic liquid quasicrystals. *Nature* 428 (6979): 157–160.

11 Talapin, D.V., Shevchenko, E.V., Bodnarchuk, M.I. et al. (2009). Quasicrystalline order in self-assembled binary nanoparticle superlattices. *Nature* 461 (7266): 964–967.

12 Xiao, C., Fujita, N., Miyasaka, K. et al. (2012). Dodecagonal tiling in mesoporous silica. *Nature* 487 (7407): 349–353.

13 Levine, D. and Steinhardt, P.J. (1986). Quasi-crystals: 1. Definition and structure. *Phys. Rev. B* 34 (2): 596–616.

14 Nagaoka, Y., Zhu, H., Eggert, D., and Chen, O. (2018). Single-component quasicrystalline nanocrystal superlattices through flexible polygon tiling rule. *Science* 362 (6421, SI): 1396.

15 Liu, L., Li, Z., Li, Y., and Mao, C. (2019). Rational design and self-assembly of two-dimensional, dodecagonal DNA quasicrystals. *J. Am. Chem. Soc.* 141 (10): 4248–4251.

16 Brown, R.D., Corcelli, S.A., and Alex Kandel, S. (2018). Structural polymorphism as the result of kinetically controlled self-assembly. *Acc. Chem. Res.* 51 (2): 465–474.

17 Shu, C.-H., Zhang, S.-Z., Wang, C.-X. et al. (2018). Diverse supramolecular structures self-assembled by a simple aryl chloride on Ag(111) and Cu(111). *Chem. Commun.* 54 (97): 13670–13673.

18 Yan, L., Kuang, G., Zhang, Q. et al. (2017). Self-assembly of a binodal metal-organic framework exhibiting a demi-regular lattice. *Faraday Discuss.* 204: 111–121.

19 Cheng, F., Wu, X.-J., Hu, Z. et al. (2018). Two-dimensional tessellation by molecular tiles constructed from halogen–halogen and halogen–metal networks. *Nat. Commun.* 9.

20 Zhang, Y.-Q., Paszkiewicz, M., Du, P. et al. (2018). Complex supramolecular interfacial tessellation through convergent multi-step reaction of a dissymmetric simple organic precursor. *Nat. Chem.* 10 (3): 296–304.

21 Blunt, M.O., Russell, J.C., Gimenez-Lopez, M.d.C. et al. (2008). Random tiling and topological defects in a two-dimensional molecular network. *Science* 322 (5904): 1077–1081.

22 Karner, C., Dellago, C., and Bianchi, E. (2019). Design of patchy rhombi: from close-packed tilings to open lattices. *Nano Lett.* 19 (11): 7806–7815.

23 Lipton-Duffin, J., Abyazisani, M., and MacLeod, J. (2018). Periodic and nonperiodic chiral self-assembled networks from 1,3,5-benzenetricarboxylic acid on Ag(111). *Chem. Commun.* 54 (60): 8316–8319.

24 Bauert, T., Merz, L., Bandera, D. et al. (2009). Building 2D crystals from 5-fold-symmetric molecules. *J. Am. Chem. Soc.* 131 (10): 3460.

25 Ren, C., Zhou, F., Qin, B. et al. (2011). Crystallographic realization of the mathematically predicted densest all-pentagon packing lattice by C-5-symmetric "sticky" fluoropentamers. *Angew. Chem. Int. Ed.* 50 (45): 10612–10615.

26 Guillermet, O., Niemi, E., Nagarajan, S. et al. (2009). Self-assembly of fivefold-symmetric molecules on a threefold-symmetric surface. *Angew. Chem. Int. Ed.* 48 (11): 1970–1973.

27 Wasio, N.A., Quardokus, R.C., Forrest, R.P. et al. (2014). Self-assembly of hydrogen-bonded two-dimensional quasicrystals. *Nature* 507 (7490): 86.

28 Wasio, N.A., Quardokus, R.C., Brown, R.D. et al. (2015). Cyclic hydrogen bonding in indole carboxylic acid clusters. *J. Phys. Chem. C* 119 (36): 21011–21017.

29 De Marchi, F., Cui, D., Lipton-Duffin, J. et al. (2015). Self-assembly of indole-2-carboxylic acid at graphite and gold surfaces. *J. Chem. Phys.* 142 (10).

30 Urgel, J.I., Ecija, D., Lyu, G. et al. (2016). Quasicrystallinity expressed in two-dimensional coordination networks. *Nat. Chem.* 8 (7): 657–662.

6

Using Self-Assembly to Control On-Surface Reactions

Zhantao Peng, Lingbo Xing, and Kai Wu

Peking University, BNLMS, College of Chemistry and Molecular Engineering, No.292 Chengfu Road, Haidian District, Beijing 100871, China

6.1 Introduction

The key issue in chemistry is how to design and steer various chemical reactions along a specific direction in an efficient and economical way. Until today, tremendous efforts to understand fundamental principles and develop applicable methodologies have been made by scientists and engineers from every walk of the chemical community and industry. As a central science [1], chemistry has been devoted to tackling great challenges the human being faces by not only innovating large-scale industrial processes such as ammonia synthesis [2] and petroleum refining [3], but also developing cutting-edge technologies for small compounds as well.

In conventional and homespun chemical synthetic processes that involve simple mixing of reactants; however, the reaction direction and rate are often determined by parameters such as temperature, concentration, and feedstock stoichiometry. These parameters are usually taken as the sole controllable variables in the scope of classical chemistry. Modern experimental techniques have accelerated the revealing of mechanisms for a variety of reactions at the molecular and atomic levels. Subsequently, sophisticated theories to uncover the underlying mechanisms of the reactions have been developed [4, 5]. A classical but elegant expression of the reaction kinetics is the Arrhenius equation [6, 7], which describes the temperature dependence of the reaction rate:

$$k = Ae^{-\frac{E}{RT}},$$

where k is the rate constant, A is the pre-exponential factor, E is the energy barrier, R is the universal gas constant, and T is the reaction temperature.

The pre-exponential factor, which contains the information of the molecular arrangement and collision, plays a crucial role in the Arrhenius equation. In the transition state theory, for example, the rate constant is given by:

$$k = \frac{k_B T}{h} \exp\left(\frac{\Delta S^{\neq}}{R}\right) \exp\left(-\frac{E}{RT}\right),$$

Supramolecular Chemistry on Surfaces: 2D Networks and 2D Structures, First Edition.
Edited by Neil R. Champness.

where k_B is the Boltzmann constant, h is the Planck constant, and $\Delta S^{\neq}$ is the change in entropy from the reactants to the transition state intermediates. Compared to the Arrhenius equation, the pre-exponential factor A can then be expressed as:

$$A = \frac{k_B T}{h} \exp\left(\frac{\Delta S^{\neq}}{R}\right).$$

The entropy change $\Delta S^{\neq}$ is the entropy difference between the elementary chemical processes. Obviously, the entropy, S, relates to the number of microstates of the thermodynamic system at a specific stage and can be generally expressed as:

$$S = k_B \ln \Omega,$$

where Ω is the number of microstates and does contain the configuration information of molecular species involved in the reaction.

Accommodation of the reactants on a surface to limit their configurations, which is termed surface confinement effect, is a generally employed strategy in the field of heterogeneous catalysis. Therefore, both surface atoms and species should be included along the reaction coordinate. However, the mechanisms behind various surface reaction systems are often difficult to elucidate. Fortunately, with the development of modern surface science [8–10], explorations of these surface reactions may shed new light on uncovering these mechanisms that involve the cleavage and formation of chemical bonds taken place at the surface in a restricted manner.

Normally it is unfeasible to directly manipulate the reacting molecules one by one in a reaction-favourable pattern on the surface. However, various non-covalent interactions between the molecules and surfaces, including van der Waals forces [11–13], hydrogen bonding (HB) [14–17], halogen bonding [18, 19], electrostatic interaction [20–22], metal–organic coordination [23–25] and the like, lead to spontaneous self-organisation of the surface species into ordered structures or patterns without any external mediation, which is termed self-assembly. To some extent, the surface self-assembly process could be fine-tuned by varying the assembly building blocks, surface molecular density, temperature, external field, and so on and so forth.

Molecular self-assembly on surface may not only possess unique and promising optical, electrical, magnetic, and electronic properties but can also be utilised to steer, in some cases, the reactant configuration and arrangement on surface by changing its adsorption site via the self-assembly confinement effect and reaction collision probability via its restricted docking orientation. In such a way, the reaction kinetics can be efficiently tweaked according to the Arrhenius equation. As a result, the same reactant on surface may dramatically react along different reaction pathways, leading to diverse products. Such a self-assembly approach on surface would come up with a novel strategy to steer on-surface reactions.

Assembly-steered on-surface reactions may fall into two categories, namely, assembly-assisted and assembly-involved reactions. The former refers to the reactions where the assembled surface structure simply serves as a reaction template for the reactants. For instance, a two-dimensional (2D) molecular assembly with a porous network can trap the reactants through selective molecular recognition or non-selective accommodation. This type of on-surface reaction is closely analogous to the reactions on a porous zeolite template and lock–key recognition in the enzyme-catalysed processes. The latter refers to the reactions where the molecular building blocks of the self-assembly participate in the reactions. These types of reactions might be far more complex due to the confinement effect invoked by the assembly of the involved reactants as the building blocks, imposing significant influence on the reaction pathway and ultimate reaction selectivity. The reactants are spatially confined by themselves so that their degree of freedom including translational, rotational, and vibrational motions may be hindered. Consequently, the reaction collision probability and corresponding selectivity are changed. This confinement effect is analogous to the cage effect [26] in solution chemistry where the reactants are encaged by solvent molecules, leading to their collision probability inside a cage that is dramatically boosted and becomes even comparable to that in their gas-phase reactions, albeit the collision probability between molecules inside different cages is essentially null. Overall, their apparent reaction rate may be drastically enhanced. The densely packed 2D molecular assembly could be envisioned as a planar cage playing a similar role as its counterpart in solution.

Surface molecular self-assembly is anticipated to function in three main aspects on the on-surface reactions. Firstly, the reaction selectivity can be tuned by the bonding directionality of the molecules that are self-assembled in advance. This is achieved by enhancing the pre-exponential factor of a favourable pathway yielding the target products while suppressing other unfavourable pathways producing by-products. Secondly, the stability of the reaction intermediates on surface can be changed by their surface assemblies, giving rise to the variations of the activation energy and pre-exponential factor, and accordingly the reaction pathway. Furthermore, the same intermediates may possess disparate properties on different surfaces. Finally, restriction of the reactive species on a specific surface site may eliminate steric hindrance because other surface sites for the reactions may be blocked during the cause of the molecular self-assembly.

In this chapter, we will focus on how the 2D molecular self-assembly controls on-surface reactions in both assembly-assisted and -involved modes. Several typical case studies relating to the above-mentioned three main aspects of controlling the on-surface reactions will be described in detail to demonstrate the capability and efficiency of such an established self-assembly strategy.

6.2 Mediating On-Surface Reaction Selectivity

Efficient control on the product selectivity is a key issue in on-surface chemistry [27]. Various measures have been reportedly taken by changing the substrate [28–30], lattice plane [31–33], molecular backbone [34, 35], functional group [36–38], coverage [39], stoichiometric ratio [40, 41], temperature [42], solvent [43], and the like. More importantly, the substrate-induced confinement effect plays a crucial role in surface reactions. Chi and coworkers [44] have reported the polymerisation of linear alkane on surfaces, as revealed by scanning tunnelling microscopy (STM) and other surface science techniques. The anisotropic Au(110) surface constrains one-dimensional diffusion of the alkanes and subsequent selective C–H activation as well as dehydrogenative C–C coupling reactions along its reconstruction troughs. Similarly, the intermolecular confinement in the surface molecular assembly would also steer reaction selectivity by mediating the self-organisation of the reaction precursors. The molecules confined in the assembly are immobilised in unique ways to strengthen their local collision probability, thus the selectivity and reaction rate are mediated.

Chen et al. [45] have reported the assembly-mediated reactions of 4,4′-bis(2,6-difluoropyridin-4-yl)-1,1′:4′,1″-terphenyl (BDFPTP) on Au(111). The BDFPTP molecules undergo two reaction pathways, i.e. dehydrocyclisation (DHC) and coupling reactions. The DHC reaction proceeds via four reaction positions from D1 to D4 (marked in blue) and coupling reaction occurs on the C position (marked in red). As a result, the DHC and coupling reactions could form four pairs of enantiomers and two isomers, respectively. Disordered (Figure 6.1c) and ordered (Figure 6.1d) self-assembly structures of BDFPTP can be obtained by controlling the annealing time at 500 K (Figure 6.1a). Surprisingly, further thermal treatments of the two samples at 520 K result in different reaction products. As shown in Figure 6.1e, the disordered regions form randomly distributed hoe-shaped products. Statistical analyses (black columns in Figure 6.1b) show the distribution of the DHC products from D1 to D4 and the coupling products with a low yield of 13.8% as well. Nonetheless, exclusive DHC products (Figure 6.1f) at positions D2 and D3 with an approximate ratio of 1 : 1 (Figure 6.1b) are observed after the thermal treatment of the ordered self-assemblies. Due to the intermolecular repulsions in the assembly domain, the endings of each rotating BDFPTP molecule are restricted and can only dock to positions D2 or D3 of another immobile neighbouring BDFPTP molecule, leaving no space for the coupling reaction. Thus, both chemo- and regio-selectivity for BDFPTP can be tuned via the pre-assembly of the molecules.

Glaser coupling has also attracted extensive attention due to its capability of forming carbon-based scaffolds such as graphyne and graphdiyne [35, 46–48]. However, the homocoupling of terminal alkynes is always accompanied by side reactions such as cyclotrimerisation, cycloaddition, and radical cyclisation and organic metalation [49–56]. Given the complexity of the reactions of the terminal alkynes, it is imperative to improve the chemo- and regio-selectivity. Gao et al. [34] have studied light-induced Glaser coupling of 1,4-diethynyl-2,5-dihexylbenzene on metal surfaces. As depicted in Figure 6.2a, the flexible alkane side chains induce the reaction precursors to assemble into ordered structures due to the π–π interaction. After the UV light irradiation, the reacting terminal alkynes in close locations carry on the Glaser coupling reaction (Figure 6.2b) to form high-yield conjugated linear polymers without by-products.

In the same way, Colazzo et al. [57] have reported light-induced metal-free homocoupling of 4-ethynylbenzoic (*para*-ethynylbenzoic) acid (PEBA, Figure 6.2c) on highly oriented pyrolytic graphite (HOPG). Figure 6.2d shows a representative STM image of a 2D ordered assembly monolayer upon drop-casting of a solution of PEBA in 7COOH on the HOPG surface. The assembly is stabilised by both intermolecular HB of the carboxylic group and twofold cyclic weak HB of the terminal alkynes. By taking advantage of the collinear displacement of neighbouring alkynes, a complete transformation of PEBA molecules into large domains of homocoupling

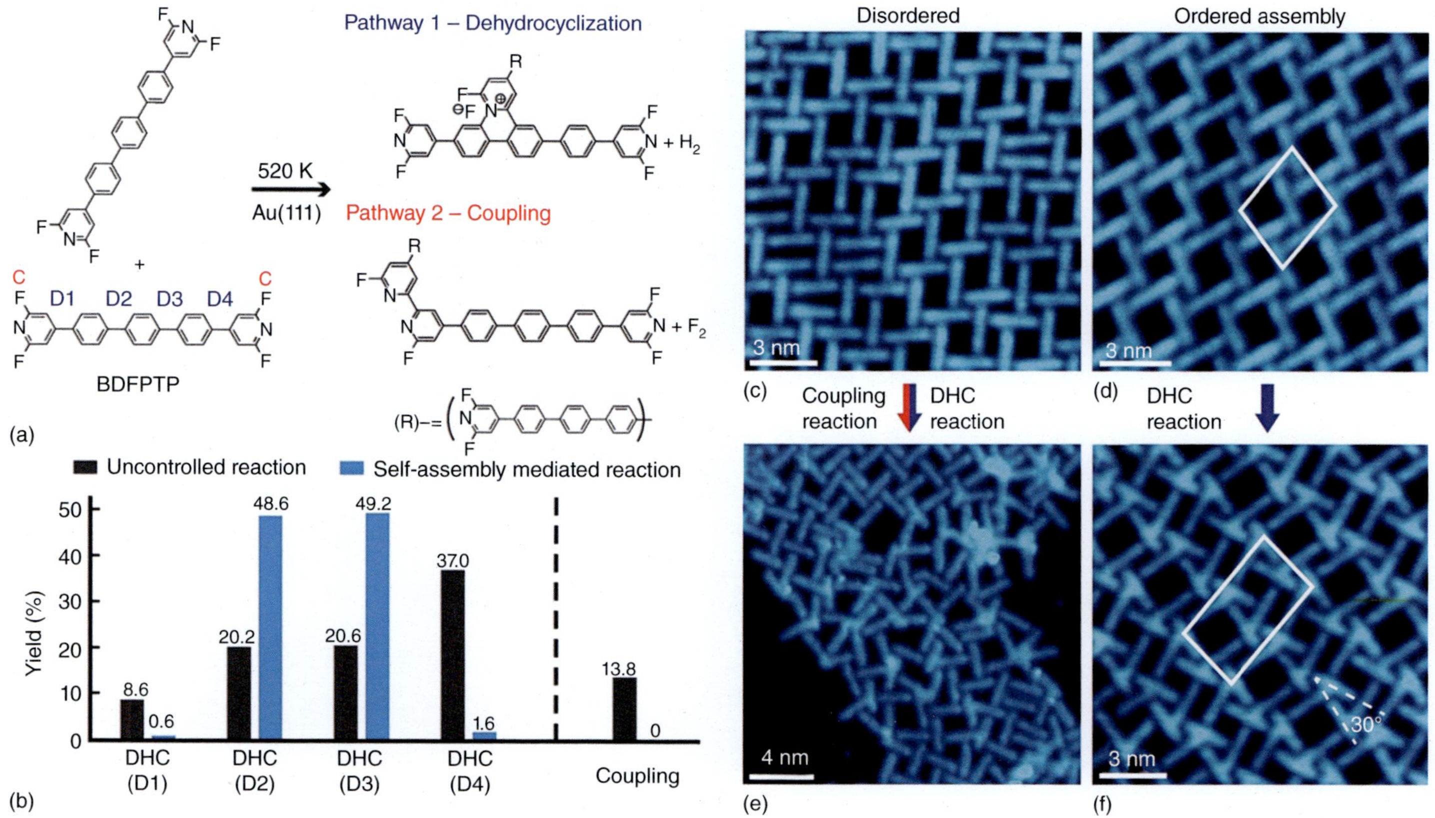

Figure 6.1 (a) Possible dehydrocyclisation and coupling reaction pathways of BDFPTP on Au(111). (b) Statistic histograms of the products from disordered (black bars) and assembled (blue bars) adlayers of the BDFPTP molecules. STM images of (c) the disordered structure and (d) the ordered assembly. (e) and (f) Reaction products from structures (c) and (d), respectively. Source: Reprinted from Chen et al. [45] with permission from Wiley-VCH Verlag GmbH, copyright 2017.

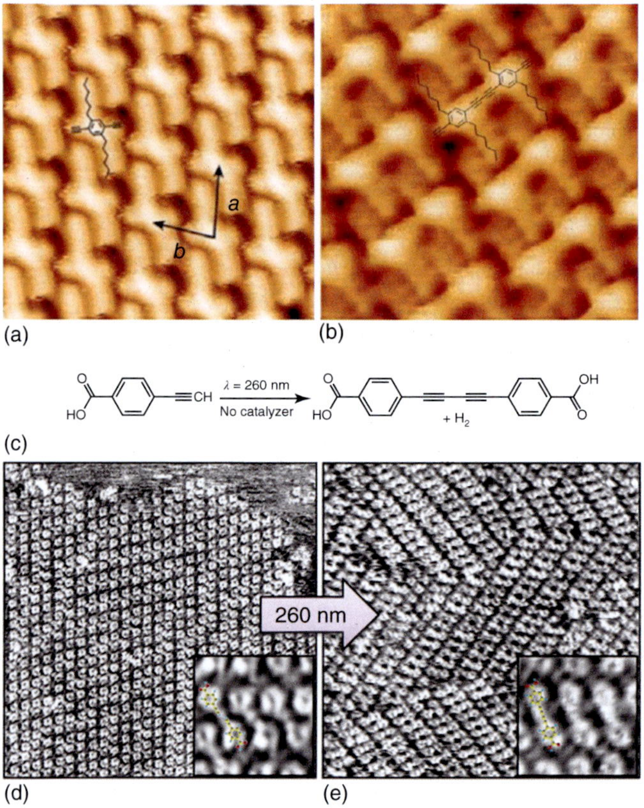

Figure 6.2 (a) STM image of the self-assembly structure of the aryl–alkyne on Ag(111) with the chemical structure superimposed. (b) STM image of the dimeric Glaser coupling product on Ag(111). Image size: 5 nm × 5 nm. Source: (a, b) Reprinted from Gao et al. [34] with permission from the American Chemical Society, copyright 2014. (c) Schematic illustration of the photochemical homocoupling of PEBA. (d) STM image of the self-assembly formed by PEBA monomers on HOPG. (e) STM image of the assembly formed by homocoupled dimers after UV-irradiation. Inset: High-resolution STM image of PEBA dimers. Source: (d, e) Adapted with permission from Colazzo et al. [57], copyright 2016 American Chemical Society.

dimers (Figure 6.2e) is achieved by UV irradiation of 260 nm wavelength. It is inevitable that the reaction that occurred in the self-assembled monolayer (SAM) shows a high selectivity.

Supramolecular chemistry deals with patterns or structures that are spontaneously formed by components without human intervention, in which molecular self-assembly plays an important role [58–63]. A variety of chiral supramolecular catalysts have been developed via self-assembly [64–70]. For instance, Figure 6.3a shows the approach to preparing a heterogeneous asymmetric catalyst based on orthogonal self-assembly by Ding et al. [71]. Ureido-4[1*H*]-ureidopyrimidone (UP) and Feringa's MonoPhos ligands are integrated into a new organic one that can form supramolecular polymers with $[Rh(cod)_2]BF_4$ by means of HB and

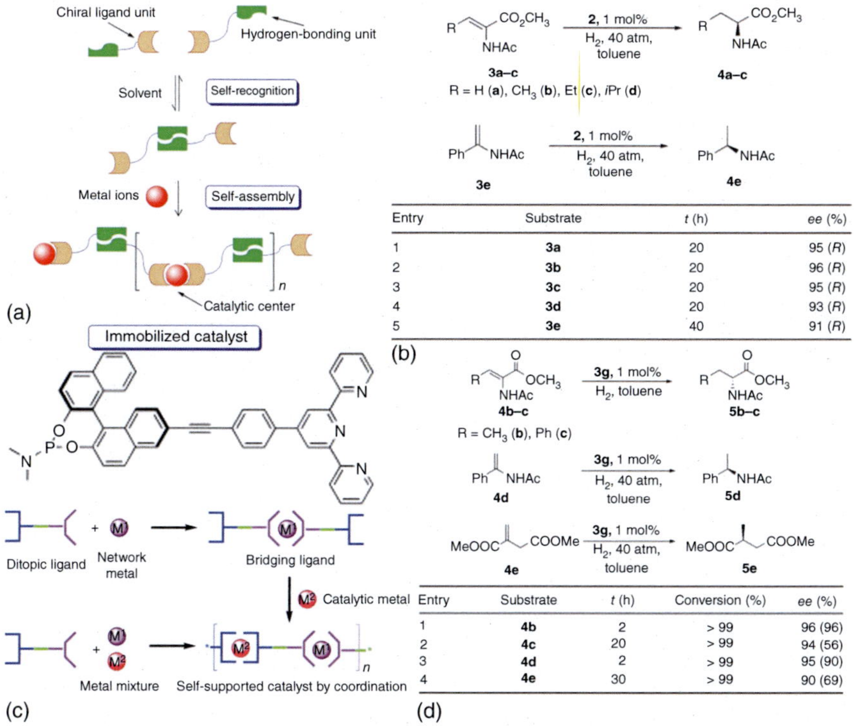

Entry	Substrate	*t* (h)	*ee* (%)
1	**3a**	20	95 (*R*)
2	**3b**	20	96 (*R*)
3	**3c**	20	95 (*R*)
4	**3d**	20	93 (*R*)
5	**3e**	40	91 (*R*)

Entry	Substrate	*t* (h)	Conversion (%)	*ee* (%)
1	**4b**	2	> 99	96 (96)
2	**4c**	20	> 99	94 (56)
3	**4d**	2	> 99	95 (90)
4	**4e**	30	> 99	90 (69)

Figure 6.3 (a) Generated supramolecular chiral catalyst via the self-assembly with noncovalent interactions. (b) Asymmetric hydrogenation of dehydro-α-amino acid and enamide derivatives with a high enantioselectivity of 91–96%. Source: (a, b) Shi et al. [71]. (c) Syntheses of self-supported catalysts via orthogonal coordination of 2,2′:6′,2″-terpyridine (tpy) unit and Feringa's MonoPhos with a single ditopic ligand. (d) Enantioselective hydrogenation of dehydroamino acid, enamide, and itaconic acid derivatives on self-supported catalysts. Source: (c,d) Yu et al. [72].

ligand-to-metal co-ordination interactions. This supramolecular polymer shows excellent asymmetric induction in the catalysis of asymmetric hydrogenation of dehydro-α-amino acid and enamide derivatives with a high enantioselectivity of 91–96% (Figure 6.3b). Similarly, the synthesised ligand containing 2,2′:6′,2″-terpyridine (tpy) and Feringa's MonoPhos could co-ordinate to Fe(II) and Rh(I) ions [72]. The chiral bimetallic self-supported catalysts would lead to a high reaction rate and an excellent enantioselectivity of 90–97% ee in the hydrogenation of α-dehydroamino acid, enamide, and itaconic acid derivatives (Figure 6.3c,d). In addition, supramolecular catalysts without metal ions made up of the co-assembly of chiral diols, arylboronic acids, and tris(pentafluorophenyl)borane possess high enantioselectivities in the Diels–Alder reactions of cyclopentadiene with different acroleins.

Obviously, the molecular self-assembly strategy can be routinely employed to steer the selectivities of both on-surface reaction and supramolecular chemistry taken place in solution.

6.3 Mediating On-Surface Reaction Pathway

Reaction pathways dictate the target structures. Efficient control of the on-surface reaction pathway has been a great challenge in surface chemistry. Therefore, a new strategy based on self-assembly has been proposed and testified in this regard. Reaction pathways in the on-surface synthesis can be scrutinised by STM imaging and scanning tunnelling spectroscopy (STS). For instance, on-surface Ullmann couplings have been revealed to occur via two disparate pathways involving organometallic intermediates and surface-anchored aromatic species [73–77], respectively.

Very recently, we have reported [78] that self-assembly of 4-bromobiphenyl (BBP) can effectively alter the pathway of its Ullmann coupling from a single-barrier process to a double-barrier one on Ag(111) (Figure 6.4a). At a low coverage of

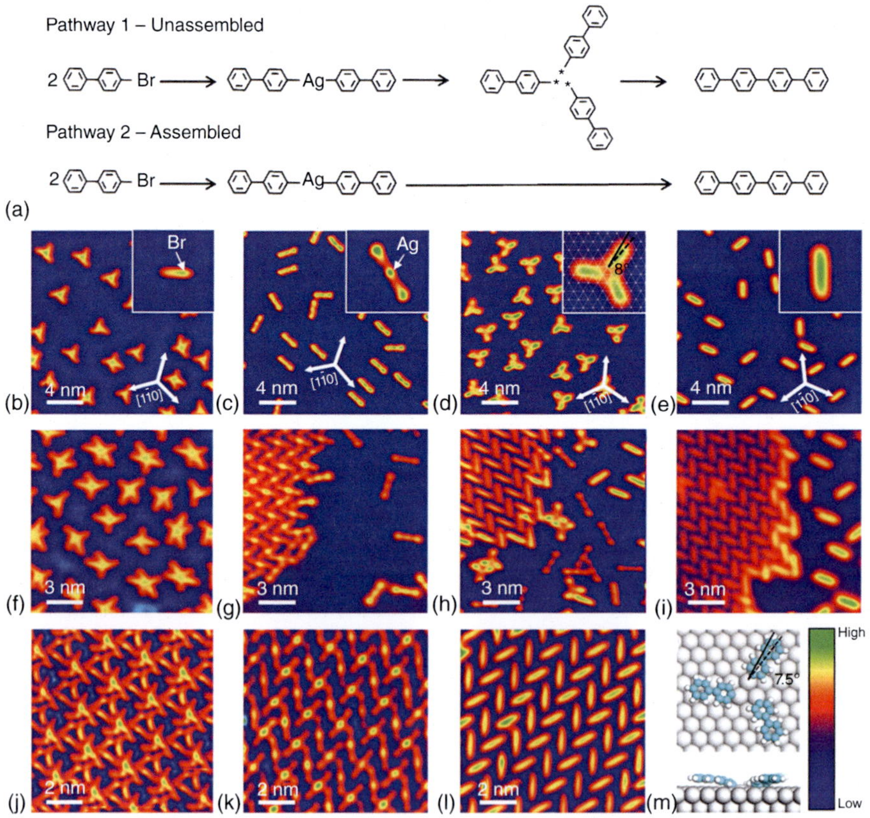

Figure 6.4 (a) Two pathways for Ullmann reactions of BBP on Ag(111). (b–e) Four reaction steps in pathway 1 at 0.1 ML, involving BBP trimers (120 K, (b)), Ag-COI (300 K, (c)), CI (400 K, (d)), and QP (430 K, (e)). Insets (b–e): corresponding high-resolution STM images of BBP, Ag-COI, CI, and QP. (f–i) Four reaction steps at 0.5 ML, involving BBP (120 K, (f)), assembled Ag-COI island and unassembled Ag-COI (300 K, (g)), QP island surrounded by CI (420 K, (h)), QP (430 K, (i)). (j–l) Three reaction steps in pathway 2 at 1.0 ML, involving BBP (120 K, (j)), Ag-COI (300 K, (k)), and QP (420 K, (l)). (m) Top and side views of the theoretically optimised configuration of CI. Source: Reprinted from Zhou et al. [78] with permission from Wiley-VCH Verlag GmbH, copyright 2017.

0.1 monolayer (ML) (Figure 6.4b–e), the BBP molecules gather into trimers by cyclic halogen bonds. Annealed at RT for 30 minutes, an isolated Ag-coordinated organometallic intermediate (Ag-COI) in a peanut-shaped feature comes into sight. However, a new clover intermediate (CI, Figure 6.4m) containing three biphenyl groups appears before the formation of the ultimate coupling product, *p*-quaterphenyl (QP). At the coverage of 1.0 ML (Figure 6.4j–l), the self-assembled BBP phases turn into ordered fishbone-like Ag-COI islands where the Ag-COIs directly convert into QPs. At a middle coverage of 0.5 ML (Figure 6.4f–i), the above-mentioned pathways co-exist. Further studies confirm the ubiquity of the assembly-steered reaction pathway for the Ullmann coupling reactions on Cu(111) and Cu(100).

The self-assembled organometallic intermediates actually apply translational and rotational restrictions on themselves. In fact, the measured pre-exponential factor in the Arrhenius equation for the single-barrier process (direct QP formation, $2 \times 10^{11}\ s^{-1}$) is much higher than that for the first reaction step in the double-barrier process (CI generation, $5 \times 10^{8}\ s^{-1}$), resulting in that pathway 2 dominates at full coverage.

According to the example described above, the molecular self-assembly is indeed an efficient strategy to steer the on-surface reaction pathway.

6.4 Mediating On-Surface Reaction Site

Once deposited onto surface, molecules may adopt different configurations, depending on the experimental parameters such as temperature and coverage. The molecules may exhibit distinct on-surface reaction activities at different surface sites that have a slight difference in energetics.

For example, in heterogeneous catalysis, only a few sites can function as active ones. These sites might be either the defects such as point defects, kinks, and edges or some specific sites such as top, bridge, and hollow sites (depending on the surface symmetry). A practical approach that can boost the catalyst performance is to increase the number of active sites or generate new active sites by surface modification or reconstruction.

The available surface sites can be generated by external species or the reactants themselves. External species may adsorb to induce surface construction that leads to exposure of specific surface sites. If designed carefully, this spontaneous process can provide an excellent way to eliminate by-products. In contrast, the surface can become dynamic upon adsorption and conversion of the reactants as the on-surface reactions proceed, which generates novel surface sites dynamically.

Normally, it is quite difficult to tune spectacular molecular distribution on surface and guide the molecular anchoring to a specific reaction site. Fortunately, some preliminary studies have shown that with the help of combined surface science techniques, such a goal could be achieved, at least partially, with the surface molecular assembly strategy.

Thiol or thiolate (R–SH or $[R{-}S]^{-}M^{+}$) has drawn great attention because of the flexibility of attachment of side or ending substitution groups. Moreover, the functional thiol group in the molecule enables its anchoring to some metal surfaces via feasible formation of the S-metal bonds. In a series of studies, Medlin et al. have

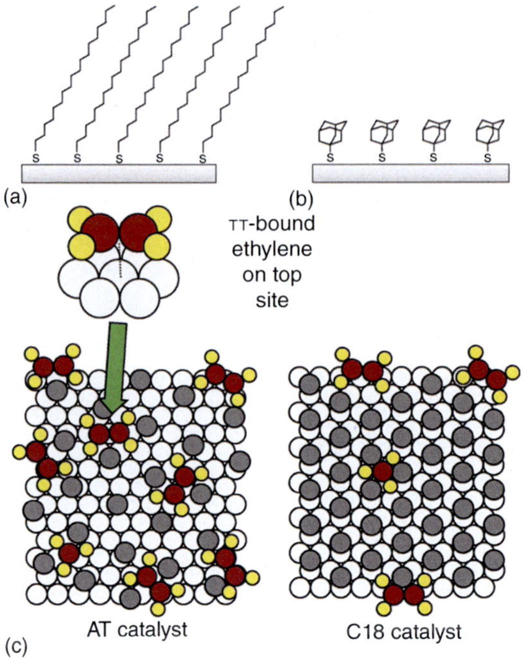

Figure 6.5 Structure of (a) 1-octadecanethiol (C18) and (b) 1-adamantanethiol (AT) used to control the density of thiolates on 5 wt% Pd/Al_2O_3 catalysts. (c) Proposed mechanisms for ethylene adsorption on both catalysts. Source: Schoenbaum et al. [80].

demonstrated that different thiolate SAMs can be prepared to preferentially expose specific types of surface sites for a desired reaction pathway while block other surface sites for undesired products [79–81]. For example, both octadecanethiol (C18, Figure 6.5a) and 1-adamantanethiol (AT, Figure 6.5b) are employed to produce SAMs on Pd/Al_2O_3 catalysts with a densely packed or a sparsely distributed monolayer, respectively [80]. The yielded surface density of the S-metal bonds strongly depends on specific van der Waals volume of their rear-ending group, a thin and long carbon chain or a sphere-like adamantane. According to the measurements of CO diffuse reflectance Fourier transform infrared spectroscopy (DRIFTS), a tool being able to reveal the available adsorption sites on a solid surface, the C18 SAMs inhibit CO adsorption on the top sites on the particle terraces, while other sites can still capture CO molecules on AT-modified surface. For a practical probe reaction of ethylene hydrogenation, both the C18- and AT-modified Pd/Al_2O_3 catalysts are used at a ratio of 10 : 1 for hydrogen to reactant feedstock in a gas-phase plug flow reactor held at 323 K. The AT-modified Pd/Al_2O_3 catalyst can convert ethylene into ethane at a reaction rate about 17 times faster than its C18-modified counterpart. The same result is retained when the AT-based thiolate surface coverage is decreased by a half. The proposed mechanism for this reaction is depicted in Figure 6.5c. Such a difference in catalytic activities seems to originate from the blocking of specific sites on the catalyst surface. The most active reaction sites for ethylene hydrogenation are the top sites on terraces that are blocked in the C18-modified catalyst, leading to a much lower catalysis efficiency compared with the AT-modified catalyst whose surface possesses plentiful top sites.

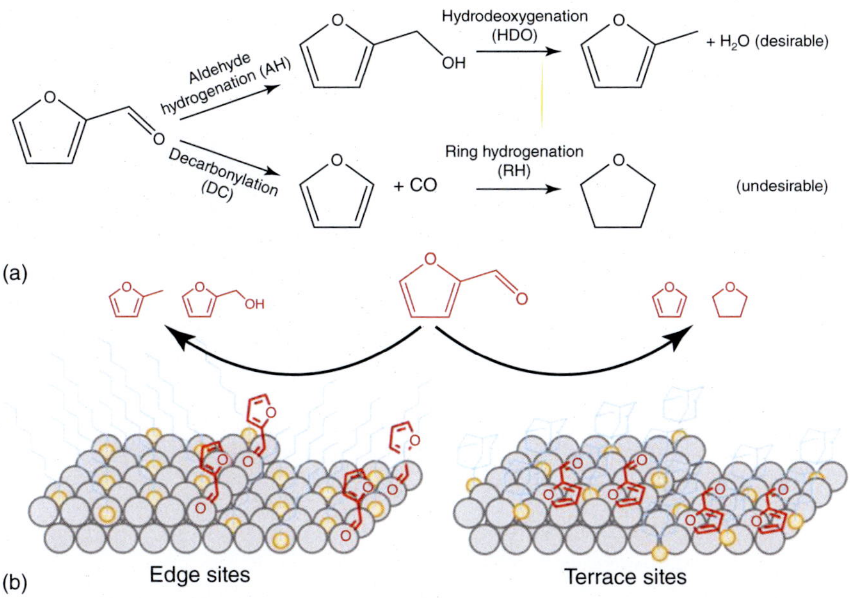

Figure 6.6 (a) Major reaction pathways for furfural hydrogenation on Pd catalysts. (b) Proposed adsorption mechanisms of furfural on C18- and AT-modified catalysts. Source: Based on Marshall et al. [83].

The selectivity of a catalyst obviously gauges the catalytic performance. With the same self-assembly strategy, Medlin et al. [82, 83] have reported another case study on tuning the reaction selectivity by blocking the unwanted sites that are responsible for the formation of by-products. Palladium catalysts are frequently used in the furfural hydrogenation process to produce methylfuran via hydrogenation of the aldehyde moiety while the undesired decarbonylated furan and CO are usually accompanied, as depicted in Figure 6.6a. Blocking the contiguous active sites on palladium catalysts with a properly densely packed SAM can help inhibit the by-product formation. According to the proposed reaction mechanism, the sites at the terrace edge are hardly affected by the thiolate SAM and are mainly responsible for the desired product, the methylfuran. Therefore, the reaction pathway towards the desired product is much less affected by the SAM modification, showing again that the reaction selectivity has been improved by this self-assembly strategy, as described in Figure 6.6b.

The thiolate can also play an important role in the formation of a surface porous metal template. Buck and co-worker [84, 85] have used 1,3,5-triazine-2,4,6-triamine (melamine) and perylene-3,4,9,10-tetracarboxylicdi-imide (PTCDI) to form a 2D honeycomb structure on the Au(111)/mica surface (Figure 6.7a,b). After being exposed to a dilute thiol solution, the bare Au substrate underlying the porous honeycomb structure is fully engrafted by densely packed thiol molecules while the backbone framework of the honeycomb structure remains intact (Figure 6.7c–e). After being treated by Cu(II) ions, a porous metallic layer of copper develops on

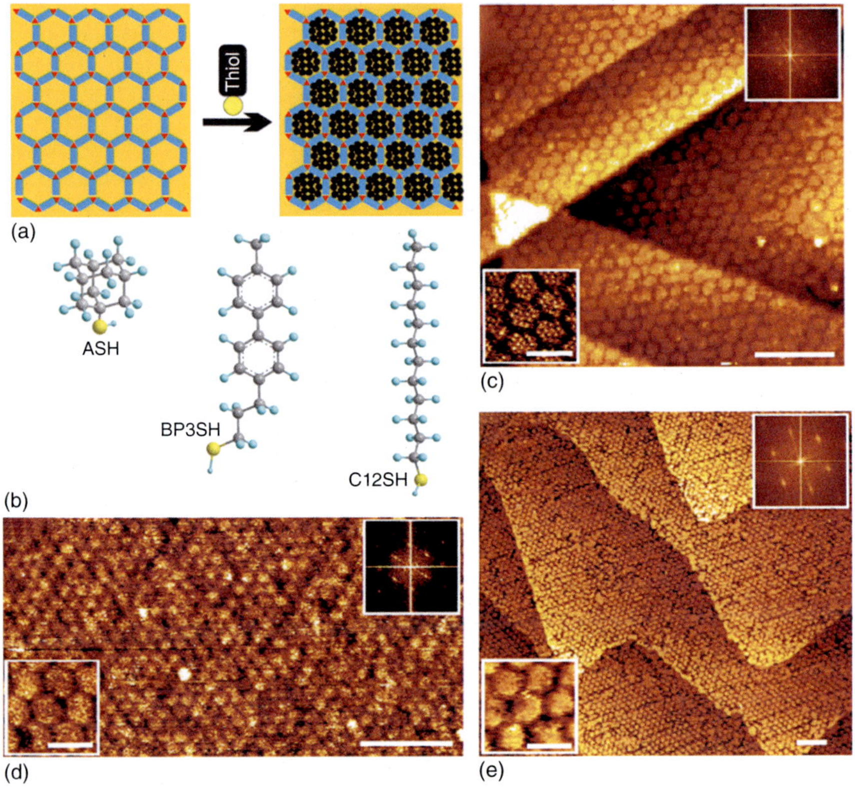

Figure 6.7 (a) Schematic diagram of the assembly filled with thiols; (b) three thiol molecules, ASH, C12SH, and BP3SH, used to fill in the assembly pores; (c–e) STM images of the assemblies filled with ASH, C12SH, and BP3SH whose molecular structures are shown in (a). Scale bars: 20 nm (large-scale images) and 5 nm (insets). Source: Reprinted from Madueno et al. [85] with permission from Nature Publishing Group, copyright 2008.

Au(111). Although intrinsically unspecific to anchor thiol-Cu cations, these sites can be regarded as new ones on the surface created by the molecular self-assembly.

Another case study on controlling the selectivity of surface oxygen reduction reaction with the self-assembly strategy is reported by Sambi and coworker [86] In this report, sub-monolayered Fe phthalocyanine (FePc) molecules are deposited on the Ag(110) surface at RT and then two energetically degenerate assemblies with distinct surface structures are formed. In these two assemblies, the FePc centres are located at the on-top (notation by R1-lower density [LD]) and short-bridge sites, respectively, as depicted in Figure 6.8a. As the FePc coverage increases slightly above 1 ML, an oblique lattice registered on the short-bridge sites is observed (oblique higher density [OHD], Figure 6.8d). A dose of 1700 L oxygen at RT results in the disappearance of the brightest spots at the FePc centres in the R1-LD assembly (Figure 6.8b), while subsequent heating at 370 K for 60 minutes leads to the recovery of the bright central protrusions in the FePc molecules (Figure 6.8c). As a contrast, no obvious change is observed when the OHD structure is exposed to the same amount of oxygen (Figure 6.8e). Based on the experimental results and

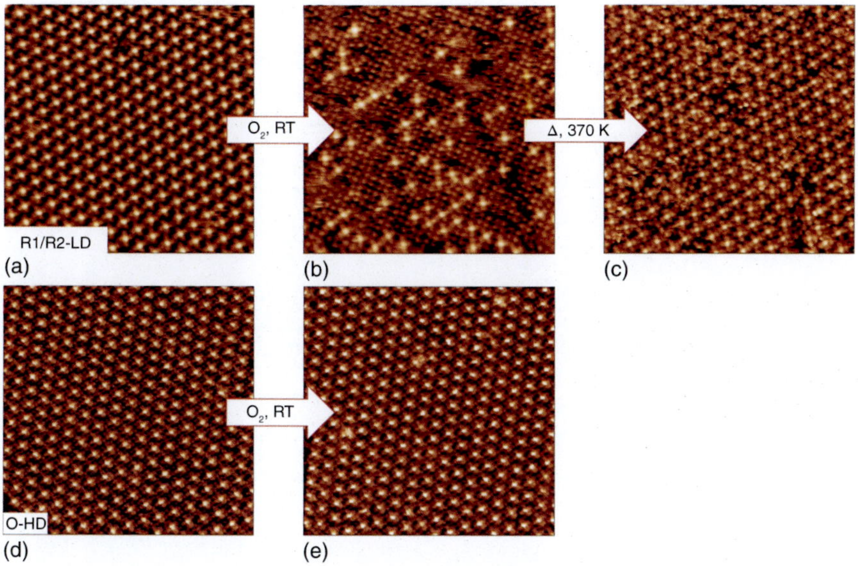

Figure 6.8 (a–c), STM images of R1/R2-LD FePc/Ag(110): (a) as-deposited, (b) oxygen-dosed, (c) annealed at 370 K; (d, e) STM images of O-HD FePc/Ag(110): (d) as-deposited and (e) oxygen-dosed. Image size: 30 nm × 30 nm. Source: Reprinted from Sedona et al. [86] with permission from Nature Publishing Group, copyright 2013.

DFT calculations, the protrusion change upon oxygen exposure is ascribed to the interaction of the substrate Ag atoms and FePc at different sites. The location of the FePc molecules at the on-top sites results in strong FePc–Ag coordination, which subsequently promotes the reduction of oxygen. The location sites for the densely packed FePc molecules swap from on-top to short-bridge sites, which explains the different reactivities of these two assemblies.

The above assembly-assisted on-surface reactions focus on directing the capability of the non-reactive species in the self-assemblies, which, to some extent, can be analogous to confined catalysts. However, there could be another analogy between the assembly-involved on-surface reaction and the autocatalysis because the reactants themselves serve as the building blocks of the self-assemblies. Zhou et al. have adopted this strategy to explore the Ullmann coupling reaction of BBP on Ag(111) (Figure 6.9a) [87].

At a full coverage, thermally deposited BBP molecules on Ag(111) held at low temperature (Figure 6.9b) are warmed up to RT to form an ordered assembly structure of the Ag-COI intermediates, as described above. Two groups of different bright protrusions at the centres of the formed Ag-COI intermediates within the assembly are noticed (Figure 6.9c). Combined STM imaging, molecular manipulation, and DFT calculations indicate that this difference in the central protrusion brightness stems from distinct adsorption sites of alternating Ag-COI arrays located as different sites. The Ag-COIs in the array with a brighter protrusion locate on the bridge sites of Ag(111) (B-type), while those in the array with a relatively dimmer protrusion sit on the hollow sites of the fcc surface (H-type). Once heated to 390 K, some

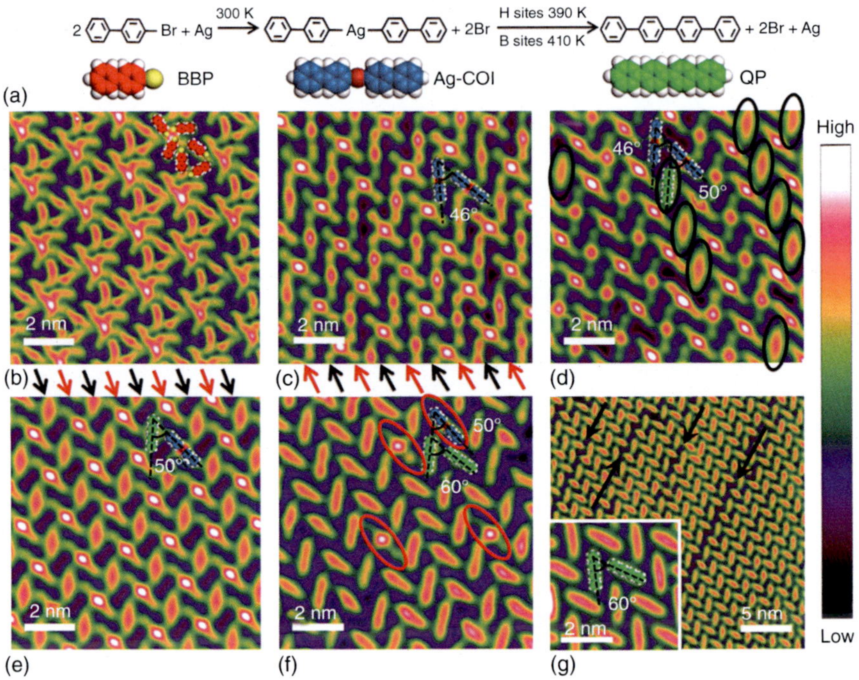

Figure 6.9 Sequential evolution of the Ullmann coupling of BBP molecules on Ag(111). (a) Scheme for the catalytic reaction process. (b) Intact BBP molecules. (c) Assembly structure of Ag-COIs. (d) Some H-type Ag-COIs (black elliptical circles) convert into the final QP product at 390 K. (e) All H-type Ag-COIs convert into the QP product at 400 K and all left B type Ag-COIs remain unreacted. (f) Most B-type Ag-COIs convert into QP at 410 K. (g) All Ag-COIs convert into the product QP at 420 K. Source: Reprinted from Zhou et al. [87] with permission from the American Chemical Society, copyright 2017.

H-type Ag-COI intermediates turn into the final coupling product, QP (Figure 6.9d). Further annealing up to 400 K leads to the full conversion of the H-type Ag-COI intermediates into alternating arrays of QP and B-type intermediates (Figure 6.9e). At a higher temperature, 410 K, the B-type intermediates start to react (Figure 6.9f) and completely turn into QP at 420 K (Figure 6.9g). It is therefore concluded that even on an atomic flat terrace, different sites have distinct catalytic activities, i.e. the hollow sites being more active than the bridge sites for the BBP Ullmann couplings on Ag(111).

6.5 Brief Summary and Perspective

In this chapter, we have overviewed some proof-of-principle case studies focusing on steering the on-surface reactions with the newly established self-assembly strategy in three main aspects: controlling the reaction selectivity, pathway, and site on surfaces. It is demonstrated that the surface molecular assemblies can not only form ordered and aesthetic structures on surfaces but also serve as an efficient strategy to control on-surface reactions to fabricate covalently bonded structures that may possess unique electronic, magnetic, and optic properties [88–92].

However, the pressure gap between the laboratory-employed ultrahigh vacuum and industrially applied high-pressure conditions as well as the materials gap between the laboratory-adopted single crystals and industrially practical supported catalysts have restricted such a self-assembly strategy from being widely applied. More efforts have to be made to minimise or even eliminate these pressure and materials gaps. If successful, it could be anticipated that any mature industrial process incorporated with such a self-assembly might have a huge economic impact on future chemistry-related industries such as chemical engineering and semiconductor industry.

Meanwhile, some novel systems could be further explored, i.e. those mimicking natural and biological functionalities [93–96]. Molecular self-assembly is actually of fundamental importance in biological systems that are full of weak intermolecular interactions. Any research ventures along this direction would undoubtedly bring the self-assembly strategy to a new horizon for the understanding of life sciences.

Acknowledgement

This work is financially supported by NSFC (21821004, 91527303, 21333001) and MOST (2017YFA0204702), China.

References

1 Brown, T.L. (2009). *Chemistry: The Central Science*. Pearson Education.
2 Haber, F. (2002). *Resonance* 7 (9): 86–94.
3 Gary, J.H., Handwerk, G.E., and Kaiser, M.J. (2007). *Petroleum Refining: Technology and Economics*. CRC Press.
4 Houston, P.L. (2012). *Chemical Kinetics and Reaction Dynamics*. Courier Corporation.
5 Chorkendorff, I. and Niemantsverdriet, J.W. (2017). *Concepts of Modern Catalysis and Kinetics*. Wiley.
6 Laidler, K.J. (1984). *J. Chem. Educ.* 61 (6): 494.
7 Logan, S. (1982). *J. Chem. Educ.* 59 (4): 279.
8 Somorjai, G.A. and Li, Y. (2010). *Introduction to Surface Chemistry and Catalysis*. Wiley.
9 Woodruff, D.P. (2016). *Modern Techniques of Surface Science*. Cambridge University Press.
10 Ertl, G. (2010). *Reactions at Solid Surfaces*, vol. 14. Wiley.
11 Shao, X., Luo, X., Hu, X., and Wu, K. (2006). *J. Phys. Chem. B* 110 (31): 15393–15402.
12 Lei, S., Tahara, K., Feng, X. et al. (2008). *J. Am. Chem. Soc.* 130 (22): 7119–7129.

13 Shen, Y.-T., Guan, L., Zhu, X.-Y. et al. (2009). *J. Phys. Chem. B* 131 (17): 6174–6180.
14 Ye, Y., Sun, W., Wang, Y. et al. (2007). *J. Phys. Chem. C* 111 (28): 10138–10141.
15 Xiao, W., Feng, X., Ruffieux, P. et al. (2008). *J. Am. Chem. Soc.* 130 (28): 8910–8912.
16 Ruben, M., Payer, D., Landa, A. et al. (2006). *J. Am. Chem. Soc.* 128 (49): 15644–15651.
17 Liang, H., Sun, W., Jin, X. et al. (2011). *Angew. Chem. Int. Ed.* 50 (33): 7562–7566.
18 Zhang, X., Li, N., Gu, G.-C. et al. (2015). *ACS Nano* 9 (12): 11909–11915.
19 Shang, J., Wang, Y., Chen, M. et al. (2015). *Nat. Chem.* 7: 389–393.
20 Liu, J., Chen, T., Deng, X. et al. (2011). *J. Am. Chem. Soc.* 133 (51): 21010–21015.
21 Geng, Y.-f., Li, P., Li, J.-z. et al. (2017). *Coord. Chem. Rev.* 337: 145–177.
22 Shao, X., Luo, X., Hu, X., and Wu, K. (2006). *J. Phys. Chem. B* 110 (3): 1288–1293.
23 Liu, J., Fu, X., Chen, Q. et al. (2016). *Chem. Commun.* 52 (88): 12944–12947.
24 Kühne, D., Klappenberger, F., Decker, R. et al. (2009). *J. Am. Chem. Soc.* 131 (11): 3881–3883.
25 Lingenfelder, M.A., Spillmann, H., Dmitriev, A. et al. (2004). *Chem. Eur. J.* 10 (8): 1913–1919.
26 Braden, D.A., Parrack, E.E., and Tyler, D.R. (2001). *Coord. Chem. Rev.* 211 (1): 279–294.
27 Dong, L., Liu, P.N., and Lin, N. (2015). *Acc. Chem. Res.* 48 (10): 2765–2774.
28 Gao, H.-Y., Franke, J.-H., Wagner, H. et al. (2013). *J. Phys. Chem. C* 117 (36): 18595–18602.
29 Cirera, B., Zhang, Y.-Q., Bjork, J. et al. (2014). *Nano Lett.* 14 (4): 1891–1897.
30 Kong, H., Yang, S., Gao, H. et al. (2017). *J. Am. Chem. Soc.* 139 (10): 3669–3675.
31 Liu, J., Chen, Q., Xiao, L. et al. (2015). *ACS Nano* 9 (6): 6305–6314.
32 Kalashnyk, N., Salomon, E., Mun, S.H. et al. (2018). *ChemPhysChem* 19 (15): 1802–1808.
33 Simonov, K.A., Vinogradov, N.A., Vinogradov, A.S. et al. (2015). *ACS Nano* 9 (9): 8997–9011.
34 Gao, H.Y., Zhong, D., Mönig, H. et al. (2014). *J. Phys. Chem. C* 118: 6272–6277.
35 Gao, H.Y., Wagner, H., Zhong, D. et al. (2013). *Angew. Chem. Int. Ed.* 52 (14): 4024–4028.
36 Wang, T., Lv, H., Fan, Q. et al. (2017). *Angew. Chem. Int. Ed.* 129 (17): 4840–4844.
37 Liu, J., Chen, Q., He, Q. et al. (2018). *Phys. Chem. Chem. Phys.* 20 (16): 11081–11088.
38 Lafferentz, L., Eberhardt, V., Dri, C. et al. (2012). *Nat. Chem.* 4 (3): 215.
39 Klaasen, H., Liu, L., Meng, X. et al. (2018). *Chem. Eur. J.*
40 Gong, Z., Yang, B., Lin, H. et al. (2016). *ACS Nano* 10 (4): 4228–4235.
41 Yu, Y., Zheng, Y., and Lei, S. (2017). *J. Phys. Chem. C* 121 (1): 593–599.
42 Fan, Q., Wang, C., Han, Y. et al. (2013). *Angew. Chem. Int. Ed.* 52 (17): 4668–4672.

43 Zhang, X., Liao, L., Wang, S. et al. (2014). *Sci. Rep.* 4: 1–5.
44 Zhong, D., Franke, J.-H., Podiyanachari, S.K. et al. (2011). *Science* 334 (6053): 213–216.
45 Chen, Q., Cramer, J.R., Liu, J. et al. (2017). *Angew. Chem. Int. Ed* 56: 5026–5030.
46 Klappenberger, F., Zhang, Y.-Q., Bjork, J. et al. (2015). *Acc. Chem. Res.* 48 (7): 2140–2150.
47 Chen, Z., Molina-Jirón, C., Klyatskaya, S. et al. (2017). *Ann. Phys. (Berl.)* 529 (11): 1700056.
48 Liu, J., Chen, Q.-W., and Wu, K. (2017). *Chin. Chem. Lett.* 28 (8): 1631–1639.
49 Eichhorn, J., Heckl, W.M., and Lackinger, M. (2013). *Chem. Commun.* 49 (28): 2900–2902.
50 Liu, J., Ruffieux, P., Feng, X. et al. (2014). *Chem. Commun.* 50 (76): 11200–11203.
51 Lin, T., Zhang, L., Bjork, J. et al. (2017). *Chemistry* 23 (62): 15588–15593.
52 Sun, Q., Zhang, C., Li, Z. et al. (2013). *J. Am. Chem. Soc.* 135 (23): 8448–8451.
53 Zhou, H., Liu, J., Du, S. et al. (2014). *J. Am. Chem. Soc.* 136 (15): 5567–5570.
54 Riss, A., Wickenburg, S., Gorman, P. et al. (2014). *Nano Lett.* 14 (5): 2251–2255.
55 Schuler, B., Fatayer, S., Mohn, F. et al. (2016). *Nat. Chem.* 8: 220–224.
56 Zhang, Y.-Q., Paszkiewicz, M., Du, P. et al. (2018). *Nat. Chem.* 10: 296.
57 Colazzo, L., Sedona, F., Moretto, A. et al. (2016). *J. Am. Chem. Soc.* 138 (32): 10151–10156.
58 Menger, F.M. (2002). *Proc. Natl. Acad. Sci. U.S.A.* 99 (8): 4818–4822.
59 De Feyter, S. and De Schryver, F.C. (2003). *Chem. Soc. Rev.* 32 (3): 139–150.
60 Chakrabarty, R., Mukherjee, P.S., and Stang, P.J. (2011). *Chem. Rev.* 111 (11): 6810–6918.
61 Cook, T.R., Zheng, Y.-R., and Stang, P.J. (2013). *Chem. Rev.* 113 (1): 734–777.
62 Stupp, S.I., LeBonheur, V., Walker, K. et al. (1997). *Science* 276 (5311): 384–389.
63 Steed, J.W. and Atwood, J.L. (2013). *Supramolecular Chemistry*. Wiley.
64 Liu, M., Zhang, L., and Wang, T. (2015). *Chem. Rev.* 115 (15): 7304–7397.
65 Ortega Lorenzo, M., Baddeley, C.J., Muryn, C., and Raval, R. (2000). *Nature* 404: 376–379.
66 Park, J., Lang, K., Abboud, K.A., and Hong, S. (2008). *J. Am. Chem. Soc.* 130 (49): 16484–16485.
67 Jiao, Y., Tang, B., Zhang, Y. et al. (2018). *Angew. Chem. Int. Ed.* 130 (21): 6185–6189.
68 Van Leeuwen, P.W. (2008). *Supramolecular Catalysis*. Wiley.
69 Brown, C.J., Toste, F.D., Bergman, R.G., and Raymond, K.N. (2015). *Chem. Rev.* 115 (9): 3012–3035.
70 Meeuwissen, J. and Reek, J.N.H. (2010). *Nat. Chem.* 2: 615.
71 Shi, L., Wang, X., Sandoval, C.A. et al. (2006). *Angew. Chem. Int. Ed.* 118 (25): 4214–4218.
72 Yu, L., Wang, Z., Wu, J. et al. (2010). *Angew. Chem. Int. Ed.* 49 (21): 3627–3630.
73 Fan, Q., Gottfried, J.M., and Zhu, J. (2015). *Acc. Chem. Res.* 48 (8): 2484–2494.
74 Basagni, A., Sedona, F., Pignedoli, C.A. et al. (2015). *J. Am. Chem. Soc.* 137 (5): 1802–1808.

75 Wang, W., Shi, X., Wang, S. et al. (2011). *J. Am. Chem. Soc.* 133 (34): 13264–13267.

76 Lewis, E.A., Murphy, C.J., Liriano, M.L., and Sykes, E.C.H. (2014). *Chem. Commun.* 50 (8): 1006–1008.

77 Lipton-Duffin, J., Miwa, J., Kondratenko, M. et al. (2010). *Proc. Natl. Acad. Sci. U.S.A.* 107 (25): 11200–11204.

78 Zhou, X., Wang, C., Zhang, Y. et al. (2017). *Angew. Chem. Int. Ed.* 56: 12852–12856.

79 Schoenbaum, C.A., Schwartz, D.K., and Medlin, J.W. (2014). *Acc. Chem. Res.* 47: 1438–1445.

80 Schoenbaum, C.A., Schwartz, D.K., and Medlin, J.W. (2013). *J. Catal.* 303: 92–99.

81 Pang, S.H., Schoenbaum, C.A., Schwartz, D.K., and Medlin, J.W. (2013). *Nat. Commun.* 4: 1–6.

82 Kumar, G., Lien, C.H., Janik, M.J., and Will Medlin, J. (2016). *ACS Catal.* 6: 5086–5094.

83 Marshall, S.T., O'Brien, M., Oetter, B. et al. (2010). *Nat. Mater.* 9: 853–858.

84 Silien, C. and Buck, M. (2008). *J. Phys. Chem. C* 112: 3881–3890.

85 Madueno, R., Räisänen, M.T., Silien, C., and Buck, M. (2008). *Nature* 454: 618–621.

86 Sedona, F., Di Marino, M., Forrer, D. et al. (2012). *Nat. Mater.* 11: 970–977.

87 Zhou, X., Bebensee, F., Yang, M. et al. (2017). *ACS Nano* 11: 9397–9404.

88 Fernández-Torrente, I., Franke, K.J., and Pascual, J.I. (2008). *Phys. Rev. Lett.* 101 (21): 217203.

89 Lobo-Checa, J. and Stöhr, M. (2009). *Science* 325 (5938): 300–303.

90 Huang, Z., Zhang, Y., He, Y. et al. (2017). *Chem. Soc. Rev.* 46 (7): 1955–1976.

91 Huang, N., Wang, P., and Jiang, D. (2016). *Nat. Rev. Mater.* 1 (10).

92 Diercks, C.S. and Yaghi, O.M. (2017). *Science* 355 (6328): eaal1585.

93 Liu, L., Zhang, L., Niu, L. et al. (2011). *ACS Nano* 5 (7): 6001.

94 Aldaye, F.A., Palmer, A.L., and Sleiman, H.F. (2008). *Science* 321 (5897): 1795–1799.

95 Mao, X.B., Wang, C.X., Wu, X.K. et al. (2011). *Proc. Natl. Acad. Sci. U.S.A.* 108 (49): 19605–19610.

96 Ciesielski, A., El Garah, M., Masiero, S., and Samori, P. (2016). *Small* 12 (1): 83–95.

7

Covalently Bonded Organic Structures via On-Surface Synthesis

Can Wang, Haiming Zhang, and Lifeng Chi

Soochow University, Institute of Functional Nano and Soft Materials (FUNSOM), Jiangsu Key Laboratory for Carbon-Based Functional Materials and Devices, 199 Ren'ai Road, Suzhou, Jiangsu 215123, China

7.1 Introduction

On-surface synthesis is the bottom-up construction of covalent bonds between molecular building blocks, which has been greatly developed during the past decade. Unlike molecular self-assembly based on weak intermolecular interactions [1], on-surface synthesis provides a new method to prepare complex molecular systems with well-defined covalent bonded structures. On two-dimensional (2D) surfaces, molecules can move and rearrange into self-assembled structures. Depending on the synthetic approach employed, the underlying surface can play a role as either the catalyst or template to initiate the reaction [2].

On catalytic metal substrates, conjugated polymers, such as one-dimensional (1D) wires and 2D networks have been synthesised via a bottom-up approach. The monomers are first deposited onto the surfaces and subsequently react with each other via homocoupling or heterocoupling. The specific polymer structures can be designed by using functional monomers. It is worth noticing that performing reactions directly on 2D surfaces is a novel approach for surface functionalisation, further used in the preparation of molecular electronic devices [3].

In solution-based chemistry, solvents always act as reaction media, which dissolve organic reactants to form a homogeneous system and allow the reaction to happen. However, the on-surface reactions occur in solvent-free conditions that can be investigated at the atomic level by scanning tunnelling microscope (STM) under ultra-high vacuum (UHV). Hitherto, it is still a challenge to transfer solution-phase reactions onto 2D surfaces. Moreover, the novel reaction environments on crystal metal surfaces bring us new chemical reactions. In this chapter, the on-surface chemical reactions are collected and classified, giving an overview of the existing repertoire of synthetic strategies for on-surface covalent bond formations [4].

Supramolecular Chemistry on Surfaces: 2D Networks and 2D Structures, First Edition.
Edited by Neil R. Champness.

7.2 Dehalogenation

7.2.1 Ullmann Coupling

In 1904, the classical Ullmann synthesis was firstly described by Ullmann and Bielecki. They introduced copper powder as catalyst to cleavage C—Br bond, further inducing the formation of diaryl [5]. As an important approach to link aromatics unit via C—C bonds, Ullmann reaction is popular used in organic synthesis. Almost after a century, Rieder's group demonstrated the feasibility of on-surface Ullmann reaction [6]. By multistep STM-tip manipulations on Cu(111) surface in a low-temperature ultra-high vacuum (LT-UHV) system (here 4 K), they managed to covalently couple individual iodobenzene molecules. As shown in Figure 7.1, the tip was located above one iodobenzene molecule and injected an electron into it. The injected electron could induce the cleavage of one C—I bond and generate a free phenyl radical. The free phenyl radicals, which are supposed to be highly reactive species in solution here, could be stabilised via both π–π interaction with the underlying substrate and σ-interaction with the Cu atoms at the step edge. In this C—C bond formation process, the STM tip had to bring two phenyl radicals into proximity, in a head-to-head fashion. Followed by applying a pulse voltage of 0.5 V, covalent coupling between two phenyl radicals was triggered to form biphenyl identified as product. This work has been highly awarded as the starting of on-surface chemistry.

However, the 'tip manipulation' approach also has limitations, such as time-consuming and inefficiency. Nowadays, the on-surface Ullmann-type coupling has been wildly investigated by chemists, and more optimised experimental approaches have been introduced for large-scale fabrication [7]. The milestone work is the thermal-induced Ullmann coupling reported by Grill et al. in 2007 [8]. They found that the C—X bond cleavage and the subsequent C—C bond coupling could be stimulated by simple thermal annealing at a proper temperature. Noteworthy, the C—Br bond cleavage occurred both on hot Au(111) surface and in the crucible before the deposit, if the temperature reached 590 K. Compared with the

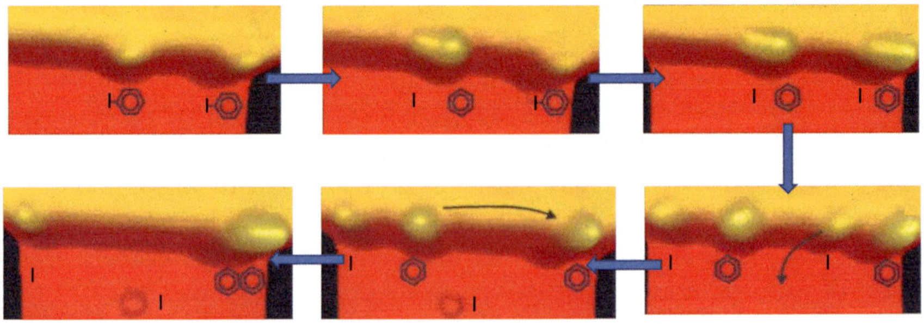

Figure 7.1 STM images showing the initial steps of the tip-induced on-surface Ullmann reaction. Source: Hia et al. [6]. Reproduced with permission of the American Physical Society.

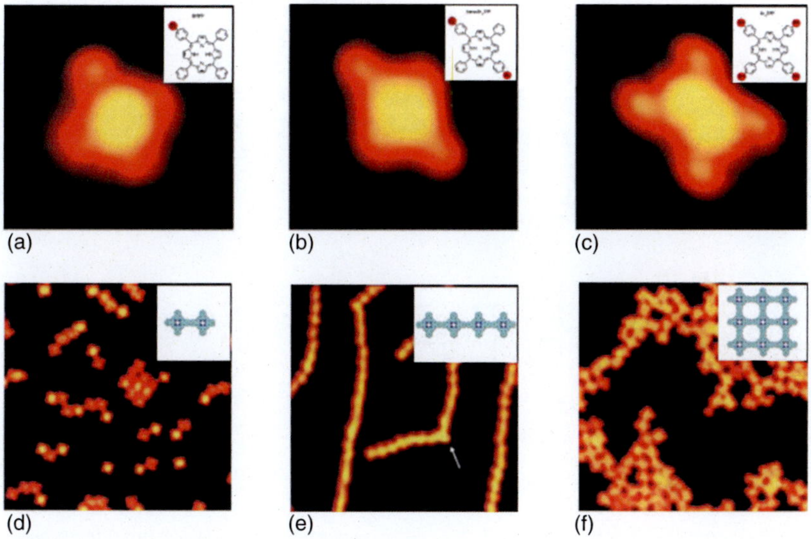

Figure 7.2 Thermal-induced Ullmann reactions of TPP derivatives. (a–c) High-resolution STM images of single intact molecules with one (a), two (b), and four (c) Br substituents ($3.5 \times 3.5\ nm^2$). (d–f) Overview STM images of the nanostructures after reaction ($30 \times 30\ nm^2$). Source: Grill et al. [8]. Reproduced with permission of the Nature.

tip-induced approach, thermal-induced Ullmann reaction is more simplified, which sets a precedent for dissemination of surface-assisted reaction. The number and position of the connection points were controlled by tailoring the Br-substitution at the tetraphenylporphyrin (TPP) core, which allowed the further control of the macromolecular architecture of the generated nanostructure. As shown in Figure 7.2, dimerisation was observed from *mono*-BrTPP, whereas *trans*-Br_2TPP led to linear chains, and Br_4TPP enabled the construction of small area 2D networks.

To facilitate further applications, the lack of long-range order in the formation of on-surface chemical synthesis has become an issue. By taking advantage of the unequal bond energy of C—X bonds with different halogen substituents, Grill's group proposed a hierarchical reaction route to obtain highly regulated 2D networks on Au(111) surface, as seen in Figure 7.3 [9]. The *trans*-Br_2I_2TPP molecules, equipped with two bromine and two iodine substituents at orthogonal terminal sites, were first coupled into a linear structure by cleavage of iodine atoms at 393 K. Then, the surface was annealed to 523 K to dissociate C—Br bonds, which led to the combination of the linear polymer chains and resulted in the final 2D conjugated networks. In the second reaction step, once the first inter-chain C—C bond was formed, the activated sites were automatically arranged for the neighbouring C—C bond formation. This 'molecular zipper' method has been proven to be more efficient, because all the activated sites are equally spaced along the TPP polymer chain. It is also reasonable to expect that the 2D covalent networks obtained through this hierarchical strategy will contain fewer defects compared with the one-step growth procedure.

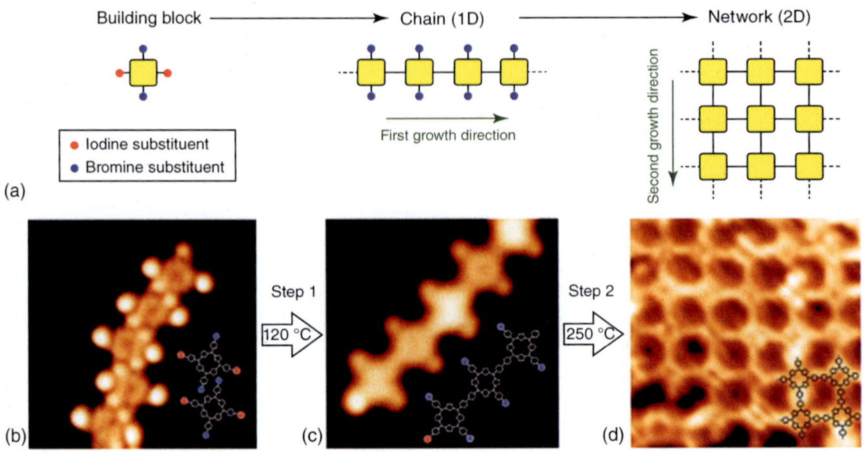

Figure 7.3 Hierarchical growths following sequential thermal activation. (a) Schematic illustration of the two-step procedure. (b–d) STM images of *trans*-Br_2I_2TPP molecules on Au(111), after deposition ($8 \times 8\,nm^2$) (b), step 1, after heating to 393 K ($8 \times 8\,nm^2$) (c), step 2, after further heating to 523 K ($10 \times 10\,nm^2$) (d). Source: Lafferentz et al. [9]. Reproduced with permission of the Nature.

The substrates not only act as static support in this on-surface Ullmann reaction process, but also actively participate as a catalyst to influence the regularities of the 2D networks. Fasel's group studied the homocoupling processes between the hexaiodo-substituted macrocycle cyclohexa-*m*-phenylene (CHP) molecules on Cu(111), Au(111), and Ag(111) substrates [10]. The resulting polyphenylene networks showed significant differences in morphology on the three substrates. As seen in Figure 7.4, 'open' branched structures, small domains of compact network clusters, and highly ordered 2D networks were obtained after thermal annealing on Cu(111), Au(111), and Ag(111), respectively. According to DFT calculation results, the mobility of CHP molecule is highest on the silver surface, and lowest on the copper surface. In the case of Ag(111), the energy barrier for molecule diffusion is significantly lower than the energy barrier for intermolecular bond formation. Hexaiodo-substituted CHP monomers have a chance to rearrange before reaction to obtain a highly ordered 2D structure. However, the reverse relation holds on Cu(111), which results in a disordered structure. Therefore, to synthesise large-scale highly-ordered 2D covalent networks, scientists must select the substrate properly and control the kinetic effect carefully.

As described above, halide precursor, in which the terminal carbon atom has been halogenated, can produce merely one unpaired electron at the carbon site to form C—C single bond with a neighboured monomer. Xu's group innovatively investigated the dehalogenation of germinal dihalide and trihalide precursors on Au(111) [11]. At room temperature, alkenyl gem-dibromides generated two unpaired electrons via debromination after deposition on the gold surface. As seen in Figure 7.5a, high-resolution STM together with noncontact atomic force microscopy (nc-AFM) images confirmed the dimer formation, and these dimers

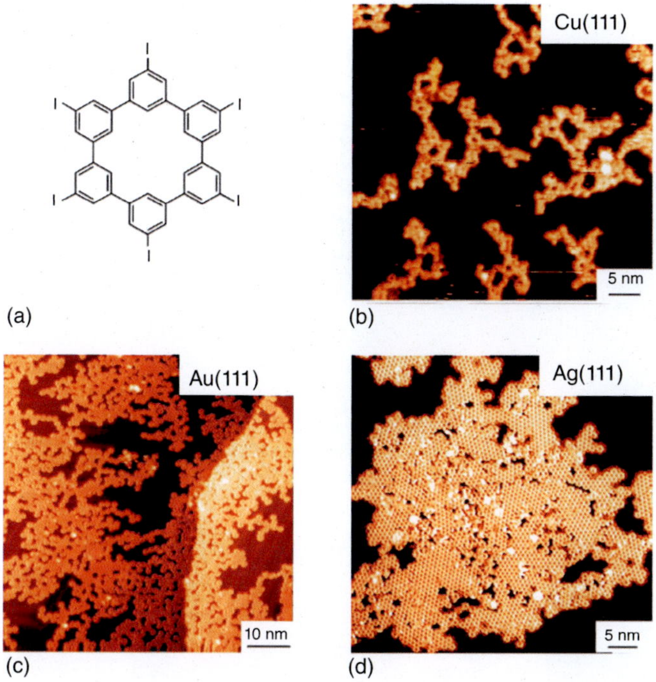

Figure 7.4 Substrate influence in Ullmann reaction. (a) Chemical structure of hexaiodo-substituted CHP molecule. (b–d) STM images of polyphenylene networks fabricated on Cu(111) (b), Au(111) (c), and Ag(111) (d). Source: Bieri et al. [10]. Reproduced with permission of the American Chemical Society.

performed isomeric configurations, trans- or cis-structure. Each two biphenyl group was linked together by three consecutive C=C double bonds to form a cumulene group, as illustrated in the DFT relaxed model. One step further, they explored the possibility of dehalogenative homocoupling of tribromomethyl-substituted precursors. Interestingly, the C≡C triple bonds could form directly from sp^3-hybridised carbon atoms at room temperature on Au(111). By functionalising the precursor with two terminal tribromomethyl groups, graphyne-like molecular wires were fabricated successfully, as seen in Figure 7.5b. Moreover, 2D-network acetylenic scaffolds were obtained using the precursor with three tribromomethyl groups, as seen in Figure 7.5c. Liu and coworkers investigated the reaction mechanism via homocoupling between trichloromethyl groups on Cu(111) surface, as seen in Figure 7.5d [12]. High-resolution nc-AFM images combined with DFT calculations, the sequential dechlorination steps involving surface-bound intermediates of benzyl, carbine, and carbyne radicals were revealed, seen in Figure 7.5e–g. The dechlorination process was exothermic, and the formation of carbine radical was energetically favourable. Compared with the reaction barriers in the gas phase, they also pointed out that Cu(111) surface had a strong catalytic effect by reducing the reaction barriers (about 2–4 eV), thereby reducing the experimental temperature to 300 K.

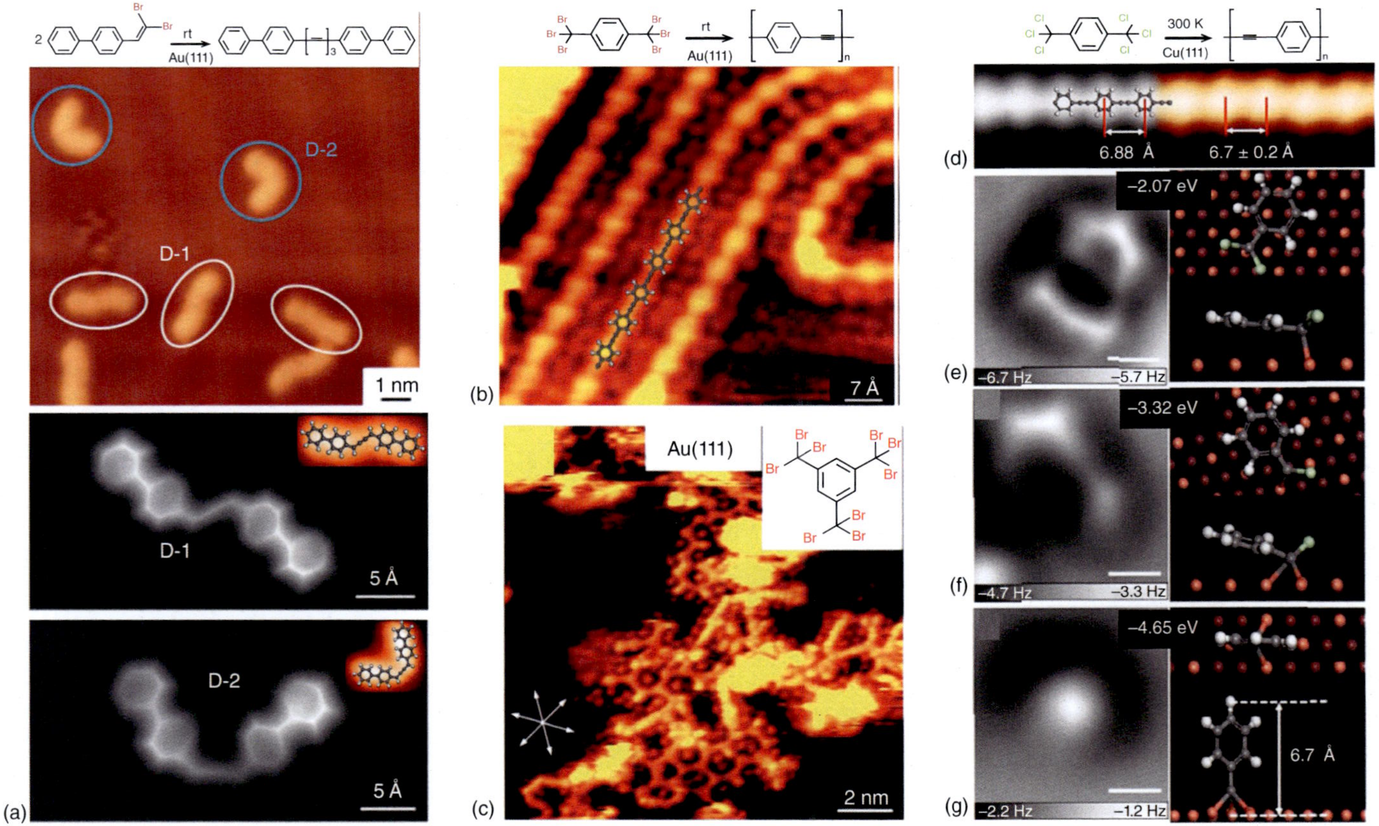

Figure 7.5 (a) STM and nc-AFM images of *trans*-cumulene (D-1) and *cis*-cumulene (D-2) formed by dehalogenative homocoupling of 4-(2,2-dibromovinyl)-1,1′-biphenyl on Au(111). Source: Sun et al. [11a]. Reproduced with permission of Wiley. (b) STM image of graphyne wires formed by annealing 1,4-bis(tribromomethyl)benzene molecules on Au(111) at about 430 K. (c) 2D molecular network obtained by annealing 1,3,5-tris(tribromomethyl)benzene molecules to approximately 500 K. Source: Sun et al. [11b]. Reproduced with permission of Wiley. (d) Simulated (left part) and experimental (right part) STM image of a poly(*p*-phenyleneethynylene) molecular wire. (e, f) The nc-AFM images and top/side view of DFT-optimised models of the three typical reaction intermediates: benzyl radical (e), carbene radical (f), and carbyne radical (g). Source: Shu et al. [12]. Reproduced with permission of the Nature.

As halogenated aromatic hydrocarbons are usually chosen as precursor monomers, Ullmann-type coupling reaction could also be utilised as a critical step towards the synthesis of graphene nanoribbons (GNR). The halogen substituents on monomers could be removed to yield biradical species, which further undergo the radical addition reactions to produce linear polymer chains, as the embryonic form of the targeted GNR [13].

7.2.2 Sonogashira Coupling

Lambert's group successfully implemented the Sonogashira cross-coupling on metal surface in vacuum condition, which had been widely used to grow carbon-chains in wet chemistry. The phenylacetylene (PA) and iodobenzene (IB) reacted on Au(111) surface via thermal annealing to yield diphenylacetylene [14]. The annealing process-induced cleavage of the iodine atom on iodobenzene, and then the produced phenyl radicals attacked the PA with the help of underlying gold atoms (acted as base). Because of the unequal adsorption energies, the two molecules preferred to self-assemble into separated domains when adsorbed on the gold surface. Only the reactants located at the domain boundaries had the ability to initiate the reaction, resulting in a quite low yield of about 10%. The most efficient approach to increase the yield was to expand the boundary area. The flat Au(111) and the rough Au(100) surfaces both have numerous nucleation centres, which lead the molecules assembled into numerous small domains, further increasing the mixed boundary area. The herringbone structures on the reconstructed Au(111) surface act as the nucleation centres, which disappear if the surface becomes roughened. As displayed in Figure 7.6a, PA adsorbed in fourfold-symmetric islands on the unreconstructed Au(100) surface, while IB only adsorbed on the reconstructed Au(100) surface [15]. As a result, the adsorbed PA and IB molecules on well-annealed Au(100) surface generated homocoupling products, respectively. However, heterocoupling between PA and IB molecules occurred on the Ar^+ roughened Au(100) surface because the roughening treatment could generate new nucleation centres to increase the area of mixed domains.

7.2.3 Heck Reaction

Unlike the independent behaviours of co-adsorbed PA and IB, aryl ethylene and aryl bromides are co-assemble on the surface, providing an advantage for the cross-coupling products. Liu's group firstly reported the heck reaction between aryl bromides and alkene on Au(111) surface [16]. Because the gold surface was ineffective to catalyze the cross-coupling reaction, Pd adatoms were dosed on the gold surface as a catalyst. As seen in Figure 7.6b–d, the on-surface Heck reaction had excellent selectivity, and no obvious homocoupling of these two reactants was observed. With a Pd adatom as a catalyst, the barrier energy of debromination was reduced, and the obtained benzene radical could also be stabilised by another Pd adatom in the exothermal process. Moreover, the isolated alkenes approached the benzene radicals via coordination interaction with the Pd adatom. DFT calculations

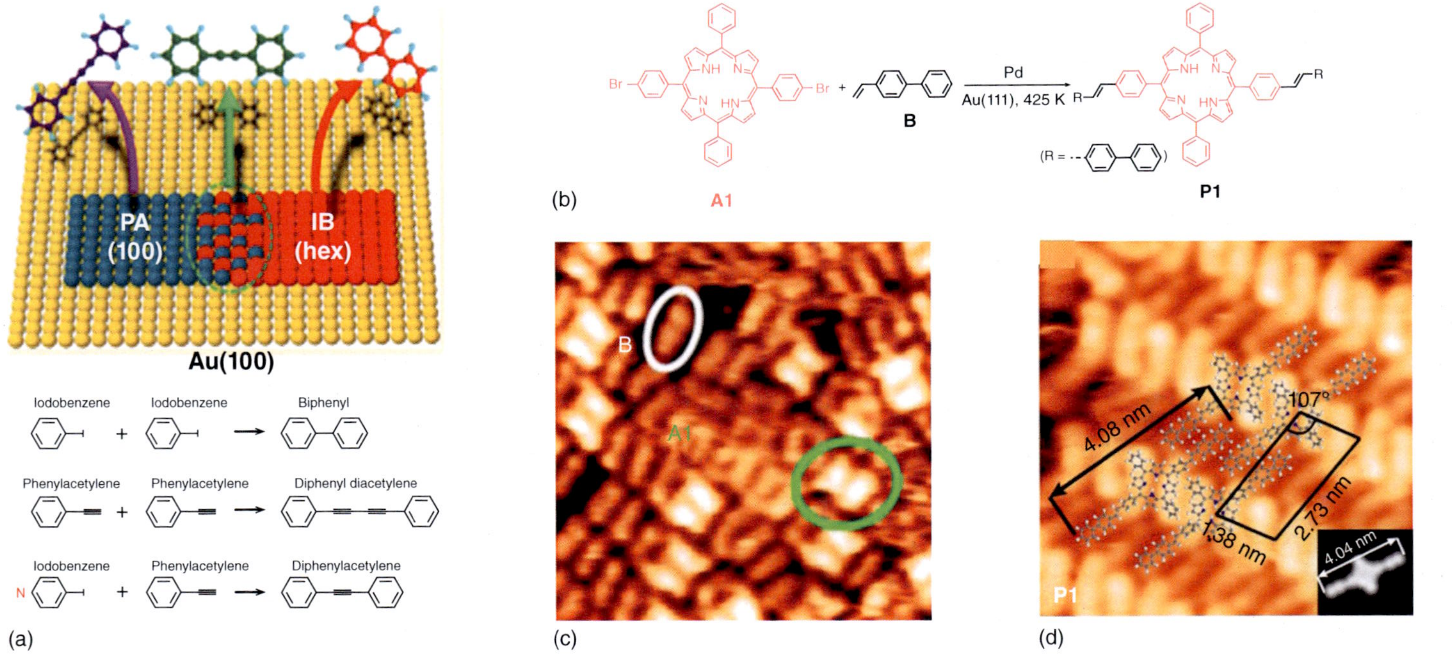

Figure 7.6 (a) Homocoupling and heterocoupling of phenylacetylene (PA) and iodobenzene (IB) molecules take place at different regions on Au(100) surface. Source: Sánchez-Sánchez et al. [15]. Reproduced with permission of the American Chemical Society. (b) Heck reaction on Au(111) with Pd adatoms as catalyst. (c) STM image of the co-adsorbed structure of aryl bromides and alkene molecules. (d) Cross-coupling products obtained by annealing at 423 K. Source: Shi et al. [16]. Reproduced with permission of the American Chemical Society.

suggested that the on-surface Heck reaction proceeded via debromination of aryl bromide, addition to alkene, and elimination of hydrogen. Noteworthy, the presence of Pd may prohibit the dehydrogenative dimerisation in Ullmann reaction.

7.3 Dehydrogenation

7.3.1 (SP^3-C) Alkane Polymerisation

As one convenient strategy, dehydrogenation via cleaving the C—H bond is widely used to generate C–C coupling structure in traditional organic chemistry. Under UHV condition, the by-product generated with dehydrogenation reaction is usually the inert hydrogen gas, which perfectly qualifies for clean on-surface processes. The selective activation of only specific C—H bond in the presence of other C—H bonds in the chosen organic building block is important and challenging for the chemical industry. The unexpected formation of covalently connected porphyrin units from simple tetra(mesityl)porphyrins (TMP) at Cu(110) surface was investigated by Amabilino, Raval and coworkers [17]. By annealing the molecules to 423 K, C–H activation of the 4-methyl group in mesityl moiety occurred, and subsequently, the homocoupling between benzyl radicals was generated, as seen in Figure 7.7.

Besides the benzylic group, the alkyl C—H bond could also be activated on catalytically-active surfaces. Chi's group succeeded in introducing alkane molecules into linear polymers on reconstructed Au(110)-(1×2) surface [18]. They found that the dehydrogenation could only occur at a more elevated temperature on

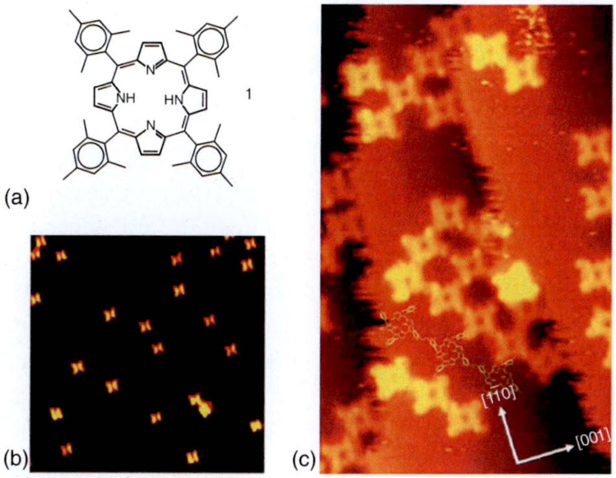

Figure 7.7 Dehydrogenation coupling between benzylic groups. (a) Molecular structure of tetra(mesityl)porphyrins (TMP). (b) STM image of TMP adsorbed on Cu(110), 27 × 27 nm^2. (c) Lines and grid structures obtained after annealing at 423 K, 12.6 × 20.6 nm^2. Source: In't Veld et al. [17]. Reproduced with permission of the Royal Society of Chemistry.

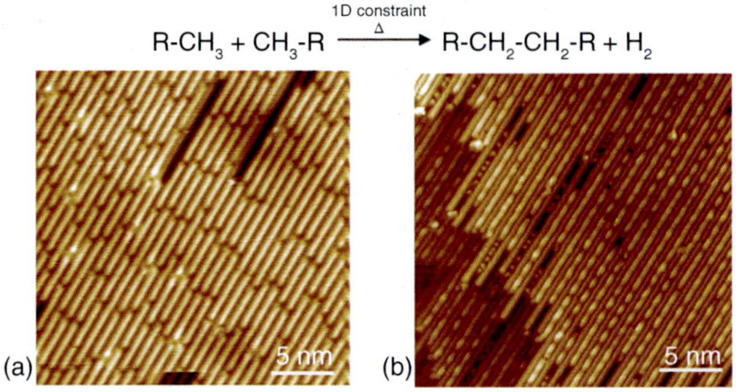

Figure 7.8 Linear alkane polymerisation on Au(100) surface. (a) Monolayer of *n*-dotriacontane on Au(110)-(1×2) after annealing at 300 K. (b) Parallel polyethylene chains formed in the Au(110)-(1×3) reconstruction grooves by heating at 440 K for 30 minutes.

Au(111) surface (470 K) than on Au(110) surface (440 K). DFT calculation shows that the grooves on the Au(110)-(1×3) surface enhanced the interaction between the molecule and substrate, thus reducing the energy barrier for dehydrogenation. The selectivity of this reaction was kinetically forced by 1D substrate structure of Au(110), and the *n*-dotriacontane ($C_{32}H_{66}$) monomers were bonded end-to-end after annealing at a mild temperature, as seen in Figure 7.8. The maximum length of the covalent bonded alkane polymer chains could reach at least 200 nm, equal to 50 monomer units, and some polymer chains could even cross the atomic step edges of the surface. This remarkable work may have a great significance to generate cheap alkanes, as raw materials in industrial products, with high selectivity under mild conditions from petroleum and natural gas.

As the surface structure was fully transformed from (1×2)-Au(110) to (1×3)-Au(110) during *n*-dotriacontane polymerisation, the real active sites of Au(110) were ambiguous. A direct comparison of the *n*-alkane reactivity on these two kinds of Au(110) surfaces became necessary. Further, (1×2)-Au(110) and (1×3)-Au(110) structures co-existed surface was prepared [19]. The thermally stable (1×2)-Au(110) structure could partly shift to (1×3)-Au(110) structure at the spot radiated by low-energy electrons. As seen in Figure 7.9b, the pre-deposited *n*-dotriacontane molecules kinked in the grooves of (1×3) structure, and this thermally unstable structure could be stabled by the widened geometry of dotriacontane molecules. When applied an annealing temperature (450 K) slightly lower than the typical triggering temperature for alkane polymerisation (470 K) was applied on this sample, the alkane molecules in the (1×3) grooves were polymerised (Figure 7.9c), but most in the (1×2) region stayed intact (Figure 7.9d). Experimental results suggested that alkane molecules adsorbed in the (1×3) region were more reactive than those adsorbed in the (1×2) region. C—H bond activation barrier of *n*-hexane on reconstructed Au(110) surface was calculated, which was 1.0 eV higher on (1×2)-Au(110) than that on (1×3)-Au(110). Such a pronounced difference was caused by the presence of the extra gold row in the grooves of the (1×3)-Au(110). As seen in Figure 7.9a, these gold atoms with lower coordination numbers could provide active sites in the polymerisation process.

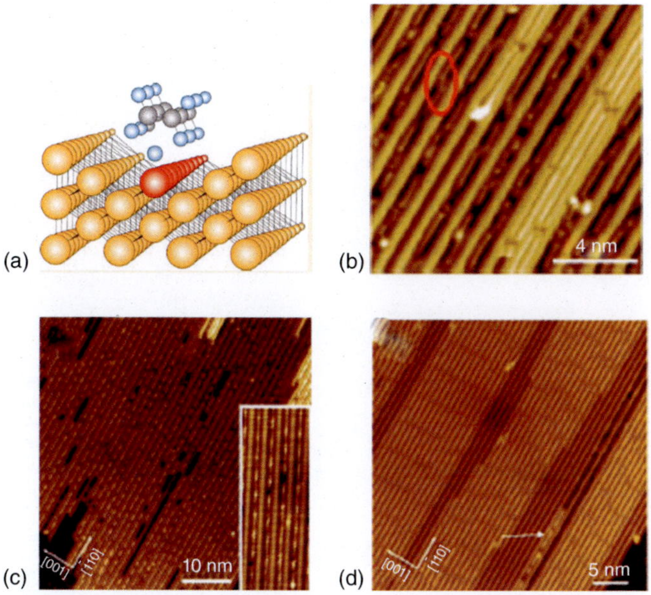

Figure 7.9 (a) High catalytic activity of extra row of gold atoms in (1×3)-Au(110) grooves. (b) STM image of the dotriacontane/(1×2)-Au(110) substrate prepared by radiation-assisted annealing, a region containing both (1×3) and (1×2) structures. (c) STM image of completely polymerised dotriacontane in the areas of the (1×3)-Au(110). (d) Typical STM image obtained in a (1×2)-Au(110) region, and only a few molecules have been polymerised (marked by the white arrow). Source: Sun et al. [19]. Reproduced with permission of the American Chemical Society.

7.3.2 (SP^2-C) Aryl and Alkene Cyclodehydrogenation

7.3.2.1 Aryl–Aryl Dehydrogenation Coupling

Harsh conditions, such as high temperature and super acid environment, are always necessary during the formation of aromatic compounds via direct aryl–aryl dehydrogenation coupling in solution [20]. When the precursors were constrained on the metal crystal surface, the catalytic effect of the underlying substrate could allow the reaction to occur at a relatively mild temperature. As seen in Figure 7.10, Otero et al. demonstrated a simple and efficient method for transferring $C_{60}H_{30}$ and $C_{57}N_3H_{30}$ to the corresponding C_{60} and $C_{57}N_3$ ball-shaped architectures on Pt(111) surface [21]. By annealing the precursor to 750 K, the ball-shaped products were generated with a high yield of 100%. Compared with less than 1% in solution preparation, the strong interaction between the precursors and the Pt surface acted as a crucial factor. The molecular bulking was reduced by both π—π bonding of the benzene rings with the underlying surface and the sterically induced molecular deformation. The strong adsorption capacity of hydrogen atoms on Pt(111) surface promotes the next step intramolecular dehydrogenation reaction. The transition from precursor to fullerene was proved to be an ordered process, in which the hydrogen atoms dissociate one by one to form radicals. As no rearrangement of C—C bonds occurred, different types of fullerene derivatives could be synthesised solely by controlling the precursor

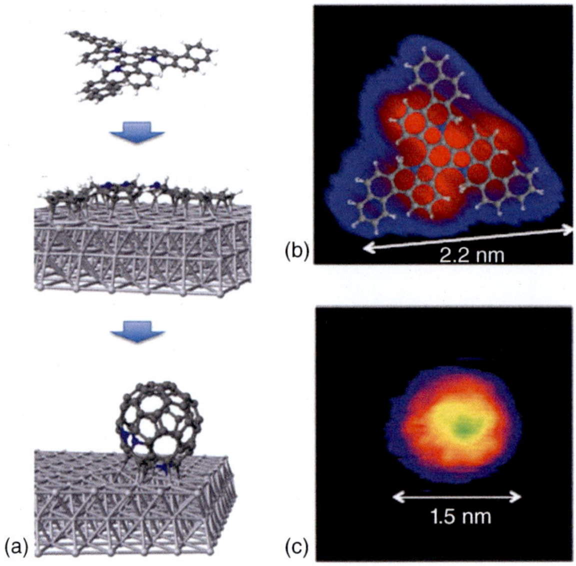

Figure 7.10 Aryl–aryl dehydrogenation coupling on surface. (a) Optimised geometrical structure of the $C_{57}H_{33}N_3$ molecule at the different stages of the process. (b) STM image of $C_{57}H_{33}N_3$ molecule deposited on the Pt(111) surface, with the atomic structures upper imposed following the calculations. (c) STM image of the obtained $C_{57}N_3$ molecule after annealing the substrate at 750 K. Source: Otero et al. [21]. Reproduced with permission of the Nature.

design. The beauty of this cyclisation process lies in the finding of adequate precursor molecules, which will spontaneously curve after dehydrogenation. For example, the large molecular precursors $C_{78}H_{48}$ and $C_{84}H_{42}$ were topologically transformed into higher fullerenes C_{78} and C_{84}, respectively [22]. It is worth noting that this fullerene formation process was not efficient on Au(111) or Cu(111) surfaces. The molecule–surface interaction was weak on these two substrates, which enhanced the molecular diffusion at room temperature and inhibited the dehydrogenation process.

Apart from the intramolecular reaction for 0D fullerenes synthesis, this aryl–aryl dehydrogenation could also occur between molecules. It has been proven that the intermolecular aryl-aryl coupling via C—H bond activation is efficient for the fabrication of 1D and 2D polymers. Han and Asao reported the fabrication of zigzag (3,1)-GNR by thermally induced polymerisation of 10,10′-dibromo-9,9′-bianthryl precursor monomers (DBPMs) on Cu(111) surface [23]. Unlike the Ullmann coupling at the brominated sites on Au(111) surface, these debrominated DBPM radicals did not react with each other. As seen in Figure 7.11a–c, the cyclodehydrogenation between the DBPM radicals was followed by the dehydrogenation of the downward-pointing anthryl tips (red crosses), which triggered a cascade of reactions (blue dashed oval). The radicals prefer polymerisation via cyclodehydrogenation rather than Ullmann-coupling mostly because of the high reactivity and strong molecule–substrate interaction on the copper surface. Another example of 1D aromatic polymer obtained through aryl–aryl dehydrogenation was the selective C–C coupling between tetraphenyl (4Ph) molecules on Cu(110) surface, as shown in Figure 7.11d [24]. Upon heating up the surface to 773 K, the aryl–aryl coupling between 4Ph only took place at the *meta*-C sites of the terminal phenyl groups. The monomers were covalently linked together into chain-like structures along the close-packed [1–10] direction of the underlying substrate. 2D conjugated polymer

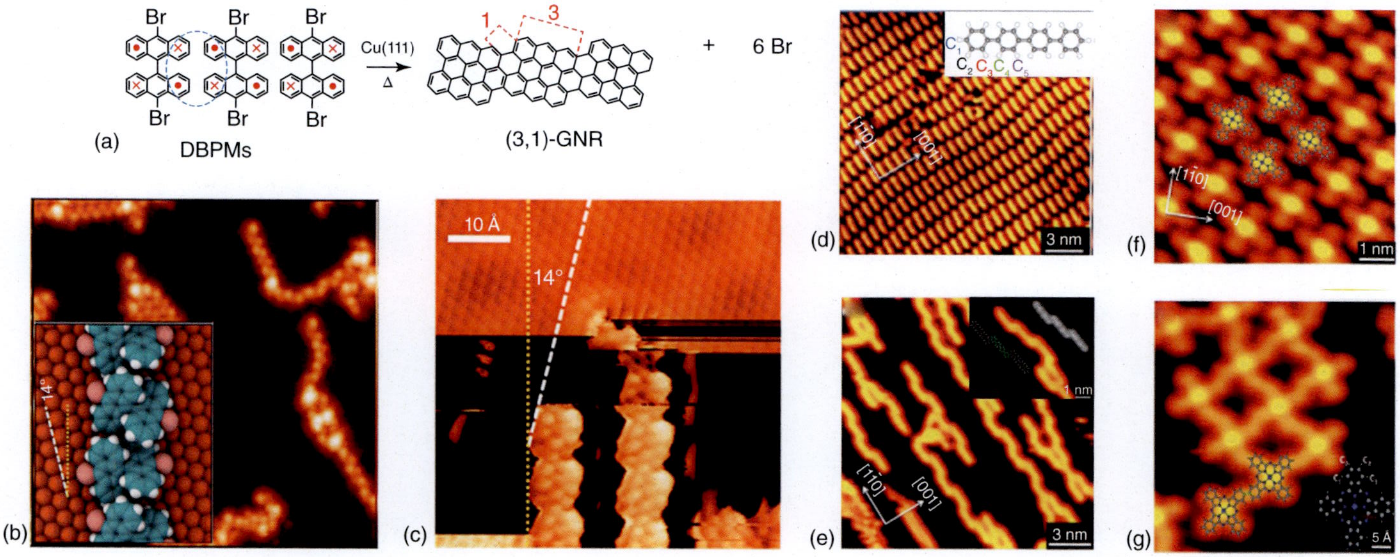

Figure 7.11 (a) Reaction scheme for the polymerisation of DBPM into (3,1)-GNR. (b) STM image of self-assembled DBPM monomers on Cu(111) surface, $32 \times 32\ nm^2$. (c) (3,1)-GNR obtained by heating at 773 K. Source: Han et al. [23]. Reproduced with permission of the American Chemical Society. (d) STM image of self-assembled 4Ph molecules on Cu(110) surface. (e) STM image of the polyphenylene chains via cross-dehydrogenation by annealing 500 K. Source: Sun et al. [24]. Reproduced with permission of the Royal Society of Chemistry. (f) STM image of the self-assembled structure of CoPc on Ag(110) surface. (g) 2D polymer structure formed by annealing at 680 K. Source: Sun et al. [25]. Reproduced with permission of the Royal Society of Chemistry.

structure formed through direct C–H activation, followed by aryl–aryl coupling of a metal–phthalocyanine (CoPc) molecule on Ag(110) surface (Figure 7.11f) [25]. Because of the smaller steric hindrance of the C2—C2 bonding than that of the C1—C1 or C1—C2 bonding, the C–C coupling reaction was selectively activated at the eight equivalent C2 sites, as seen in Figure 7.11e,g. Amabilino, Raval and coworkers introduced the 'pick-mix-and-link' strategy to open up the hetero-coupling between a wide range of π-functional molecules, such as porphyrin, pentacene, and perylene [26]. Complex multicomponent macromolecules were directly built on a copper surface, displaying diverse compositions, structures, and topologies. Interestingly, some specific products, such as block copolymers, were controllably generated by exploiting the differences in C—H bond reactivity. The experimental results also demonstrated that the symmetry and geometry of the molecules and the surfaces together determined the outcome of the covalent bond-forming reactions.

7.3.2.2 Bottom-Up Fabrication of Graphene Nanoribbons (GNRs)

Opening the band gap of graphene is a prerequisite for graphene application in semiconductor devices, and attractive for the fabrication of nanoscale electronic devices [27]. Scientists have explored several methods to fabricate GNRs, e.g. lithographic method, unzipping carbon nanotubes, and cutting with catalytic particles, summarised as 'top-down' approaches [28]. However, these approaches have disadvantages in controlling the width and edge structures with atomic precision. In 2010, Fasel and Müllen first applied a 'bottom-up' approach, i.e. molecular precursor-based growth, to build prototypical armchair ribbon of width $n = 7$ (7-AGNR), as seen in Figure 7.12a–c [29]. This 7-AGNR was formed out of 10,10′-dibromo-9,9′-bianthryl (DBBA) monomers on Au(111) surface via a two-step reaction: Ullmann coupling and subsequent cyclodehydrogenation. The first step, an Ullmann-type oligomerisation between DBBA monomers, occurred at 473 K to generate linear polyethylene chains. The second step, increasing annealing temperature to 673 K led to the intramolecular cyclodehydrogenation, which resulted in the formation of planar GNRs. Using this 'bottom-up' approach, the synthesis of GNRs on catalytic metal surfaces has bloomed quickly. Various precursors have been designed to generate GNRs with fine-tuned widths and band gaps [32]. 5-AGNR (3p+2 type) was synthesised both by Liljeroth's and Chi's groups with different precursors [33]. 6-AGNR (3p type) and 13-AGNR (3p+1 type) were obtained by Basagni's and Crommie's group respectively [34]. An innovative pathway was proposed for generating 6-AGNR: 4,4″-dibromo-*p*-terphenyl (DBTP) as precursor firstly forming extended poly(*p*-phenylene) wires by dehalogenation, and subsequent intermolecular cyclodehydrogenation took place along the poly(*p*-phenylene) wires. 13-AGNR showed a smaller band gap than 7-AGNR, which confirmed that Increasing the width of the nanoribbon resulted in its electrical properties closer to graphene. Besides controlling the width, different edge structures can also tune the properties of GNRs. To prepare zigzag-edge structures (ZGNRs) is still a high challenge in solution-mediate fabrication because of its relatively unstable nature,

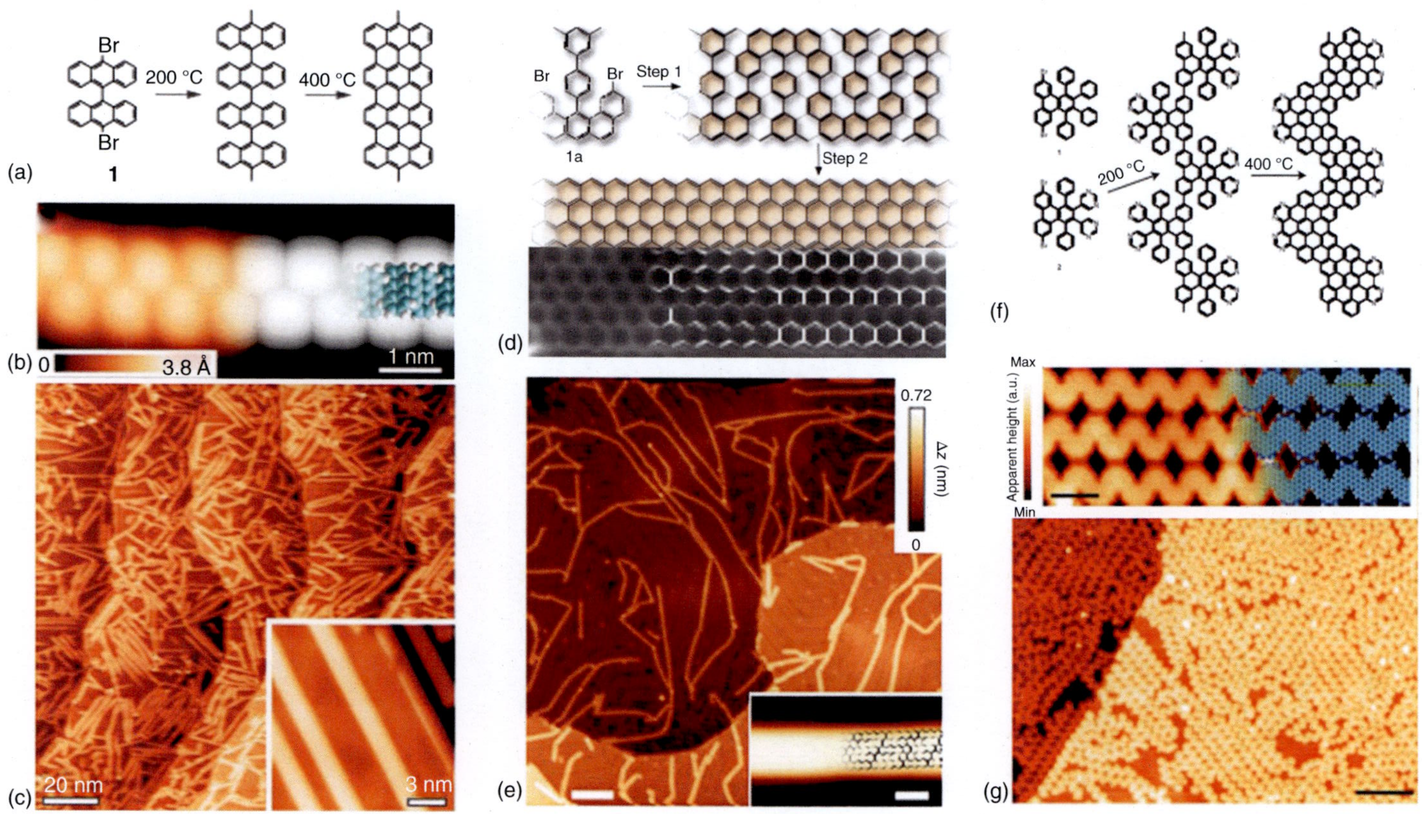

Figure 7.12 (a) Schematic illustration of the bottom-up fabrication of armchair GNRs on Au(111) surface. (b) STM image of a polyanthrylene chain after dehalogenation at 200 °C, DFT simulation shown in the right side. (c) STM image of the straight 7-AGNRs after cyclodehydrogenation at 400 °C. Figure inset is the high-resolution STM image. Source: (b, c) Cai et al. [29]. Reproduced with permission of the Nature. (d) Schematic illustration of the synthesis strategy of zigzag GNRs on Au(111) surface, and high-resolution nc-AFM image of the ZGNR structure. (e) Large scale and high-resolution (inset) STM images of the ZGNRs. Source: Ruffieux et al. [30]. Reproduced with permission of the Nature. (f) Schematic illustration of the synthetic strategy of N-doped GNRs from nitrogen-substituted monomers on Au(111) surface. (g) Large scale and high-resolution (upper) STM images of the N-doped GNRs. Source: Cai et al. [31]. Reproduced with permission of the Nature.

especially in an oxidative environment. Remarkably, under the extremely stable UHV condition on Au(111) substrate, Fasel, Müllen and coworkers obtained the first example of full zigzag edge GNR, named 6-ZGNR [30]. Specifically designed U-shaped monomer, a dihalogenated precursor with additional methyl groups at the periphery, was crucial for building up the full zigzag edge by surface-assisted coupling and subsequent graphitisation of the corresponding polyphenylene intermediates, as seen in Figure 7.12d,e. ZGNRs were predicted to show unique electronic properties including profound lower band gap and the spin-polarised localised edge states. The dI/dV spectrums revealed the energy splitting, and acquired the corresponding states that were highly localised at the zigzag edges. As a result, the strong chemical reactivity of the pristine 6-ZGNR caused strong interaction with the underlying metal surface that impeded the application of ZGNR at ambient conditions. The first principle method revealed that hydrogen passivation of edges of GNRs can modify their electronic properties. This provides a simple way for band gap engineering of graphene as the relative stability of sp^2 and sp^3-like bonds at the edges depends on the chemical potential of hydrogen gas [35]. Some other hybrid-edge structures were also obtained with cleverly designed procurers, such as chevron-type [36] and acene-type [37]. Furthermore, doped GNRs were explored to modify their electronic properties. N-substituted precursors were first introduced to synthesise N-doped GNRs by Fasel's group, as seen in Figure 7.12f,g [31]. They further combined N-substituted precursors with pristine hydrocarbon precursors to fabricate partially doped GNR heterostructures, which showed similar behaviour as traditional p–n junctions. Subsequently, B-doped GNRs were obtained by Fischer's group, featuring a regular pattern of B atoms along the central backbone of the nanoribbon [38]. Trigonal-planar B atoms along the backbone of the GNR shared an empty p-orbital with the extended π-band for dopant functionality, leading to a predominance of the local density of states. Substituting carbon atoms with B/N atoms could only upshift/downshift the band gaps relative to the corresponding undoped GNRs, instead of affecting the magnitude of the band gaps [39]. Gao's group successfully tailored the band gap via adjusting S-doped GNRs structure [40]. By applying different sulfur configurations from cyclodehydrogenated isomers, the related heterostructures achieved varied band gap magnitudes. The mechanism of this adjustment was a sequence of multiple heterojunctions, resulting in a sequence of quantum dots.

7.3.2.3 Homo-Coupling of Terminal Alkene

On-surface coupling of alkenes was first reported by Xu's group: The 4-vinyl-1,1′-biophenyl (VBP) molecule with a terminal alkenyl group connected on a biphenyl backbone were homo-coupled to form diene compounds on Cu(110) and Cu(100) substrates. Interestingly, on Cu(110) surface, the molecules could perform high selectivity to give unique *trans*-diene moiety with a quite high yield (>80%). As reported, the alkene molecules normally dimerised into multiple products with different kinds of butane moieties in solution synthesis [41]. However, on-surface synthesis would present an efficient route for increasing the yield of products.

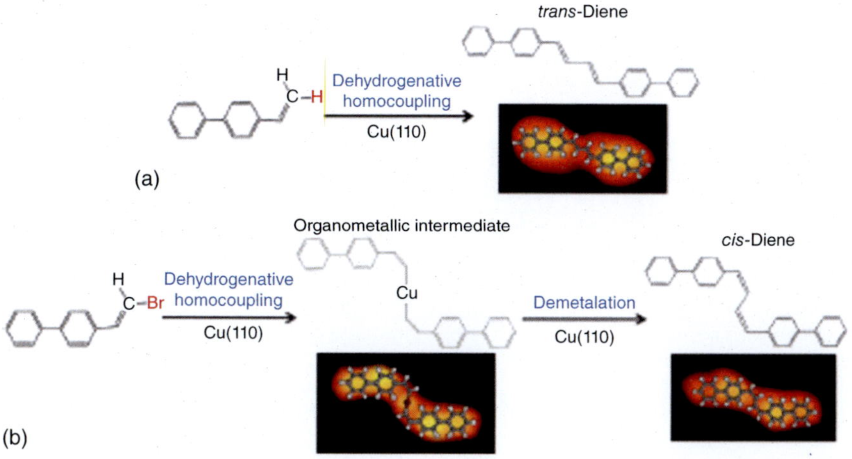

Figure 7.13 Stereoselective synthesis of diene moieties on Cu(110) surface. (a) *trans*-Diene formation by dehydrogenative homocoupling of VBP. (b) *cis*-Diene formation by dehydrogenative homo-coupling of BVBP on Cu(110). The BVBP molecule had a terminal alkenyl bromide group functionalised on a biphenyl backbone. Source: Sun et al. [42]. Reproduced with permission of the Royal Society of Chemistry.

Subsequently, they demonstrated that the stereo selective synthesis of a specific *cis*-diene moiety could be successfully achieved via dehalogenation homo-coupling of the monomer with a terminal alkenyl bromide group functionalised, as seen in Figure 7.13 [42].

7.3.3 (SP^1-C) Alkyne – Glaser Coupling

The homocoupling of terminal alkynes (Glaser coupling reaction) has been widely employed to build 1D and 2D nanostructures on various noble metal surfaces. The first attempt was done by Barth's group using 1,3,5-tris-(4-ethynylphenyl)benzene (Ext-TEB) molecules as precursors [43]. The molecules were annealed at 600 K on Ag(111) surface and small covalently linked 2D polymers up to 10 nm were produced by the terminal alkyne C–H activation, as shown in Figure 7.14a. The chemical nature and the lattice structure of the underlying surface were demonstrated to effectively influence the reaction pathway. In the case of Cu(111) surface, different reactions, such as homocoupling, trimerisation, and cross-coupling, were found to proceed concurrently [47]. Unlike the versatile reaction pathways on Cu surface, cyclotrimerisation was the dominant reaction pathway on Au surface. In Figure 7.14b, Fasel's group reported that the Ext-TEB molecules combined into a highly uniformed polyphenylene structure on Au(111) [44]. To avoid the side products, Fuchs's group introduced hexyl groups into the aromatic ring to enhance steric hindrance [45]. Noteworthy, long linear oligomers generated by homocoupling between terminal alkynes could perform high yield on Ag(111), as shown in Figure 7.14c. Compared with the products obtained on Au(111), 'Glaser coupling'

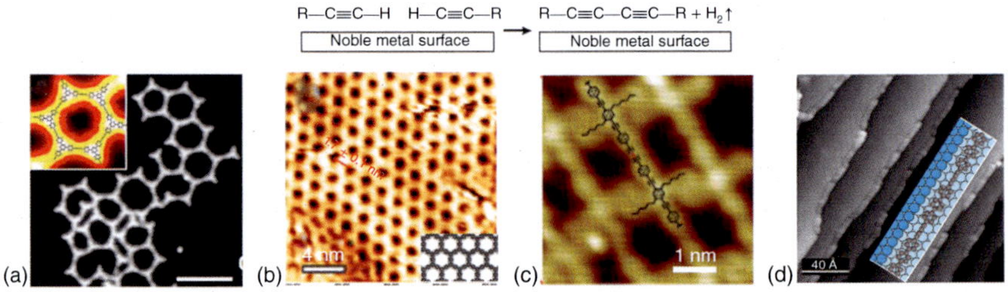

Figure 7.14 Glaser coupling reaction on surfaces. (a) Small size covalently linked 2D polymers obtained on Ag(111) at 300 K from Ext-TEB molecules. Source: Zhang et al. [43]. Reproduced with permission of the Nature. (b) High-quality polyphenylene nanostructures obtained by cyclotrimerisation of Ext-TEB on Au(111) at 433 K. Source: Liu et al. [44]. Reproduced with permission of the Royal Society of Chemistry. (c) Terminal alkanyl functionalised arylalkynes oligomerisation on Ag(111) at 398 K. Source: Gao et al. [45]. Reproduced with permission of Wiley. (d) Polymerisation of DETP at the step edge of Ag(877) at 450 K. Source: Cirera et al. [46]. Reproduced with permission of the American Chemical Society.

showed more efficiency and selectivity on Ag(111). As a template, step edge could also be used to increase the reaction selectivity. 4,4′-Diethynyl-1,1′:4′,1″-terphenyl (DETP) molecules at the step edge of a Ag(877) surface formed extended graphydyne wires over 30 nm long, as seen in Figure 7.14d [46]. However, irregularly branched networks were obtained on Au(111) due to thermo activated side reactions.

7.3.4 Hierarchical Dehydrogenation of X—H Bonds (X = N and C)

Besides C—H bond, N—H bond is also used as the hydrogen-rich organic reactants. Chi's group successfully realised the hierarchical dehydrogenation of aromatic amine (4,4″-diamino-*p*-terphenyl, DATP) on Cu(111) surface [48]. By gradually increasing the substrate temperature, the scissions of N–H and C–H occurred one by one, as shown in Figure 7.15. The reaction pathways were monitored by STM and nc-AFM, which demonstrated the formation of corresponding metal-organic supramolecular structures during each step of the dehydrogenations. Three DATP molecules lost the first H atom on amino groups to form a triangle structure at 340 K. Here, each resulting imino group coordinated with one Cu atom as the interlinking of the triangles. Further increasing the substrate temperature to 360 K, the second H on the amino group was detached. Each N atom coordinated with two Cu atoms, which could connect several triangles together as a larger annular coordination complex. The structure evolution of the triangles took place via annealing at 400 K, resulting in a smaller triangle size and brighter protrusions at the vertex. This elongated shape of the protrusions referred to the coexistence of the C—Cu—C bonds and the N—Cu—N bonds. DFT calculations suggested that the energy barrier of N—H bond cleavage from the amino group (1.04 eV) was significantly lower than the energy barrier of C—H bond cleavage from the adjacent phenyl group (1.67 eV), so that the N—Cu—N and C—Cu—C bonds hierarchically formed with gradually increasing temperature.

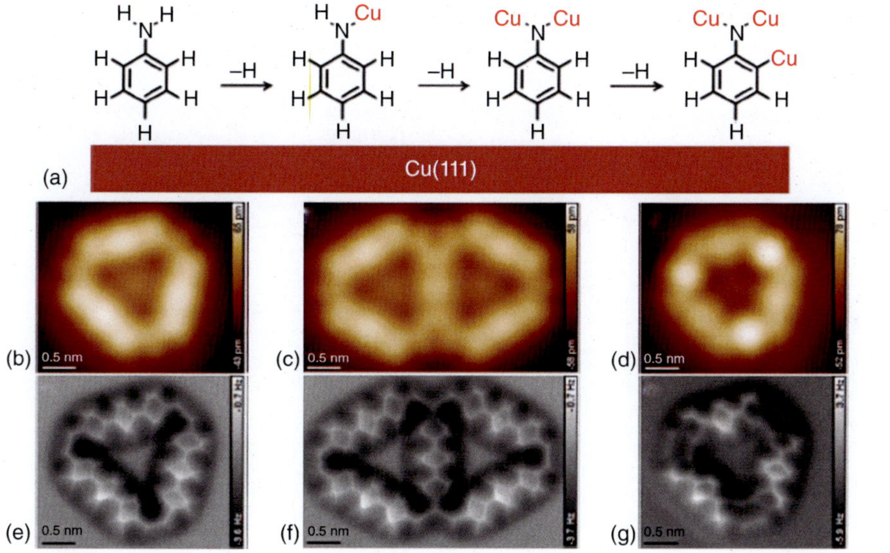

Figure 7.15 (a) Proposed reaction steps of dehydrogenation of DATP molecules. (b) STM and (e) nc-AFM images of triangle structure obtained by annealing at 340 K. (c) STM and (f) nc-AFM images of diamond-shaped product obtained by annealing at 360 K. (d) STM and (g) nc-AFM images of the product by annealing at 400 K, formed by the *ortho* C–H activations. Source: Li et al. [48]. Reproduced with permission of the American Chemical Society.

7.4 Dehydration Reaction

7.4.1 Schiff-Base Reaction

Schiff-base reaction (dehydration between an aldehyde and an amine) is a common approach to obtain imine. Linderoth's group firstly reported the on-surface synthesis of imines [49]. Under UHV condition, the condensation reaction between benzylic aldehyde and primary amine could initiate at room temperature according to the excellent catalytic effect of the underlying Au(111) surface, as seen in Figure 7.16a. The products were checked by comparing the self-assembled structure with that synthesised in solution, and the same behaviour gave strong evidence for on-surface Schiff-base reaction. DFT calculations were conducted to simulate the reaction pathways. The amino group first underwent a nucleophilic addition to the carbonyl group to form the tetrahedral hemiaminal intermediate. Then the phenol hydroxyl group donated a proton to the hydroxyl group of the hemiaminal to eliminate water. Finally, the iminium group donated a proton to the phenoxy group to recover the phenol hydroxyl group, and the C=N bond formed as well. Interestingly, the phenol hydroxyl group acted as either an internal proton donor or acceptor to catalyse this reaction. Larger domains of diimines were observed via increasing the surface temperature to 400 K. The improved reaction efficiency was related to the increased surface mobility of the reactants in the co-assembly

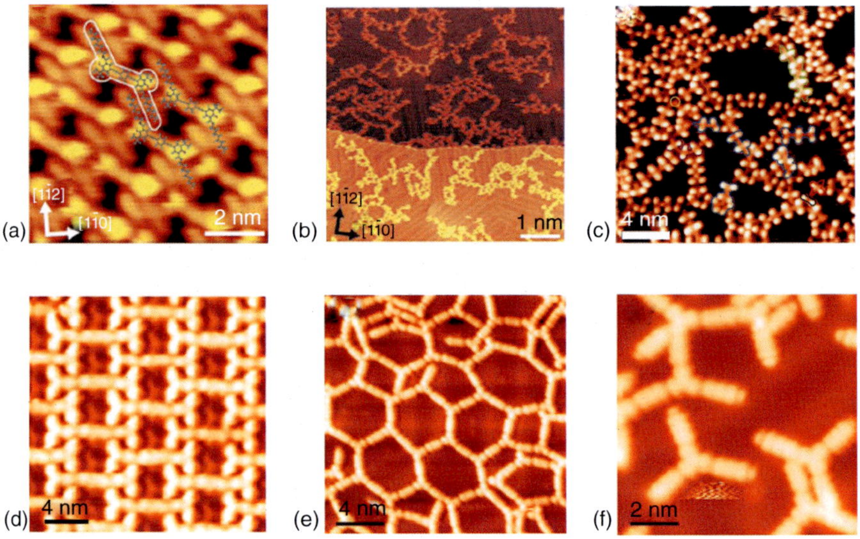

Figure 7.16 (a) Imine formation via on-surface condensation reaction of benzylic aldehydes and primary amines on Au(111) at room temperature. Source: Weigelt et al. [49]. Reproduced with permission of Wiley. (b) Branched polymer structure obtained by condensation of trialdehydes and 1,6-diaminohexanes on Au(111). Source: Weigelt et al. [50]. Reproduced with permission of Wiley. (c) Condensation products via annealing tetraketone and tetraamine (4 : 1) to 510 K on Ag(111). Source: Jiang et al. [51]. Reproduced with permission of the American Chemical Society. (d–f) Products obtained by adjusting stoichiometric ratios of TFPB and DATP on Au(111), 2 : 1 (d), 2 : 3 (e), 1 : 3 (f). Source: Gong et al. [52]. Reproduced with permission of the American Chemical Society.

process, and possibly the molecular conformation was also switched at elevated temperature. As depicted in Figure 7.16b, the same group reported the branched polymer containing 3–10 nm pores by using trialdehydes and alky diamines as building blocks [50]. However, the high flexibility of the diamine linkers on Au(111) led to the less ordered networks. One step further, Barth's group investigated the synthesis of 1D pyrene-fused pyrazaacenes oligomers on Ag(111) surface via polymerisation of 2,7-di-tertbutylpyrene-4,5,9,10-tetraketone (tetraketone) and 2,11-ditertbutylquinoxalino [2′,3′,9,10]phenanthro[4,5-*abc*]phenazine-6,7,15,16-tetraamine (tetraamine) (see Figure 7.16c) [51]. The experimental results demonstrated that the length of the linear polymeric structures could be controlled by the ratio between the tetraketone and tetraamine precursors. The high-quality products via Schiff-base condensation were reported by Chi's group. Desirable oligomer products and regular 2D covalently bonded networks were synthesised through the rational adjustment of stoichiometric proportions of 1,3,5-tris(4-formylphenyl)benzene (TFPB) and 4,4′-DATP on Au(111) surface [52]. Bowtie-like species were obtained by annealing the co-deposited sample to 400 K with a TFPB/DATP stoichiometric ratio 2 : 1 (Figure 7.16d). Upon decreasing the co-deposition ratio to 2 : 3, aldehyde–amine covalent 2D networks were formed (Figure 7.16f), and statistical analysis suggested that pentagons and hexagons were the dominant species. Further decreasing the co-deposition ratio to 1 : 3, fork-like isolated products appeared (Figure 7.16e). In this case, the aldehyde species were chemically saturated, leading to identical tetramers with amino group terminations. Monte Carlo simulations investigated the underlying mechanisms: the selective products balanced by the coupling rate of precursors together with the mobility of monomers on the substrate.

The Schiff-base reaction could also take place in an ambient environment under mild conditions [53]. Wan's group demonstrated a self-limiting solid–vapour interface reaction strategy to fabricate highly ordered covalent frameworks on HOPG [54]. As shown in Figure 7.17a,b, Schiff-base coupling reaction was excited at the solid–vapour interface by introducing terephthaldicarboxaldehyde (TPA) via vaporisation to the surface preloaded with 1,3,5-tris(4-aminophenyl)benzene (TAPB). Following this strategy, high-quality honeycomb 2D polymers with imine linkage were obtained in large scale. Samorì and Lehn studied the thermodynamic control during the Schiff-base reaction at solid/liquid interface [55]. By in situ addition of 1,2-diaminoethane (B2) on top of a pre-existing monolayer of 4-(hexadecyloxy)benzaldehyde (A), A_2B2 compound was formed. Upon subsequent addition of 1,6-diaminoethane (B6), the longer diamine B6 could replace the shorter B2, and the product transformed into new a A_2B6 compound (see Figure 7.17c–e). Noteworthy, a surface-confined 2D network via Schiff-base reaction was prepared on single-layer graphene grown on copper foil by Liu's group [56]. With the well-developed transfer technique for graphene, the covalent networks can be transferred onto various substrates for electrical device fabrication.

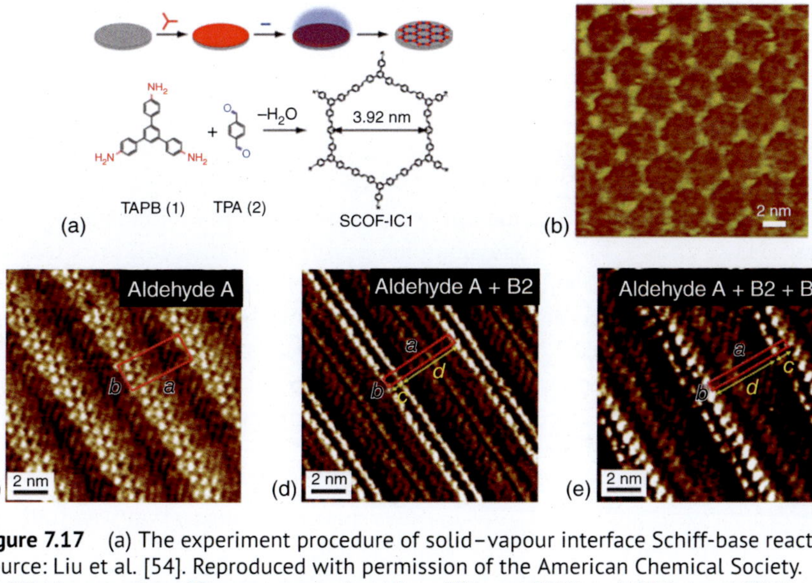

Figure 7.17 (a) The experiment procedure of solid–vapour interface Schiff-base reactions. Source: Liu et al. [54]. Reproduced with permission of the American Chemical Society. (b) STM image of the 2D network obtained from TPA and TAPP on HOPG. (c–e) STM images of self-assembled nanopatterns of A (c), A_2B2 (d) and A_2B6 (e) compounds on HOPG. Source: Ciesielski et al. [55]. Reproduced with permission of the Nature.

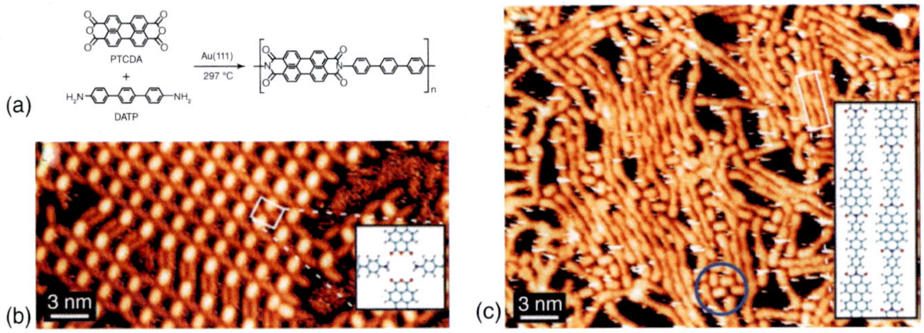

Figure 7.18 (a) Imidisation condensation reaction pathway between PTCDA and DATP on Au(111). (b) STM image of co-adsorbed structure via hydrogen bonds. (c) STM image of parallel polyimide strands formed by annealing at 570 K. Source: Treier et al. [57]. Reproduced with permission of the American Chemical Society.

7.4.2 Imidisation Condensation Reaction

The imidisation via rear side nucleophilic attack from amine to a carbonyl group proceeds in nonplanar manner. It was considered as this reaction could be strongly hindered by confining the two reactants onto 2D surface. However, the investigation reported by Fasel's group extended our knowledge [57]. Imide bond formation was successfully achieved with the assistance of gold surface. As shown in Figure 7.18, 4,4′-DATP and 3,4,9,10-perylenetetracarboxylicdianhydride (PTCDA) were co-deposited on Au(111) surface, and subsequently annealed the surface to 570 K to produce the polyimide structures. The high transfer ability of the amine molecules could promote the attack to anhydride.

7.4.3 Boronic Acid Condensation

Condensation of boronic acid has been wildly applied for the fabrication of 3D covalent organic framework in solution. 2D extended covalent organic framework synthesised via boronic acid condensation was first reported by Abel's group [58]. On Ag(111) surface, 1,4-phenylenediboronic acid (PDBA) molecules reacted in a self-condensation manner to form porous boroxine polymers at room temperature (Figure 7.19a). If PDBA co-deposited with 2,3,6,7,10,11-hexahydroxy-triphenylene (HHTP) on hot Ag(111) of 420 K, more thermo-stabled corresponding boronic esters could be produced. Under the participation of HHTP, the bimolecular condensation followed a more kinetic favourable pathway, and no homocoupling products were observed. Beside the optimal structure of hexagons, five-, seven-, and eight-membered rings were also present in the obtained framework (Figure 7.19b). These defects were probably due to the fact that the condensation was initiated before HHTP molecule fluxing to the right position.

The self-assembled structure of boronic acids, stabilised by hydrogen-bonds, is an important precursor structure for polycondensation. Lackinger's group investigated the polycondensation of *para*-diboronic acids at liquid/HOPG interface [59].

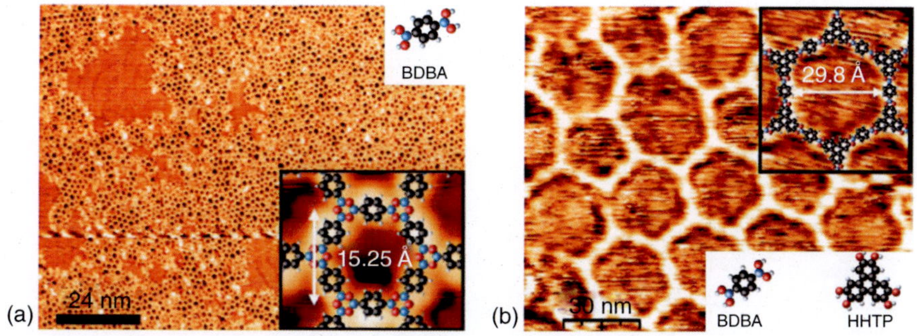

Figure 7.19 (a) Conversion of PDBA into boroxines network at room temperature on Ag(111) surface. (b) Porous network obtained via dehydration between BDBA and HHTP at 420 K Ag(111) surface. Source: Zwaneveld et al. [58]. Reproduced with permission of the American Chemical Society.

Using precursor molecules containing a different number of benzene rings, the lattice parameters and the pore size of the produced films were successfully tuned. In the presence of water, the dehydration was carried out under slightly reversible reaction conditions. The subtle balance between kinetic control and thermodynamic control of the polycondensation was favourable to form high-quality 2D structures with few defects, as seen in Figure 7.20.

7.4.4 Decarboxylative Polymerisation

On-surface chemical reactions based on the condensation of carboxyl or carbonyl function groups stand for another promising approach to build 2D covalent frameworks. Fuchs's group reported the metal-catalysed polymerisation of 2,6-naphthalenedicarboxylic acid (NDCA) to form poly-2,6-naphthalenes at various surfaces [60]. The polymeric binaphthyl-Cu was first formed at elevated temperature as an intermediate structure. Further increasing substrate temperature could lead to reductive elimination to fabricate polynaphthalenes. Via this decarboxylative homocoupling approach, the chain length of the obtained polymer could reach over 50 nm, as seen in Figure 7.21. Compared to Au(111), Ag(111), and Cu(110) surfaces, Cu(111) was demonstrated to be the most effective substrate to conduct such polymerisations. The 6-ethynyl-2-naphthoic acid (changing one carboxylic group of NDCA into ethinyl group) polymerised in a two-step domino reaction mechanism on Au(111) surface [61]. The ethinyl function group first reacted via Glaser coupling to form dimers, and subsequent dehydrogenative coupling between the acid moieties provided the corresponding polymeric bisacylperoxide. The reductive McMurry reaction of aldehydes was subsequently reported by the same research group. With 2,5-dihexylterephthalaldehyde (DTA) sublimated on Au(111) surface, the carbonyl groups connected via deoxygenative homocoupling to form C—C double bonds via thermal treatment [62]. Unfortunately, the yield of target products, *para*-polyphenylene vinylene derivatives, was quite poor, which could reflect the low reaction selectivity under the applied conditions.

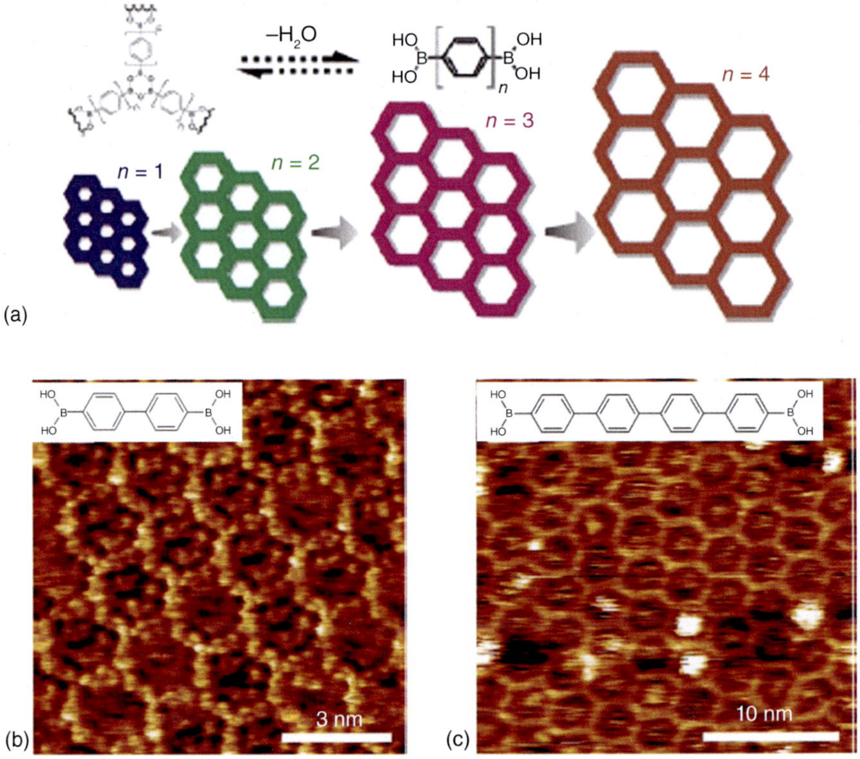

Figure 7.20 (a) General reaction scheme of diboronic acid self-condensation into hexagonal 2D porous films. (b, c) STM image of 2D networks with different pore size derived from polycondensation of smaller biphenyldiboronic acid (b) and bigger quaterphenyldiboronic acid (c). Source: Dienstmaier et al. [59]. Reproduced with permission of the American Chemical Society.

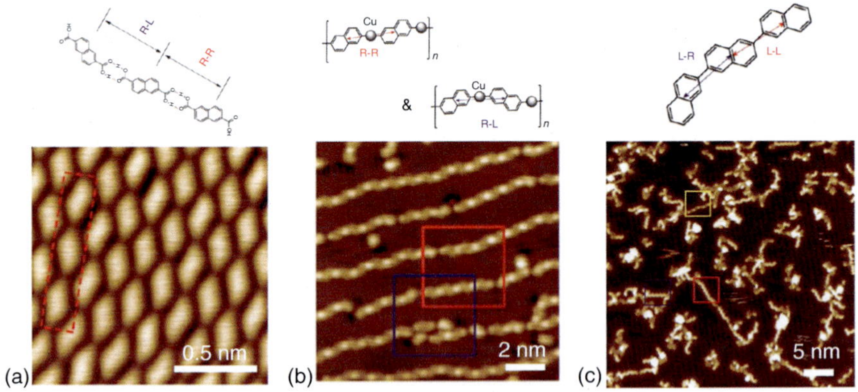

Figure 7.21 (a) NDCA self-assembly structure on Cu(111). (b) C–Cu–C organometallic intermediates by annealing surface at 433 K. (c) Final polymerisation products obtained after annealing at 464 K. Source: Gao et al. [60]. Reproduced with permission of the American Chemical Society.

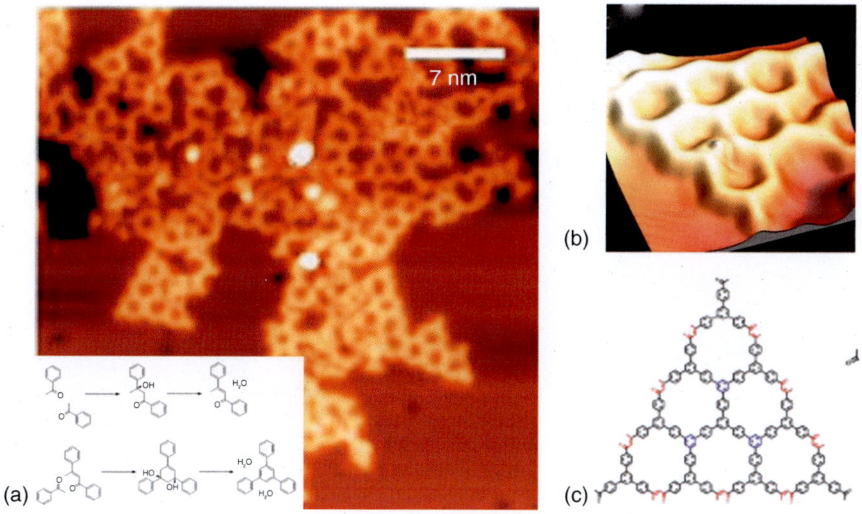

Figure 7.22 (a) STM image of 2D network formation via cyclocondensation of TAPB on Ag(111) at 590 K. Inset is the possible reaction route. (b) High resolution STM image of representative reaction product. (c) Red and blue denote the product of dimerisation coupling and cyclotrimerisation coupling of acetyls, respectively. Source: Yang et al. [63]. Reproduced with permission of the American Chemical Society.

7.4.5 Dimerisation and Cyclotrimerisation of Acetyls

The additional phenyl rings formed between precursors can also be the linkages in 2D conjugated frameworks. Chi's group first reported the cyclotrimerisation of trifunctional acetyl compounds on Ag(111) surface, as seen in Figure 7.22 [63]. According to DFT calculations and XPS measurements, a plausible reaction mechanism was explored. This thermal-induced reaction required enough dehydrogenated molecules to generate radicals. Then, these radicals attacked the aldehyde group of the adjacent molecule, and subsequent dissociation of a hydroxyl led to a dimer. A third molecule, dehydrogenated at methyl group, attacked the dimer to produce the final dimerisation product by sublimation of two water molecules. It is worth noticing that this reaction would happen on other noble metal surfaces. Ag(111) was proved to be the best choice because of the proper adsorption energy of molecules.

7.5 Other Reactions

7.5.1 Click Reaction

7.5.1.1 Azide–Alkyne Cycloaddition

With Cu(I) ions as catalysts, Huisgen 1,3-dipolar cycloaddition between a terminal alkyne and an azide could produce 1,4-disubstituted-1,2,3-triazole as the only product (CuAAC reaction). However, a mixture of 1,5- and 1,4-triazole regioisomers

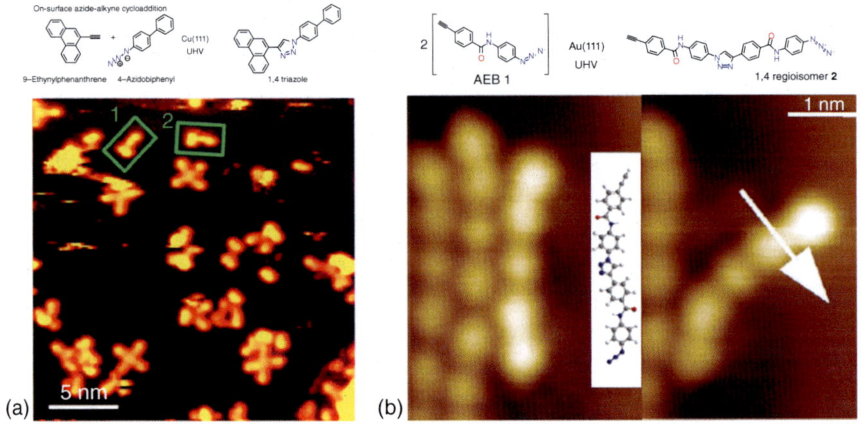

Figure 7.23 (a) 1,4-Triazole products obtained on Cu(111) at room temperature, marked in green. Source: Bebensee et al. [64]. Reproduced with permission of the American Chemical Society. (b) Linear oligomers formed via Azide–alkyne cycloaddition on Au(111) at room temperature. Source: Díaz Arado et al. [65]. Reproduced with permission of the American Chemical Society.

would be produced without Cu(I) catalysis. This CuAAC reaction is of great interest for the on-surface reaction because of low energy initiation and high product yield. Since copper was proved to be a good catalyst, Gothelf, Linderoth and coworkers implemented this coupling reaction between 9-ethynylphenanthrene and 4-azidobiphenyl on Cu(111) [64]. As seen in Figure 7.23a, the corresponding 1,4-triazole product was obtained directly after the deposition of these two precursors onto the copper surface at room temperature. However, the product yield was as low as only 2% under this condition. Further, Fuchs's group indicated that the Cu atoms were not an essential agent, meaning that this on-surface 1,3-dipolar cycloaddition could realise without the catalysis of copper. The *N*-(4-azidophenyl)-4-ethynylbenzamide (AEB) molecules were chosen as building blocks, and the 1,4-triazole-connected dimers/trimers were successfully obtained on Au(111) surface at room temperature, as shown in Figure 7.23b [65]. The DFT calculation results indicated that the gold surface only acted as a 2D-restraint rather than a catalyst in this reaction. According to the low mobility of AEB molecules on Au(111) surface, it's difficult to obtain long polymer chains.

7.5.1.2 Diels–Alder Reaction

Diels–Alder reaction, involving a conjugated diene and a substituted alkene (the 'dienophile'), is quite commonly used in chemical synthesis. A theoretical study from Sheik's group predicted that the barrier height for certain Diels–Alder reactions could be reduced via applying an extra electric field [66]. Further, Aragonès et al. proved this conjecture via a model system [67]. As seen in Figure 7.24, the diene (surface tethered furan derivative) molecules were attached to the STM gold tip through Au—S bond formation, while the non-polar dienophile (norbornylogous bridge with a terminal double bond) molecules bonded to Au surface via two CH_2S—Au bonds

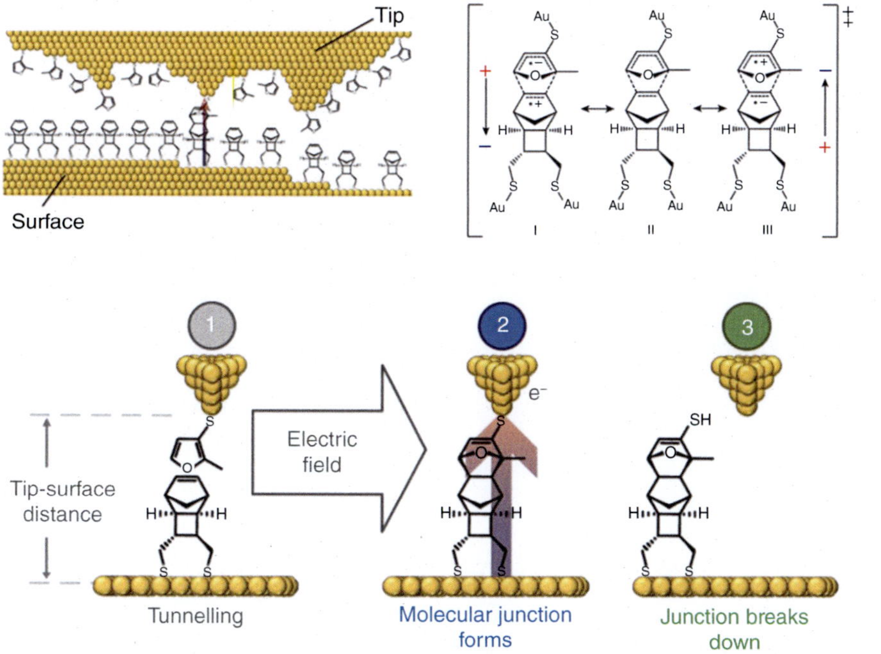

Figure 7.24 Electrostatic catalysis of a Diels–Alder reaction. The stages encountered during the external electric field induced reaction. Source: Aragonès et al. [67]. Reproduced with permission of the Nature.

in an edge-on manner. By applying an external electric field oriented from STM tip to surface ($E_{\text{surface}} < 0$), the Diels–Alder reaction was accelerated by a fivefold increase. The ability to manipulate chemical reactions with electric fields may offer new approaches to heterogeneous catalysis.

7.5.2 Bergman-Like Reaction

Bergman reaction is one rearrangement reaction of the enediyne group. The on-surface investigation of this intramolecular cycloaromatisation was greatly improved with the help of nc-AFM, which provided a possibility to investigate the internal covalent bond configuration. Compared with the unique C^1–C^6 cyclisation (six-membered rings) in solution, C^1–C^4 cyclisation (four-membered rings) and C^1–C^5 cyclisation (five-membered rings) could also be investigated on surface. Fischer and Crommie first reported the Bergman-like cyclisation in individual oligo-(phenylene-1,2-ethynylenes) molecule on Ag(100) surface [68]. By annealing the sample above 363 K, additional six-, five-, or four-membered ring in molecular structure was captured via STM and nc-AFM. The molecular mass conservation theory suggested that the intermolecular rearrangement is exclusively an isomerisation process. The explanation of this isomerisation process was given by DFT calculations, as the reactants could be thermally excited to produce different intermediate diradicals. According to a better understanding of the reaction mechanism, 1D

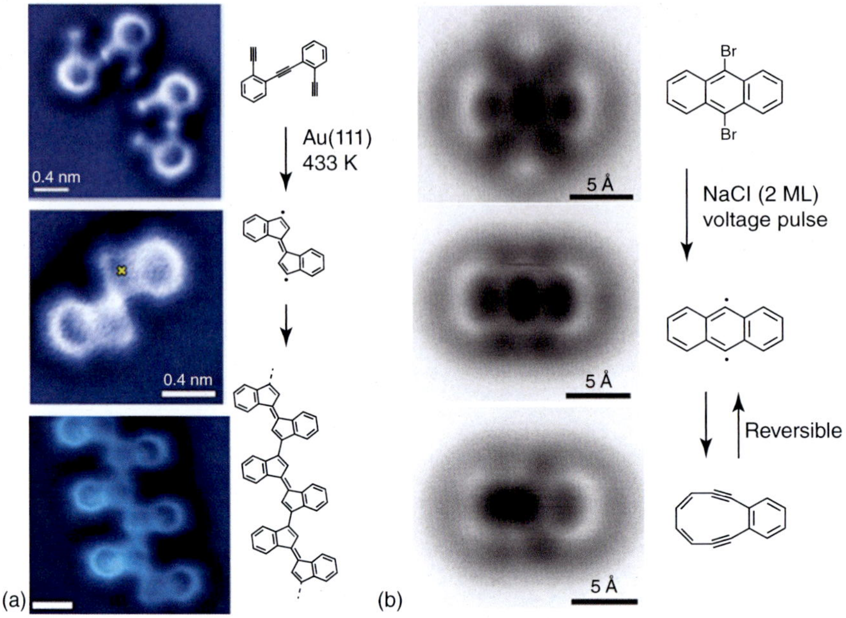

Figure 7.25 On-surface Bergman-like cyclisation reactions of various enediyne precursors. (a) On Au(111), Source: Riss et al. [69]. Reproduced with permission of the American Chemical Society. (b) On NaCl@Cu(111), Source: Schuler et al. [70]. Reproduced with permission of the Nature.

chains of the oligoacetylene derivative were realised via a smart design of the diyne building block [69]. As displayed in Figure 7.25a, the electronic structure and chemical structure of the obtained 1D polymer were characterised by UHV–STM and nc-AFM, respectively. This oligo-(E)-1,1′-bi(indenylidene) chains were obtained on Au(111) surface through the cooperation of C^1–C^5 thermal enediyne cyclisations and step-growth polymerisation of the (E)-1,1′-bi(indenylidene) diradical intermediates. Noteworthy, Gross's group demonstrated the reversible Bergman cyclisation via atomic manipulation, as seen in Figure 7.25b [70]. 9,10-Dibromoanthracene (DBA) molecules were deposited on a Cu(111) surface partly covered by two monolayers of NaCl. The two C—Br bonds of one DBA were cleaved sequentially by applying two voltage pulses (3.3 V) to obtain diradical. The final products (diyne) consisting of fused six- and ten-membered rings were obtained by applying another voltage pulse above the diradical. The high-resolution nc-AFM image suggested that the diyne was formed by homolytic cleavage of the C—C bond. The underlying ultrathin NaCl film facilitated the stabilisation of diradical and diyne, and the transformation between these two was found to be reversible.

7.5.3 N-Heterocyclic Carbenes Formation and Dimerisation

Gade's group investigated the thermal-induced C—C coupling of N-heterocyclic carbenes (NHC) on Cu(111) surface [71]. As seen in Figure 7.26a–c, 1,3,8,10-tetraazaperopyrene (TAPP) molecules self-assembled as porous 2D network with a

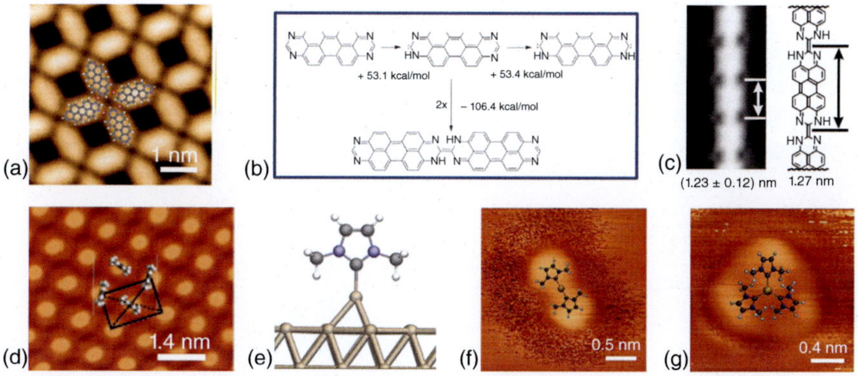

Figure 7.26 (a) STM image of TAPP molecules self-assemble structure on Cu(111) surface. (b) The formation route of TAPP polymer chain. (c) High-resolution STM image and schematic model of a TAPP chain. The distance between two monomers was determined to be (1.23 ± 0.12) nm determined by STM image and 1.27 nm indicated by the model. Source: Matena et al. [71]. Reproduced with permission of Wiley. (d) STM image of NHC self-assembled on Au(111) surface. (e) Geometric structures of one NHC molecule with an additional Au adatom as determined from DFT optimisation. (f) The STM image of the dimer with a superimposed geometric structure from the DFT calculations. (g) The STM image of the trimer with a superimposed geometric structure from the DFT calculations. Source: Wang et al. [72]. Reproduced with permission of the Nature.

sub-monolayer coverage and then evolved into 1D polymeric chains by annealing at 523 K. According to the mechanism of Wanz-lick-type dimerisation, the carbine intermediate radicals were obtained from the tautomerisation of the *N*-heterocyclic end units (Figure 7.26b). The width of the obtained polymer chains was related to the width of a single TAPP molecule and the orientation of the chains mismatched with the substrate lattice. These evidences indicated that the C–C coupling reactions led to the covalent linked polymers. One step further, Fuchs and Glorius explained the high mobility of NHCs on Au(111) surface with ballbot-type motion mechanism [72]. As shown in Figure 7.26e, the NHC molecule rode on a gold atom pulled out of the underlying surface, and moved as a complex across the surface. Together with a large desorption barrier on gold, the high surface mobility of NHC enabled the highly-ordered self-assembled monolayer (Figure 7.26d). In addition, dimeric and hitherto unknown trimeric NHC–Au complexes could also be obtained following an equilibration after the initial formation of the monolayer (Figure 7.26f,g).

7.5.4 σ-Bond Metathesis

Fuchs and Studer disclosed the intermolecular reaction between alkynyl silane and carboxylic acid to give terminal alkyne along with silyl ester via σ-bond metathesis [73]. This σ-bond metathesis reaction worked efficiently on Ag(111) surface at room temperature, and also could occur on Au(111) surface by thermal treatment. The co-deposited 2,6-NDCA and 4,9-bis((trimethylsilyl)ethynyl)anthracene (BTEA) self-assembled into two separated phases on substrates. As shown in

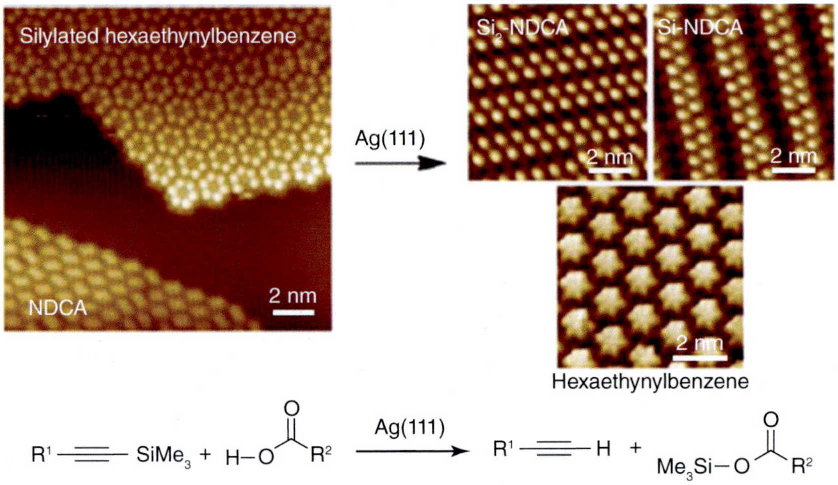

Figure 7.27 The σ-bond metathesis of silylated alkynes with aromatic carboxylic acids on Ag(111) surface to generate the corresponding terminal alkynes and silyl esters. Source: Gao et al. [73]. Reproduced with permission of the American Chemical Society.

Figure 7.27, the exchange between silyl groups on BTEA and the hydrogen atoms of the carboxylic acid group on NDCA occurred on Ag(111) to generate desilylated-anthracene-bisalkyne, Si-NDCA, and Si_2-NDCA molecules. Remarkably, a hexaethybenzene self-assembled structure cannot be obtained from the traditional 'protecting group free' approach because of high instability, but can be formed via this approach. Furthermore, the influence of central conjugated π-system in this metathesis reaction was also investigated, finding that aromatic tetrayne had a much higher reactivity compared with nonaromatic tetrayne [74].

7.5.5 Diacetylene Polymerisation

Polymeric diacetylene (PDA) and its derivatives have been widely used in optical and electrical devices as conductive materials. The highly ordered self-assembled diacetylene monolayer could provide a chance for a direct transformation from monomer to polymer on surfaces. The diacetylene polymerisation on HOPG could be induced by both STM-tip and UV light, which have been extensively investigated at atomic lever under ambient or UHV conditions [75]. The central diacetylene units polymerised to form covalently linked 1D polymer, undergoing a mechanism attributed to a chain reaction initiated by the formation of diradical. As shown in Figure 7.28a, Okawa and Aono used an STM tip to induce the polymerisation of 10,12-nonacosadiynoic acid (NCDA) [75b]. First, an artificial defect was created on the NCDA monolayer by the tip by applying a positive pulse of +5 V. Then, the tip glided from top to bottom while applying a negative bias voltage of −4 V, and polydiacetylene nanowire formed between the initial position of the tip and the artificial defect (Figure 7.28a). The UV light-driven polymerisation was introduced by De Feyter, De Schryver and coworkers [76]. On HOPG surface, the

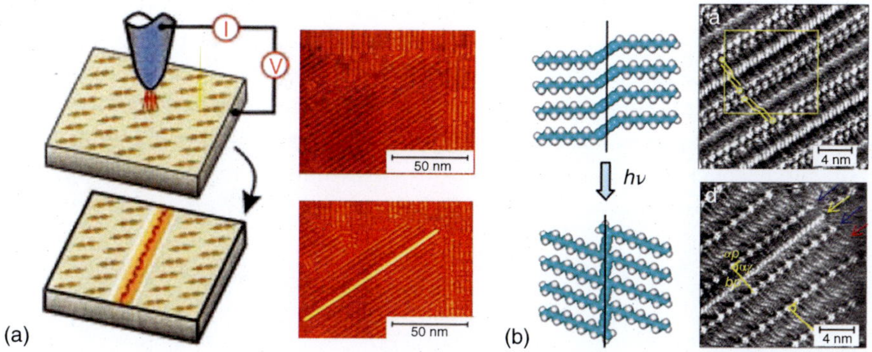

Figure 7.28 (a) Diacetylene polymerisation initiated by a STM tip at designated positions. STM images of the same area before and after manipulations. Source: Okawa et al. [75b]. Reproduced with permission of the American Chemical Society. (b) UV-light induced diacetylene polymerisation. STM images of the molecular structures before and after 30 minutes UV light irradiation. Source: Miura et al. [76]. Reproduced with permission of the American Chemical Society.

well-ordered self-assembled monolayer of diacetylene derivatives was prepared by the Langmuir–Blodgett technique with the horizontal lifting method. After UV light irradiation of 30 minutes, some bright lines, which were assigned to the polymerised diacetylene rows, were observed with the same orientation in the same domain (Figure 7.28b).

7.6 Conclusion and Perspectives

In this chapter, we summarise the typical on-surface reactions reported in the last 10 years. Although still in its infancy stage, surface assistant reactions have attracted scientist's attention due to the different reaction pathways compared with the reaction in solution. Particularly, on-surface chemistry provides promising approaches towards 'precise chemistry' via precisely selecting the active site on reactant. The low-temperature system can freeze the molecules, which makes it possible to in-situ observe the intermediate states during the reaction process. Moreover, the specific molecular structure can be designed by tip manipulation. The substrates, single-crystal metal surfaces, not only serve as a carrier for the molecular precursor, but also play an important role as catalysts. The same precursors may follow different reaction paths on different substrates, leading to various products. Meanwhile, different forms of energy can be used to stimulate on-surface reactions, as light, temperature, and electrical field, the on-surface reaction process can be followed in real space by STM and nc-AFM, from precursors to intermediates and final products. especially, the development of nc-AFM makes it possible to depict the structures of products in the chemical bond resolution. Together with DFT calculations, we gain insight into reaction pathways, activation barriers, and reaction sites.

By learning and capitalizing on the previous works, directions for future research in on-surface synthesis can emerge. One direction is to explore the basic laws of physics and chemistry in chemical reactions at the atomic level. Another direction is to design and synthesize nanomaterials with properties for further applications in nano-electronics, such as novel functional molecules, conductive polymers, vertical artificial structures, and so on. Finally, it should be noted that the rapid improvement of experimental technology might open an entirely new dimension for on-surface chemistry, towards harnessing structural and functional complexity in molecule-based materials.

References

1 a) Mamdouh, W., Uji-i, H., Ladislaw, J.S. et al. (2006). *J. Am. Chem. Soc.* 128: 317–325. b) Tahara, K., Furukawa, S., Uji-i, H. et al. (2006). *J. Am. Chem. Soc.* 128: 16613–16625. c) Wang, D., Wan, L.-J., and Bai, C.-L. (2010). *Mat. Sci. Eng., R* 70: 169–187. d) Mali Kunal, S. and De Feyter, S. (2013). *Philos. Trans. R. Soc. A* 371: 20120304. e) Tahara, K., Nakatani, K., Iritani, K. et al. (2016). *ACS Nano* 10: 2113–2120.

2 a) Sakamoto, J., van Heijst, J., Lukin, O., and Schluter, A.D. (2009). *Angew. Chem. Int. Ed. Engl.* 48: 1030–1069. b) Liu, X.-H., Guan, C.-Z., Wang, D., and Wan, L.-J. (2014). *Adv. Mater.* 26: 6912–6920.

3 a) Barth, J.V., Costantini, G., and Kern, K. (2005). *Nature* 437: 671. b) Lafferentz, L., Ample, F., Yu, H. et al. (2009). *Science* 323: 1193. c) Ciesielski, A., Palma, C.-A., Bonini, M., and Samorì, P. (2010). *Adv. Mater.* 22: 3506–3520. d) Bazarnik, M., Bugenhagen, B., Elsebach, M. et al. (2016). *Nano Lett.* 16: 577–582.

4 a) El Garah, M., MacLeod, J.M., and Rosei, F. (2013). *Surf. Sci.* 613: 6–14. b) Dong, L., Liu, P.N., and Lin, N. (2015). *Acc. Chem. Res.* 48: 2765–2774. c) Fan, Q., Gottfried, J.M., and Zhu, J. (2015). *Acc. Chem. Res.* 48: 2484–2494. d) Lindner, R. and Kühnle, A. (2015). *ChemPhysChem* 16: 1582–1592. e) Held, P.A., Fuchs, H., and Studer, A. (2017). *Chem. Eur. J.* 23: 5874–5892. f) Shen, Q., Gao, H.-Y., and Fuchs, H. (2017). *Nano Today* 13: 77–96.

5 Ullmann, F. and Bielecki, J. (1901). *Ber. Dtsch. Chem. Ges.* 34: 2174–2185.

6 Hla, S.-W., Bartels, L., Meyer, G., and Rieder, K.-H. (2000). *Phys. Rev. Lett.* 85: 2777–2780.

7 Lackinger, M. (2017). *Chem. Commun.* 53: 7872–7885.

8 Grill, L., Dyer, M., Lafferentz, L. et al. (2007). *Nat. Nanotechnol.* 2: 687.

9 Lafferentz, L., Eberhardt, V., Dri, C. et al. (2012). *Nat. Chem.* 4: 215.

10 Bieri, M., Nguyen, M.-T., Gröning, O. et al. (2010). *J. Am. Chem. Soc.* 132: 16669–16676.

11 a) Sun, Q., Tran, B.V., Cai, L. et al. (2017). *Angew. Chem. Int. Ed.* 56: 12165–12169. b) Sun, Q., Yu, X., Bao, M. et al. (2018). *Angew. Chem. Int. Ed.* 57: 4035–4038.

12 Shu, C.-H., Liu, M.-X., Zha, Z.-Q. et al. (2018). *Nat. Commun.* 9: 2322.

13 Talirz, L., Ruffieux, P., and Fasel, R. (2016). *Adv. Mater.* 28: 6222–6231.

14 Kanuru, V.K., Kyriakou, G., Beaumont, S.K. et al. (2010). *J. Am. Chem. Soc.* 132: 8081–8086.

15 Sánchez-Sánchez, C., Yubero, F., González-Elipe, A.R. et al. (2014). *J. Chem. Phys. C* 118: 11677–11684.
16 Shi, K.-J., Shu, C.-H., Wang, C.-X. et al. (2017). *Org. Lett.* 19: 2801–2804.
17 Int'Veld, M., Iavicoli, P., Haq, S. et al. (2008). *Chem. Commun.*: 1536–1538.
18 Zhong, D., Franke, J.-H., Podiyanachari, S.K. et al. (2011). *Science* 334: 213.
19 Sun, K., Chen, A., Liu, M. et al. (2018). *J. Am. Chem. Soc.* 140: 4820–4825.
20 a) Scott, L.T., Boorum, M.M., McMahon, B.J. et al. (2002). *Science* 295: 1500. b) Gómez-Lor, B. and Echavarren, A.M. (2004). *Org. Lett.* 6: 2993–2996.
21 Otero, G., Biddau, G., Sánchez-Sánchez, C. et al. (2008). *Nature* 454: 865.
22 Amsharov, K., Abdurakhmanova, N., Stepanow, S. et al. (2010). *Angew. Chem. Int. Ed.* 49: 9392–9396.
23 Han, P., Akagi, K., Federici Canova, F. et al. (2014). *ACS Nano* 8: 9181–9187.
24 Sun, Q., Zhang, C., Kong, H. et al. (2014). *Chem. Commun.* 50: 11825–11828.
25 Sun, Q., Zhang, C., Cai, L. et al. (2015). *Chem. Commun.* 51: 2836–2839.
26 Haq, S., Hanke, F., Sharp, J. et al. (2014). *ACS Nano* 8: 8856–8870.
27 a) Jensen, S.A., Ulbricht, R., Narita, A. et al. (2013). *Nano Lett.* 13: 5925–5930. b) Abbas, A.N., Liu, G., Narita, A. et al. (2014). *J. Am. Chem. Soc.* 136: 7555–7558.
28 a) Yang, X., Dou, X., Rouhanipour, A. et al. (2008). *J. Am. Chem. Soc.* 130: 4216–4217. b) Campos, L.C., Manfrinato, V.R., Sanchez-Yamagishi, J.D. et al. (2009). *Nano Lett.* 9: 2600–2604. c) Kosynkin, D.V., Higginbotham, A.L., Sinitskii, A. et al. (2009). *Nature* 458: 872. d) Wang, X. and Dai, H. (2010). *Nat. Chem.* 2: 661. e) Bonaccorso, F., Lombardo, A., Hasan, T. et al. (2012). *Mater. Today* 15: 564–589. f) Narita, A., Verzhbitskiy, I.A., Frederickx, W. et al. (2014). *ACS Nano* 8: 11622–11630. g) Hwang, W.S., Zhao, P., Tahy, K. et al. (2015). *APL Mater.* 3: 011101.
29 Cai, J., Ruffieux, P., Jaafar, R. et al. (2010). *Nature* 466: 470.
30 Ruffieux, P., Wang, S., Yang, B. et al. (2016). *Nature* 531: 489.
31 Cai, J., Pignedoli, C.A., Talirz, L. et al. (2014). *Nat. Nanotechnol.* 9: 896.
32 Deng, X.Q., Zhang, Z.H., Tang, G.P. et al. (2014). *RSC Adv.* 4: 58941–58948.
33 a) Kimouche, A., Ervasti, M.M., Drost, R. et al. (2015). *Nat. Commun.* 6: 10177. b) Zhang, H., Lin, H., Sun, K. et al. (2015). *J. Am. Chem. Soc.* 137: 4022–4025.
34 a) Chen, Y.-C., de Oteyza, D.G., Pedramrazi, Z. et al. (2013). *ACS Nano* 7: 6123–6128. b) Basagni, A., Sedona, F., Pignedoli, C.A. et al. (2015). *J. Am. Chem. Soc.* 137: 1802–1808.
35 Lu, Y.H., Wu, R.Q., Shen, L. et al. *Appl. Phys. Lett.* 94 (12): 122111.
36 Linden, S., Zhong, D., Timmer, A. et al. (2012). *Phys. Rev. Lett.* 108: 216801.
37 Sakaguchi, H., Song, S., Kojima, T., and Nakae, T. (2016). *Nat. Chem.* 9: 57.
38 Cloke, R.R., Marangoni, T., Nguyen, G.D. et al. (2015). *J. Am. Chem. Soc.* 137: 8872–8875.
39 Liang, L. and Meunier, V. (2015). *J. Chem. Phys. C* 119: 775–783.
40 Zhang, Y.-F., Zhang, Y., Li, G. et al. (2017). *Nano Res.* 10: 3377–3384.
41 Sun, Q., Cai, L., Ding, Y. et al. (2015). *Angew. Chem. Int. Ed.* 54: 4549–4552.
42 Sun, Q., Cai, L., Ma, H. et al. (2016). *Chem. Commun.* 52: 6009–6012.
43 Zhang, Y.-Q., Kepčija, N., Kleinschrodt, M. et al. (2012). *Nat. Commun.* 3: 1286.

44 Liu, J., Ruffieux, P., Feng, X. et al. (2014). *Chem. Commun.* 50: 11200–11203.

45 Gao, H.-Y., Wagner, H., Zhong, D. et al. (2013). *Angew. Chem. Int. Ed.* 52: 4024–4028.

46 Cirera, B., Zhang, Y.-Q., Björk, J. et al. (2014). *Nano Lett.* 14: 1891–1897.

47 Eichhorn, J., Heckl, W.M., and Lackinger, M. (2013). *Chem. Commun.* 49: 2900–2902.

48 Li, Q., Yang, B., Björk, J. et al. (2018). *J. Am. Chem. Soc.* 140: 6076–6082.

49 Weigelt, S., Busse, C., Bombis, C. et al. (2007). *Angew. Chem. Int. Ed.* 46: 9227–9230.

50 Weigelt, S., Busse, C., Bombis, C. et al. (2008). *Angew. Chem. Int. Ed.* 47: 4406–4410.

51 Jiang, L., Papageorgiou, A.C., Oh, S.C. et al. (2016). *ACS Nano* 10: 1033–1041.

52 Gong, Z., Yang, B., Lin, H. et al. (2016). *ACS Nano* 10: 4228–4235.

53 a Yu, Y., Sun, J., and Lei, S. (2015). *J. Chem. Phys. C* 119: 16777–16784. b Hu, Y., Goodeal, N., Chen, Y. et al. (2016). *Chem. Commun.* 52: 9941–9944. c Bilbao, N., Yu, Y., Verstraete, L. et al. (2018). *Chem. Commun.* 54: 9905–9908.

54 Liu, X.-H., Guan, C.-Z., Ding, S.-Y. et al. (2013). *J. Am. Chem. Soc.* 135: 10470–10474.

55 Ciesielski, A., El Garah, M., Haar, S. et al. (2014). *Nat. Chem.* 6: 1017.

56 Xu, L., Zhou, X., Tian, W.Q. et al. (2014). *Angew. Chem. Int. Ed.* 53: 9564–9568.

57 Treier, M., Richardson, N.V., and Fasel, R. (2008). *J. Am. Chem. Soc.* 130: 14054–14055.

58 Zwaneveld, N.A.A., Pawlak, R., Abel, M. et al. (2008). *J. Am. Chem. Soc.* 130: 6678–6679.

59 Dienstmaier, J.F., Medina, D.D., Dogru, M. et al. (2012). *ACS Nano* 6: 7234–7242.

60 Gao, H.-Y., Held, P.A., Knor, M. et al. (2014). *J. Am. Chem. Soc.* 136: 9658–9663.

61 Held, P.A., Gao, H.-Y., Liu, L. et al. (2016). *Angew. Chem. Int. Ed.* 55: 9777–9782.

62 Arado, O.D., Mönig, H., Franke, J.-H. et al. (2015). *Chem. Commun.* 51: 4887–4890.

63 Yang, B., Björk, J., Lin, H. et al. (2015). *J. Am. Chem. Soc.* 137: 4904–4907.

64 Bebensee, F., Bombis, C., Vadapoo, S.-R. et al. (2013). *J. Am. Chem. Soc.* 135: 2136–2139.

65 Díaz Arado, O., Mönig, H., Wagner, H. et al. (2013). *ACS Nano* 7: 8509–8515.

66 Meir, R., Chen, H., Lai, W., and Shaik, S. (2010). *ChemPhysChem* 11: 301–310.

67 Aragonès, A.C., Haworth, N.L., Darwish, N. et al. (2016). *Nature* 531: 88.

68 de Oteyza, D.G., Gorman, P., Chen, Y.-C. et al. (2013). *Science* 340: 1434.

69 Riss, A., Wickenburg, S., Gorman, P. et al. (2014). *Nano Lett.* 14: 2251–2255.

70 Schuler, B., Fatayer, S., Mohn, F. et al. (2016). *Nat. Chem.* 8: 220.

71 Matena, M., Riehm, T., Stöhr, M. et al. (2008). *Angew. Chem. Int. Ed.* 47: 2414–2417.

72 Wang, G., Rühling, A., Amirjalayer, S. et al. (2016). *Nat. Chem.* 9: 152.

73 Gao, H.-Y., Held, P.A., Amirjalayer, S. et al. (2017). *J. Am. Chem. Soc.* 139: 7012–7019.

74 Meng, X., Liu, L., García, F. et al. (2018). *J. Chem. Phys. C* 122: 6230–6235.

75 a) Okawa, Y. and Aono, M. (2001). *Nature* 409: 683. b) Okawa, Y. and Aono, M. (2001). *J. Chem. Phys.* 115: 2317–2322. c) Akai-Kasaya, M., Shimizu, K., Watanabe, Y. et al. (2003). *Phys. Rev. Lett.* 91: 255501. d) Sullivan, S.P., Schnieders, A., Mbugua, S.K., and Beebe, T.P. (2005). *Langmuir* 21: 1322–1327. e) Takajo, D., Okawa, Y., Hasegawa, T., and Aono, M. (2007). *Langmuir* 23: 5247–5250. f) Okawa, Y., Mandal, S.K., Hu, C. et al. (2011). *J. Am. Chem. Soc.* 133: 8227–8233. g) Deshpande, A., Sham, C.-H., Alaboson, J.M.P. et al. (2012). *J. Am. Chem. Soc.* 134: 16759–16764.

76 Miura, A., De Feyter, S., Abdel-Mottaleb, M.M.S. et al. (2003). *Langmuir* 19: 6474–6482.

8

Hybrid Organic-2D TMD Heterointerfaces: Towards Devices Using 2D Materials

Yu L. Huang[1,2] *and Andrew T. S. Wee*[2]

[1] *Institute of Materials Research and Engineering (IMRE), A*STAR (Agency for Science, Technology and Research), 2 Fusionopolis Way, Innovis 138634, Singapore*
[2] *National University of Singapore, Department of Physics, 2 Science Drive 3, Singapore 117542, Singapore*

8.1 Introduction

The emergence of two-dimensional (2D) atomically layered materials with novel and intriguing properties has led to a rapid burgeoning in the fields of nanoscience and nanotechnology over last decade [1–3]. Being atomically thin, 2D materials are of special interest in flexible, transparent electronics and optoelectronics applications. As the first and the most famous member of the 2D family, graphene – the monolayer counterpart of graphite, is considered to be one of the most promising candidate materials for post-Si electronics due to its extraordinary electrical-transport properties [1, 4–6]. For instance, graphene displays a remarkable high carrier mobility, with predicted values exceeding $10^6\ cm^2/VS$ at 2 K [7], and device-derived values exceeding $10^5\ cm^2/VS$ at room temperature [8]. However, pristine graphene with zero bandgap has limited applications in many electronic devices, e.g. field-effect transistors (FETs) which require high on/off switching ratios, and digital electronics where the conventional complementary logic operations are based on controllable stable doping processes [9]. 2D transition metal dichalcogenides (TMDs) with graphene-like laminar packing structure but large semiconducting bandgap, have attracted enormous attention recently for their chemical versatility [3, 10, 11]. Distinct from graphene, TMD materials have diverse electronic bandgaps ranging from visible to near-infrared photon wavelength, which are tunable with their crystal stoichiometric composition and geometrical symmetry [11–13]. Furthermore, 2D TMDs exhibit exotic properties such as indirect-to-direct bandgap crossover with decreasing layer number, field-induced transport with high on–off ratios, strong photovoltaic responses, large spin-orbit coupling and interesting valleytronics phenomena, enabling a wide range of applications in a variety of nanoscale and flexible devices [3, 10–13].

Organic semiconducting materials, on the other hand, have exhibited great success in flexible electronic devices, with an extensive number of available molecules with tunable electronic and optical properties. The most successful organic devices

Supramolecular Chemistry on Surfaces: 2D Networks and 2D Structures, First Edition.
Edited by Neil R. Champness.

include organic light-emitting diodes (OLEDs), which have been used in cell-phone displays and other applications, and organic field-effect transistors (OFETs), which have also been widely studied [14, 15]. However, the organic device applications greatly suffer from poor mobility, with the best reported value of 43 cm^2/VS in rubrene single crystal [16, 17].

Since both semiconducting 2D TMD materials and organic molecules have wide potential applications in next generation flexible electronic devices, it is natural to marry the two fields to construct organic-2D TMD hybrid structures, with the aim of leveraging the benefits from both [18, 19]. A number of recent studies have reported a variety of organic-2D TMD hybrid devices with promising applications such as photovoltaics, p–n junctions, chemical sensors, catalysts, and so on, with even better functionalities that are not present in either material alone. For example, a high performance photodetector constructed by the combination of organic dye molecules and monolayer MoS_2 was fabricated, benefited from the high optical absorption efficiency of the organic layer and the relatively high charge mobility in MoS_2 [20]. Organics have also been used to efficiently passivate and/or repair defects [21–24], and modify the carrier tunnelling barriers [25, 26] and concentrations in the 2D TMDs [27, 28]. In this way, the integration of 2D TMDs with selected organics is a promising and controllable approach to improve the functionalities of TMDs without structural damage. As a result, the 2D TMDs exhibit significantly enhanced charge mobility [22] and better photoluminescence (PL) yields [29–32] after appropriate organic functionalisation. Central to the realisation of all these applications is a fundamental understanding of the organic-2D TMD heterointerface.

In this chapter, we will review recent literature studies on the fundamental properties of organic/2D-TMD hybrid structures and their applications. We begin with a brief discussion on their atomic structures in Section 8.2, including both pristine 2D TMD crystals and organic-2D TMD heterointerfaces. In Section 8.3, we focus our attention on the surface functionalisation of 2D TMDs via covalently bonding with the organics. We then discuss the fundamental electronic properties in Section 8.4, focusing on the impact of the interfacial charge transfer and electronic screening on the energy level alignment (ELA) at the van der Waals (vdW) hybrid interfaces. Lastly, we highlight elegant examples of the device applications based on organic/2D-TMD p–n heterojunctions in Section 8.5, and conclude with a forward looking perspective on future scientific challenges and device developments in this field in Section 8.6.

8.2 Atomic Structures

8.2.1 Pristine 2D TMDs

TMDs in general have the chemical composition of MX_2, where M is a transition metal from group IV, V, or VI, and X is a chalcogen (S, Se, or Te), as highlighted in Figure 8.1a [33]. Compared to the plethora of organic molecules that can be synthesised, there are only about 40 different layered TMD crystals available. Typically,

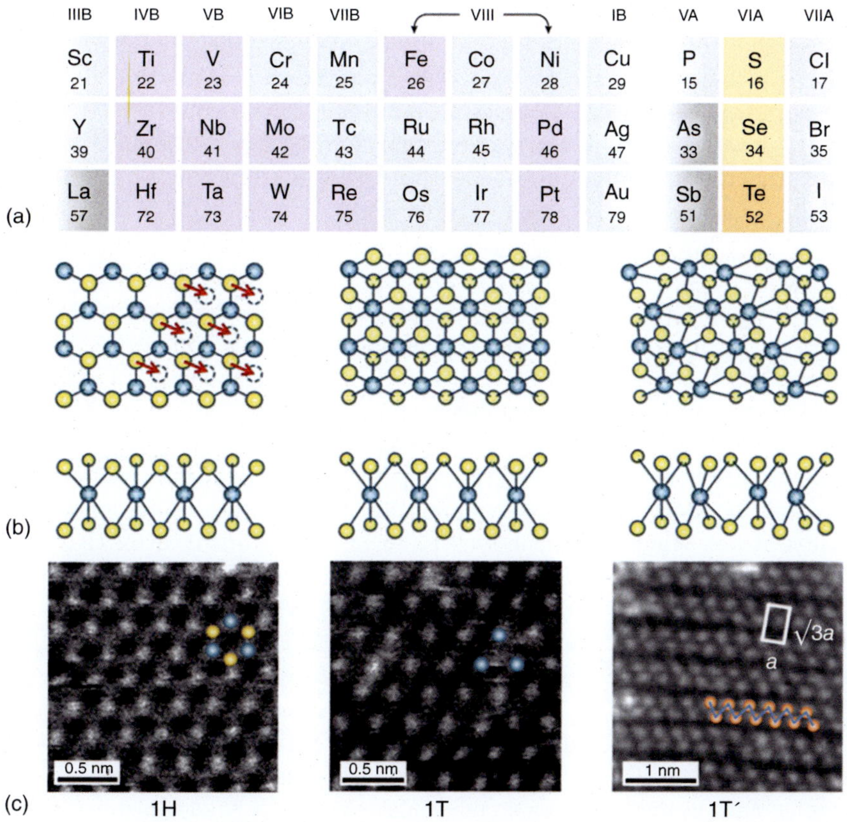

Figure 8.1 Atomic structures of monolayer TMDs. (a) Overview of the transition metals (highlighted in purple) and the three chalcogen elements (highlighted in yellow and orange) that can form about more than 40 different layered TMD compounds. (b) Top-view (top) and side-view (bottom) of the three most common phases, 1H, 1T, and 1T′ from left side to right side. The blue and yellow balls represent metal (M) and (X) atoms. (c) Dark-field scanning transmission electron microscopy (STEM) images of single-layer MX_2 (typically MoS_2 and WSe_2) corresponding to the 1H, 1T, and 1T′ phases. Source: (a) Zhou et al. [33], (b) Eda et al. [34], (c) From Chhowalla et al. [35] © 2017 Nature Publishing Group.

an MX_2 monolayer contains three layers of atoms, X–M–X, where a hexagonally packed metal layer is sandwiched between two chalcogen layers (Figure 8.1b). The intralayer M—X bonds are covalent in nature, such that the oxidation states of M and X are +4 and −2, respectively. The adjacent layers are held weakly together via vdW coupling to form the bulk crystal, and thereby allows the exfoliation into 2D atomic layers with thickness of a single unit cell and high chemical stability. In an individual TMD monolayer, two most common phases exist, namely hexagonal (1H) and tetragonal (1T), where the M atoms are coordinated with the basal X layers in trigonal and octahedral prismatic symmetries respectively (Figure 8.1b). The digit here, typically 1, 2, and 3, indicates the number of layers per repeat unit along the direction perpendicular to the (0001) plane (i.e. z direction or *c* axis) in the bulk [12].

In addition to the hexagonal 1H and 1T phases, another commonly observed phase is 1T′, which is formed by Jahn–Teller distortion with metastability (Figure 8.1c). Interestingly, the properties of TMDs in different phases are significantly different from each other. For instance, MoS_2 is semiconducting in the 1H and 1T′ phases with different band structures, and becomes metallic in the 1T phase. The chemical and physical properties of TMDs have been reviewed in detail elsewhere [11, 12, 35–39].

8.2.2 Organic/2D TMD Interfaces

Pristine semiconducting 2D TMDs are generally chemically inert and their surfaces are atomically smooth and free of dangling bonds. On such surfaces, organic molecules interact with the underlying TMDs predominantly via relatively weak vdW interactions, similar to graphene and boron nitride (BN) [40–43]. The self-assembly of organic molecules on 2D TMDs is thus driven by dominant intermolecular interactions to form high-quality crystalline films and hence well-defined vdW organic-2D TMD interfaces with atomically sharp interfaces. For example, the formation of well-ordered perylene-3,4,9,10-tetracarboxylic dianhydride (PTCDA) and fluorinated fullerene ($C_{60}F_{48}$) supramolecular arrays have been reported on tungsten diselenide (WSe_2) monolayers [44, 45], which will be discussed in Section 8.4. In contrast, TMD metallic phases and surface defects (e.g. vacancies, edges, and grain boundaries) with high chemical reactivity can facilitate stronger covalent coupling [46], and even chemical reactions. Understanding the organic-2D TMD interactions at interfaces is key to control the atomic configurations as well as electronic and chemical properties for the rational bottom-up design of devices.

A prototypical example of organic molecules preferentially adsorbed on intrinsically patterned TMD surface is illustrated in Figure 8.2. The 1H/1T phase engineering in platinum diselenide ($PtSe_2$) monolayer was achieved in vacuum by controlling the synthesis conditions [47, 48]. Figure 8.2a shows the formation of a homogeneous 1T phase on a Pt(111) substrate by direct selenisation at 270 °C under Se-rich conditions. After annealing at 400 °C, the 1T monolayer transforms into a triangle tiling pattern constructed by alternating1T and 1H domains (Figure 8.2b) due to the loss of some Se atoms at elevated annealing temperatures. The transformation is reversible. By adding Se atoms and annealing the 1T/1H patterned sample, it can revert to a defect-free 1T structure. Unlike intrinsic MoS_2 which favours the semiconducting 1H phase, the monolayer $PtSe_2$ is stable in the semiconducting 1T form and metastable in the metallic 1H configuration. The ordered 1T/1H titling patterns with different electron affinities makes it a suitable template for selective adsorption of different molecular species. Figure 8.2e,f show that pentacene molecules preferentially bind to the metastable 1H domains due to the relatively higher binding energy, forming well-ordered triangle pentacene/1H-$PtSe_2$ nanotructures [47]. This observation is consistent with another theoretical study on the pentacene/MoS_2 interface, which predicted a larger charge transfer on the metallic TMD phase. That is, when pentacene is lying flat on the surface, each molecule is able to contribute as large as ~0.437e to the metallic 1T-MoS_2, while only ~0.015e to the 1H phase [49]. Furthermore, by adsorbing different molecular species to different phase domains,

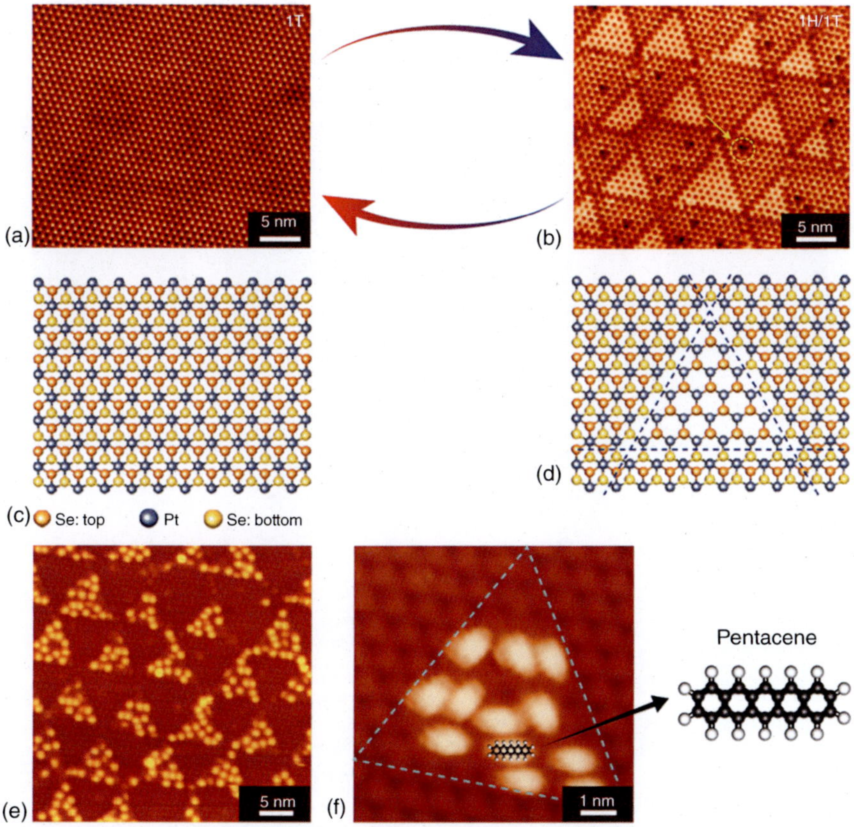

Figure 8.2 1H/1T tilting pattern in $PtSe_2$ monolayer and the selective adsorption. STM images of (a) pure 1T phase and (b) triangular tilting phase formed by alternating 1H and 1T areas, where the corresponding atomic models are given in (c) and (d), respectively. Reversible structural transition between (a) and (b) can be obtained by controlling the Se density. (e) Ordered triangular molecular clusters are formed due to the selective adsorption of pentacene on the 1H areas in the 1H/1T patterned surface. (f) A high-resolution STM image of a pentacene cluster. Source: From Lin et al. [47] © 2017 Nature Publishing Group.

the preferential adsorption indicates a possibility of dual or even multiple functionalisation, which could be used in catalysis or other applications.

Another example shown in Figure 8.3 is the selective adsorption of dibenzothiophene (DBT) on the corner sites on MoS_2 nanoclusters [50]. Single-layered MoS_2 nanoclusters with various cluster sizes were synthesised on a single-crystalline Au(111) substrate by exposing Mo clusters into an H_2S atmosphere at elevated temperature for sulfidation [50]. The size of the triangle MoS_2 nanocluster is defined by the number n of Mo atoms on the side of the triangle (Figure 8.3a,c). Depending on the nanocluster size, two different terminated edges can be found, namely S-edge for $n < 6$ (Figure 8.3a,b), and Mo-edge fully sulfided with S_2 dimers for $n > 6$ (Figure 8.3c,d) [51, 52]. The nanoclusters with $n = 6$ can be either S-edged or Mo-edged with equal probability. To explore the catalytic properties,

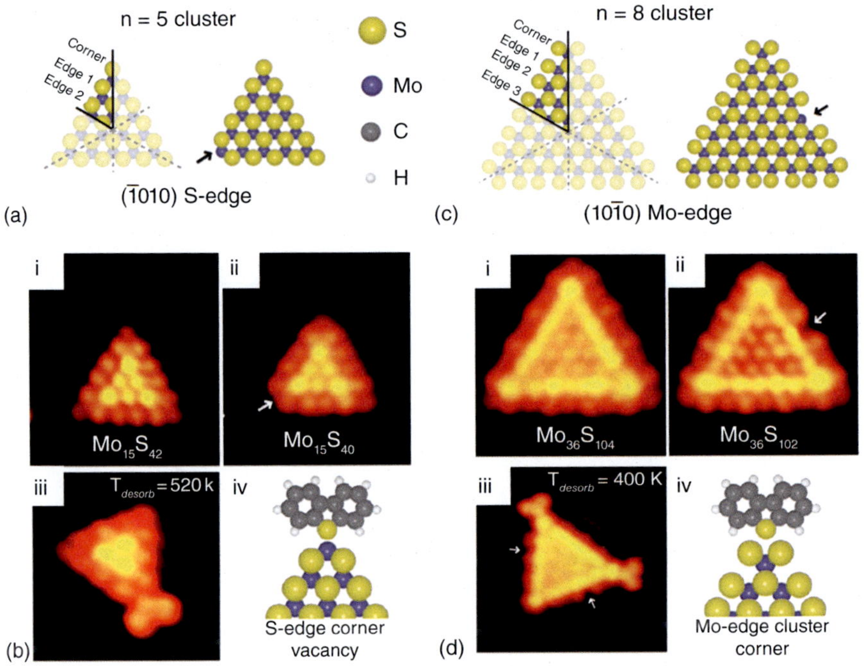

Figure 8.3 The selective adsorption of DBT molecules on MoS_2 nanoclusters. (a) The atomic models of an S-edge-cluster (left) and an S vacancy formed at the corner (right), where *n* indicates the number of Mo atoms at a nanocluster edge. (b) STM images of (i) a $Mo_{15}S_{42}$ nanocluster with S-edge, (ii) an S vacancy created at the corner, (iii) the covalent adsorption of a DBT molecule to the corner S-vacancy site, and (iv) the corresponding atomic model. (c) The atomic configurations illustrate a Mo-edge-cluster (left) and a created S vacancy at the edge (right). (d) STM images of (i) a $Mo_{36}S_{104}$ nanocluster with Mo-edge, (ii) an S vacancy created at the edge, (iii) a DBT molecule physisorbed to the corner site of the Mo-edge-cluster, and (iv) the corresponding model. Source: (a–d) From Tuxen et al. [50] © 2010 American Chemical Society.

the MoS_2 nanoclusters were subsequently exposed to atomic hydrogen and then DBT molecules at room temperature [50]. Firstly, the hydrogen can react with the sulfur atoms, yielding H_2S and leaving sulfur vacancies at the edges of the MoS_2 nanoclusters. As indicated by arrows, vacancies were created at the corner sites of the S-edged nanoclusters (Figure 8.3b(ii)), and some edge sites of the Mo-edges ones (Figure 8.3d(ii)). Subsequently, it was observed that the DBT molecules preferentially absorb at the corner S-vacancy sites of the S-edge-cluster with the formation of a strong covalent bond (Figure 8.3b(iii)), where the Mo–S distance is only ~2 Å as depicted in Figure 8.3b(iv). The strong binding interaction is consistent with the high desorption temperature of 520 K [50]. For the Mo-edge-clusters, DBT also absorbs at the corners but via relatively weak interactions, which can be removed by annealing at 400 K (Figure 8.3d(iv)). The adsorption of DBT molecule is neither observed on edge sites nor edge vacancies due to their steric hindrance [50]. The unique site-selective catalytic behaviours thus indicate that the atomic TMD surface structures can significantly determine their catalytic properties, enabling new opportunities for enhancing surface catalysis, e.g. hydrodesulfuization of fossil fuels and other technological applications in sensors.

8.3 Surface Functionalisation of 2D TMDs by Organics

As discussed in Section 8.2.2, the chemical functionalisation of 2D TMDs can be facilitated via covalent bonding at defect sites and metastable metallic phases. The covalent bonds are generally stronger than non-covalent ones, and thus such covalent functionalisation potentially has better stability and larger impact on TMD properties. In this section, we discuss the importance of the covalently surface functionalisation via defect and phase engineering to improve the advanced properties in the organic-2D TMD hybrid structures for device applications.

8.3.1 Defect Engineering

As a promising functionalisation method, covalent functionalisation can be achieved by grafting various thiol molecules with different functional groups onto the MoS_2 surface defect sites [23, 53–55]. As shown in Figure 8.4a, a prototypical thiol molecule containing a carbon-bonded sulfhydryl is depicted by an R–SH formula, where R represents an alkyl or other organic substituent. Thiolation of the MoS_2 flakes was achieved in solution as the molecular ligand conjugated to the vacancy sites via the S atom with the dissociation of H atoms. In general, the optical bandgap of such covalently modified MoS_2 flakes is almost retained regardless of the 'R' functional group, as revealed by the negligible shifts in the PL peak positions [55, 56]. However, the electronic band positions can be tuned by the different substituent group as well as the chain length through the formation of interfacial dipole. For example, the energy diagram in Figure 8.4b shows that the energy positions of the functionalised MoS_2 monolayer vary significantly with the molecular tail groups [55]. As determined by photoelectron spectroscopy in air, the upward shift of the valence band (VB) position can reach as large as ~0.5 eV in the MoS_2 flakes thiolated by the S2 and S5 molecules. Such energy level shifts are induced by the functional polar thiol molecules, which are negatively charged after H dissociation and thereby resulting in an electron donating effect. In principle, the magnitude as well as the direction of the interfacial dipole vary with the substituent compound and the chain length, and therefore allows the capability to tune the ELAs by careful selection of the functional molecules. More discussion on the ELAs at the organic-2D TMD interfaces will be found in Section 8.4.

The selective adsorption of thiol molecules at S vacancies can also be further used to repair the vacancies in MoS_2 [22, 57, 58]. An elegant example shown in Figure 8.4c is that (3-mercaptopropyl)trimethoxysilane (MPS) molecules repair S vacancies via two reactions [22]. The first reaction is the initial adsorption of MPS onto an S vacancy with an S—H bond dissociation to form a thiolate surface intermediate, the same as the functionalisation process in Figure 8.4a. The second step is the S–C dissociation of the thiolate intermediate leaving the lone S atom

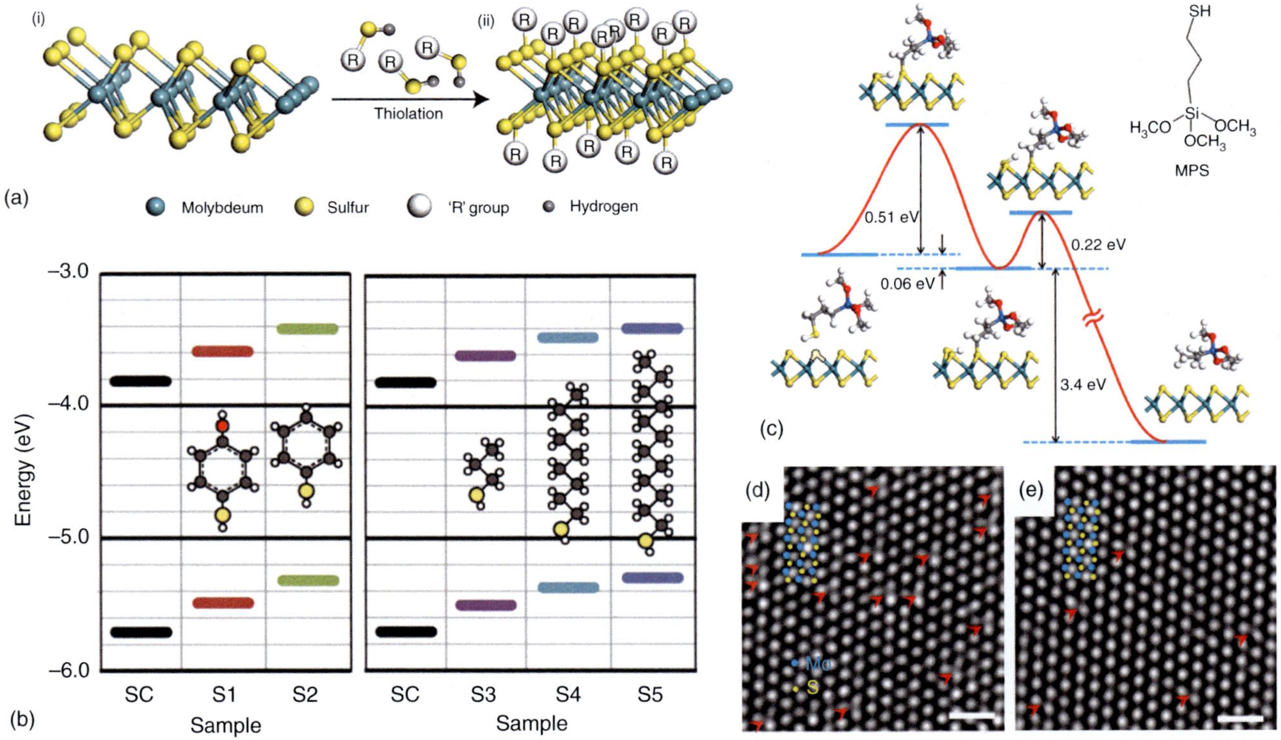

Figure 8.4 Surface engineering with vacancies. (a) Functionalisation of MoS_2 monolayers by thiolation, where the thiol molecules with 'R' group bound to S vacancy sites. (b) Energy diagram shows that the MoS_2 energy levels change by different thiol molecules with various substituent groups and chain lengths. The band gap size remains almost constant. (c) Atomic plots of the initial, transient, and final states involved in the reactions between an S vacancy and a MPS molecule. Comparing the high-resolution transmission electron microscopy (TEM) images of as-exfoliated (d) and MPS-treated (e) monolayer MoS_2 samples, a significant reduction of S vacancy density is observed. Source: (a and b) Nguyen et al. [55], (c–e) From Yu et al. [22] © 2014 Nature Publishing Group.

in the vacancy site, which can be driven by thermal annealing [22] or electric field [57]. The repair can be directly supported by comparing the vacancy densities of the samples before and after thiol treatment. For instance, the high-resolution scanning transmission electron microscopy (STEM) images recorded at the MoS_2 samples before (Figure 8.4d) and after (Figure 8.4e) the MPS treatment revealed that the density of sulfur vacancies was dramatically reduced, from $\sim 6.5 \times 10^{13}$ cm^{-2} for the as-exfoliated samples to $\sim 1.6 \times 10^{13}$ cm^{-2} for the top-side treated sample. Furthermore, a better conductivity and higher carrier mobility (μ) were obtained from the MPS treated MoS_2 flake. In the double-side treated MoS_2 sample, $\mu = 81$ cm^2/VS obtained by Yu et al. is the highest (field effect) mobility achieved at room temperature so far [22]. It further confirms the improvement of the sample quality because the transport properties in 2D materials are significantly limited by the defects, where vacancies are one of the most common types for charge trapping or electron scattering.

It is worth noting the use of chalcogen vacancies for covalent functionalisation of 2D TMDs is inherently limited by the vacancy densities. Although vacancies can be created by atomic hydrogen (Figure 8.3) [50] or thermal annealing [47], it is highly desired to develop strategies for the formation of uniform, covalently bonded organic-2D TMD hybrid structures, e.g. by phase engineering as discussed in Section 8.3.2.

8.3.2 Phase Engineering

As described in Figure 8.2, the electron-rich TMD metallic phase, 1H-$PtSe_2$, are comparatively favoured for molecule adsorption. Many other studies have also revealed that the metallic 1T-MoS_2 and WS_2 are promising for covalent functionalisation with selected organic reactants [46, 59–62]. As shown in Figure 8.5a, single-layered MoS_2 can be converted to 1T-phase by Butylithium (BuLi) treatment, due to charge transfer from the butyl group to the MoS_2 flake [46]. Subsequently, the 1T MoS_2 can be easily functionalised by 2-iodoacetamide (R–I) solution, where the functional groups (R) are covalently bonded to the 1T surface (Figure 8.5b). The optoelectronic properties of the functionalised materials were dramatically changed. As revealed by the PL spectra in Figure 8.5c, two peaks arose in the functionalised 1T MoS_2 (Fct-1T MoS_2) (red curve), indicating its semiconducting nature. The peak at ~1.9 eV originates from the pristine 2H-MoS_2, while the new one at ~1.6 eV is assigned to the modified band structures due to the covalently organic functionalisation. The intensity of the PL peak at 1.6 eV increases with the concentration of the functionalised proportion (Figure 8.5d). These phenomena are in contrast to the PL quenching in the non-functionalised 1T metallic phase (blue curve). Furthermore, the 1T phase can be restored to 2H phase by annealing (e.g. 300–350 °C) with the anchoring of the organic functional groups, and thus finally resulting in functionalised 2H TMDs [46]. This methodology to covalently functionalise 2D-TMDs by phase engineering is versatile, and has been successfully applied to $MoSe_2$, WS_2, and WSe_2 using various organic halides as reactants [46, 59–65].

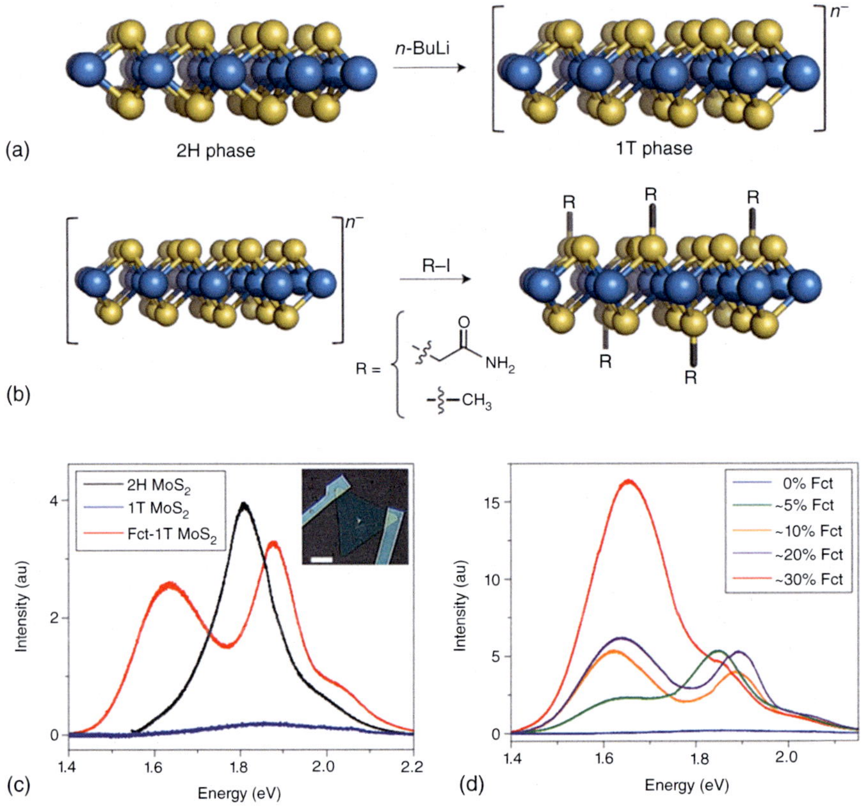

Figure 8.5 Covalent functionalisation through phase engineering. (a) A general schematic shows the conversion of 2H MoS_2 nanosheets to the 1T phase via butyllithium (BuLi) treatment. (b) The negatively charged 1T MoS_2 is subsequently functionalised by 2-iodoacetamide or iodomethane (R–I) solution. (c) PL spectra measured from different single-layer chemical vapour deposition (CVD) MoS_2 samples (inset): pristine 2H phase (black), non-functionalised 1T phase (blue) and functionalised 1T phase (red). (d) The PL intensity increases with the proportion of the functionalised 1T-MoS_2. Source: From Voiry et al. [46] © 2014 Nature Publishing Group.

8.4 Fundamental Electronic Properties

In semiconductor devices, the interface is of paramount importance as it participates in any device actions [66]. This is even more true in organic/2D-TMD nanodevices, because the interface makes up most of the active layers for carrier and energy transports due to the reduced dimensionality. In the discussion of a heterostructure, it is essential to understand the energy band diagram, or ELA at the interface, as it determines charge transfer barriers and directions between the two materials in contact. In this section, we discuss the current understanding of the ELAs at the organic/2D-TMD interfaces, and the physical processes determining the alignments.

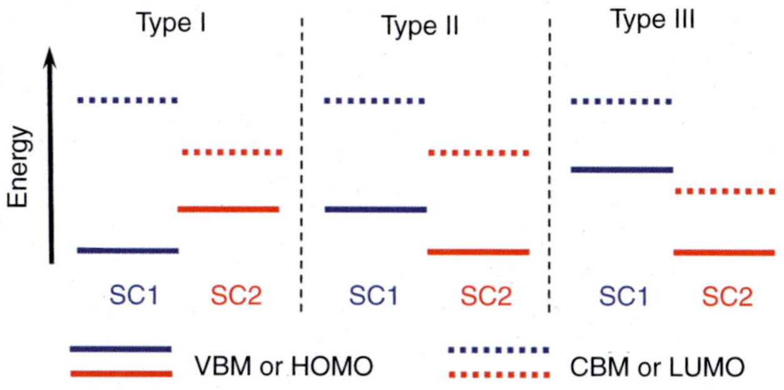

Figure 8.6 Schematic of three types of ELAs at the interface between semiconductor 1 (SC1) and semiconductor 2 (SC2): (i) type I – straddling gap, (ii) type II – staggered gap, and (iii) type III – broken gap.

8.4.1 Energy Level Alignment

The ELA is referred to a diagram depicting the relative positions of the quasiparticle energy levels of the materials at both sides of a heterointerface. In principle, the ELA at a heterointerface between semiconductor 1 (SC1) and semiconductor 2 (SC2) can be categorised into one of the three types: (i) straddling gap, (ii) staggered gap, and (iii) broken gap, as shown in Figure 8.6. The semiconductor 1 or 2 can be organic or inorganic, either in bulk form or atomically thin film. In the case of organic materials with discrete frontier orbitals, the highest occupied molecular orbital (HOMO) and the lowest unoccupied molecular orbital (LUMO) are considered instead of the valence band maximum (VBM) and conduction band minimum (CBM) in inorganic materials. The ELA at the interface is the key for device performance and engineering. It is known that (i) electrons usually transfer to lower energy levels while holes immigrate to higher energy levels among the interlayers; and (ii) energy transfer takes place at a straddling type I configuration through the hole–electron recombination. Based on these working principles, type I alignments are useful in OLEDs to facilitate energy transfer [67, 68]; while type II alignments are chosen to separate electrons and holes across the functional layers in photodetectors [20], photovoltaics [69], photodiodes [70], and so on. Therefore, it is possible to engineer the functionalities of a modern multilayer device by tuning the ELA, which are determined by the intrinsic electronic structures of the selected functional layers and also their stacking configurations.

To understand the formation of the final ELA after two materials come into contact, we need to distinguish it from the energy level lineup (ELL) before contact, which can be quite different. In principle, the initial ELL usually uses the vacuum level as a reference (Figure 8.7a), while the final ELA is established via subtle interfacial interactions, including charge transfer, chemical bonding and electronic screening at the interface (Figure 8.7b,c) [72], to equilibrate the Fermi level at both sides. We first discuss the interface charge transfer which is driven by the Fermi energy (or electrochemical potential) difference, taking a type III ELL as a prototypical model. As illustrated in Figure 8.7a, the Fermi level lies at the midline between VBM (HOMO) and CBM (LUMO) in an isolated semiconductor under

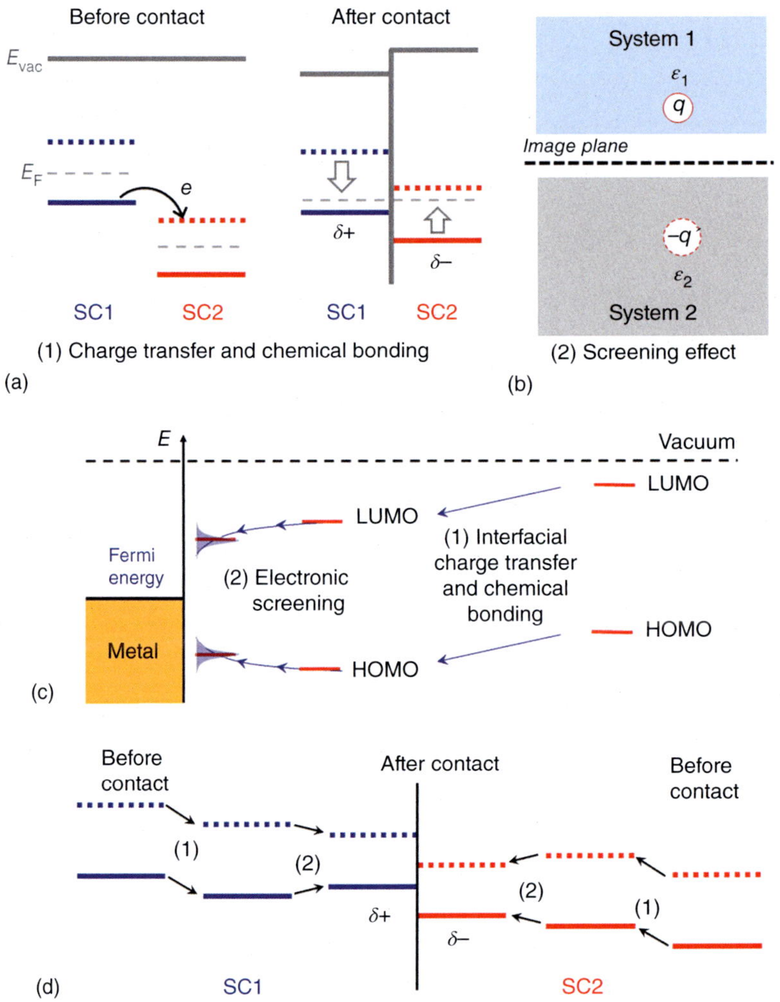

Figure 8.7 Schematics showing energy level alignments at heterointerfaces. (a) Charge transfer driven by the Fermi energy difference between SC1 and SC2. (b) Electronic screening effect can be depicted by a simplistic image charge model. (c) When a molecule is physisorbed on a metal surface, (1) the frontier molecular levels shift with charge transfer and/or chemical bonds, and (2) a reduction of the HOMO–LUMO gap induces by electronic screening. Souce: Quek and Khoo [71]. (d) Schematic shows the ELA at a 2D SC1–SC2 interface determining by the subtle interplay between charge transfer and electronic screening, which are interdependent.

thermodynamic equilibrium, which is quite different from each other in SC1 and SC2 before contact. To equilibrate the Fermi level, electrons could flow from the highest filled states of SC1 to the lowest unfilled states of SC2. As a result of the charge transfer, SC1 becomes positively charged and the energy levels in SC1 shift downwards, while SC2 is equally negatively charged and the energy levels shift upwards. In particular, the Fermi level in the electron donor (SC1) is brought down and becomes relatively closer to the VBM or HOMO, while it is raised up towards CBM or LUMO in the electron acceptor (SC2). Thus, the Fermi levels finally align at both sides when a new thermodynamic equilibrium is established. The initial ELL can quantitatively determine the direction and the amount of charge transfer. Furthermore, to some extent, covalent chemical bonds between the two materials are formed by the hybridisation of the extended electronic states at the same energy levels, which tend to shift the energy levels upward or downward (e.g. Figure 8.4b) analogous to the effect of the charge transfer. The chemical bonding can also result in new electronic states or new properties at the heterointerfaces. For example, as have been discussed in Figure 8.5, semiconducting nature instead of metallic one, is observed in the organic-functionalised 1T-MoS_2, which is not yet fully understood. Here, we only consider the energy level shifts brought in by the covalent bonding for a simplistic modelling.

In addition to charge transfer and chemical bonding, electronic screening is another important factor that affects the final ELA. The screening effect from the dielectric environment usually results in a noticeable reduction in the band gap of a semiconductor. To understand how the electronic screening renormalises the energy levels, we need to recall their definitions. Fundamentally, the VBM or HOMO level represents the ionisation energy required to remove an electron from a system, and the CBM or LUMO represents the affinity energy needed to add an electron to a system. When the two systems are in contact with each other, the added holes or electrons in the former will result in a polarisation of the latter. That is, the electrons in system 2 rearrange to screen the effect of the added positive or negative charge in system 1. In a classical image charge model, the screening effect of the charge q can be captured by placing an image charge $-q'$ in the system 2 at a certain distance from the surface, as shown in Figure 8.7b. The Coulomb interaction between the charge q and its image charge q' further raises the VBM or HOMO and lowers the CBM or LUMO in system 1, resulting in a reduction of the band gap. The electronic screening effect has been well illustrated in the organic–metal interface, as depicted in Figure 8.7c. When organic molecules, e.g. benzene, adsorb on a metallic substrate, the reduction of the HOMO–LUMO gap can be as large as several electronvolts [73–75]. The charge transfer (and/or chemical bonding), screening effect and the ELA are interdependent quantities. In practice charge transfer and chemical bonds can bring in additional (dynamical) screening and further narrow the semiconductor gap [76, 77].

A proposed ELA occurring at the organic-2D TMD interface is illustrated in Figure 8.7d, with the considerations of interfacial charge transfer and/or chemical bonding and electronic screening processes. Compared to traditional bulk semiconductor heterostructures, the impact of electronic screening on the ELA at the low dimensional organic/2D-TMD heterostructures is expected to be different. As both the semiconducting organic molecules and 2D TMD layers have smaller electronic polarisability because of their reduced dimensions, they will have relatively smaller degree of screening of each other compared to bulk materials. However, the electronic screening from the supported substrate, if any, cannot be neglected [44]. It has been found that both the TMD quasiparticle band gap and optical-exciton binding energy can be significantly reduced, as large as hundreds of million electronvolts, by electronic screening from the substrates [78, 79]. Furthermore, it is worth noting that band bending effect that is commonly observed in ELA diagrams for bulk heterostructures has to be omitted [68, 72, 80]. This is because band bending takes place over long length scales of several nanometres, whereas the organic/2D-TMD interfaces considered here are atomically thin.

In the following, we will focus on a few prototype studies showing how the charge transfer (Section 8.4.2) and dielectric screening (Section 8.4.3) affect the ELAs in the vdW organic-2D TMD heretointerfaces, for a comprehensive understanding.

8.4.2 Interfacial Charge Transfer

Surface adsorption of strong organic acceptors or donors has been revealed to be an effective approach to tune the energy levels and the carrier densities of graphene [81, 82]. For example, graphene can be p-doped by tetracyanoquinodimethane (TCNQ) [83, 84], tetrafluoro-tetracyanoquinodimethane (F_4-TCNQ) [81], and tetracyanoethylene (TCNE) [83], and n-doped by 2-(2-methoxyphenyl)-1,3-dimethyl-2,3-dihydro-1H-benzoimidazole (*o*-MeO-DMBI) [85], benzyl viologen (BV) [86], tetrathiafulvalene (TTF) [87], and so on. Recently, researchers have also begun to study how the electronic structures of 2D TMDs be modified by molecular dopants without the formation of covalent bonding, through both theoretical calculations and experimental measurements. It has been found that, similarly, F_4-TCNQ, TCNQ, TCNE, and $C_{60}F_{48}$ can be used as strong electron acceptors [25, 32], while BV, TTF, and dimethyl-*p*-phenylenediamine (DMPD) can act as effective donors [25, 26, 28]. Among them, BV is a strong electron donor, and is able to donate about 0.25*e* per molecule to semiconducting MoS_2 [26]. $C_{60}F_{48}$ is a strong electron acceptor, which can withdraw about 0.29 e per molecule from a WSe_2 monolayer [45]. Such large charge transfer in turn brings in significant changes in the ELA at the organic-2D TMD interfaces as well as the device transport properties [28].

Systematical studies on the impact of the charge transfer at organic-2D TMD heterointerfaces on the supramolecular arrangements, electronic structures, as well as device performances have been carried out by our group. Figure 8.8 shows that the vdW organic-2D TMD heterostructure is comprised of a single-layer tungsten diselenide (SL-WSe_2), a 2D-TMD semiconductor with a moderate bandgap of ~2 eV [89], and a strong molecular acceptor, $C_{60}F_{48,}$ supported by a graphite substrate. A liquid–solid surface phase transformation with the increasing molecular coverage was observed. A loosely packed liquid phase is formed at dilute coverages, e.g. 0.05 ML in Figure 8.8c, where each $C_{60}F_{48}$ molecule is isolated from each other with intermolecular separations much larger than the molecular size (~1 nm). From the corresponding fast Fourier transform (FFT) image in the inset, a blurred ring indicates a certain distribution of the intermolecular separations for the $C_{60}F_{48}$ molecules in the liquid phase. A distribution peak at ~3.7 nm was revealed from the histogram of three nearest molecule–molecule separation [88]. When the coverage is higher than 0.1 ML, e.g. 0.5 ML in Figure 8.8d, the molecules assemble into a closed-packed solid phase with a lattice constant of 1.22 nm [88]. The formation

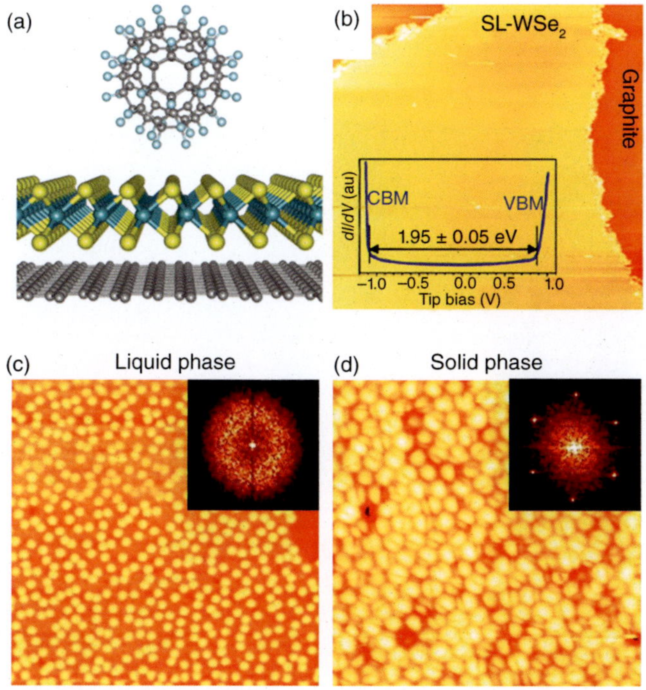

Figure 8.8 Liquid–solid phase transformation. (a) A schematic model of the adsorption of $C_{60}F_{48}$ on SL-WSe_2/graphite. Grey: carbon; turquoise: fluorine; yellow: selenium; cyan: tungsten. (b) A typical STM image of a triangular SL-WSe_2 flake on graphite. The inserted STS spectrum reveals a 1.95 eV bandgap for the clean SL-WSe_2. As the $C_{60}F_{48}$ coverage increases, a phase transformation is observed, i.e. from (c) liquid phase (0.01 ML) to (d) solid phase (0.5 ML). Insets in the upper right corners show the corresponding fast Fourier transform (FFT) images (Image size: (c), 80×80 nm^2; (d), 20×20 nm^2; $V_{tip} = -3.1$ V.) Source: (a) Song et al. [88], (b–d) From Song et al. [45] © 2017 American Chemical Society.

of the liquid phase indicates the presence of repulsive intermolecular interactions, without which the molecules would aggregate into the close-packed configuration due to attractive vdW forces. For a non-polar molecule such as $C_{60}F_{48}$, the most possible repulsive force is intermolecular dipole–dipole interaction arising from the interfacial charge transfer, which has been observed for many other atoms and molecules adsorbed on metal [90] and graphene surfaces [91].

As a powerful approach to understand the origin of the charge transfer, differential charge density (DCD) calculations across the organic-TMD heterointerface were also performed based on density functional theory (DFT). The DCD is defined as the difference between the charge density for the combined system and that for the isolated subsystems, i.e. $\Delta\rho = \rho_{\text{combined}} - \rho_{\text{TMD}} - \rho_{\text{organic}}$. Figure 8.9a shows the side view of the DCD isosurface of the $C_{60}F_{48}$/SL-WSe_2/graphene hybrid system, where blue (red) colour represents the electron (hole) accumulation regions. It is clear that significant charge rearrangement takes place at the interface and the $C_{60}F_{48}$ molecule becomes negatively charged. Interestingly, the WSe_2 interlayer becomes polarised, with electrons accumulating on the top Se layer facing the

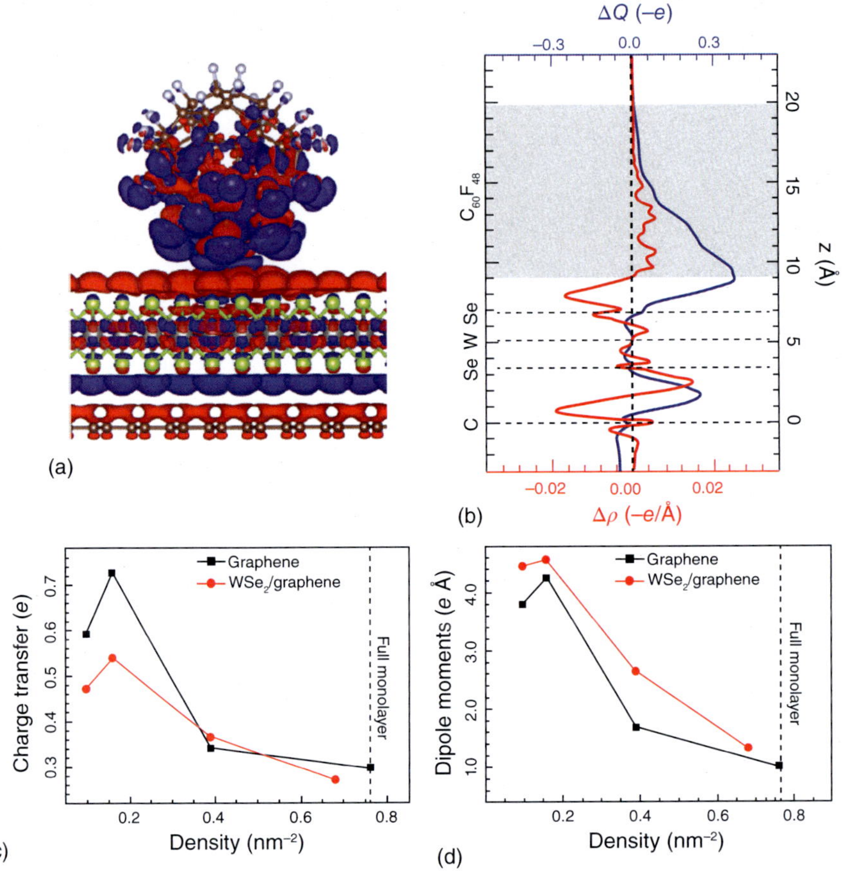

Figure 8.9 DFT calculations of the charge transfer and the dipole moment at the heterointerface. (a) Side view of the isosurface of the differential charge density for the $C_{60}F_{48}$/SL-WSe_2/graphene heterostructure. Blue (red) represents electron accumulation (depletion) region. (b) The corresponding plots of the plane-averaged $\Delta\rho(z)$ (red) and $\Delta Q(z)$ (blue) for the $C_{60}F_{48}$/SL-WSe_2/graphene system. (c) The total amount of the charge transfer and (d) the dipole moment per molecule as a function of the molecular density on both SL-WSe_2/graphene (red) and graphene (black) surfaces. Source: (a, b) Song et al. [45]. (c, d) Song et al. [88].

molecule, and holes accumulating on the bottom Se layer facing graphite. A more quantitative picture is given in Figure 8.9b, where the plane integration of the DCD at position z, $\Delta\rho(z)$ (red), and the cumulative charges transferred from the bottom to the top, $\Delta Q(z)$ (blue), are provided. The total amount of ΔQ obtained by the $C_{60}F_{48}$ molecule adsorbed on the SL-WSe_2/graphene substrate is as large as $-0.38e$ in the model, where about two-thirds of the negative charge comes from graphite and only one-third from the SL-WSe_2 interlayer [45]. Comparative calculations further confirm that $C_{60}F_{48}$ is also a strong hole dopant for isolated SL-WSe_2 and isolated graphene, which is able to withdraw -0.29 and $-0.40e$ from the supported substrates, respectively [45].

Another important factor that influences the amount of charge transfer is the molecular density, namely the number of molecules per area on the surface. The plots in Figure 8.9c,d reveal that the charge transfer as well as the interface dipole moment per molecule decreases with increasing molecular density in both the $C_{60}F_{48}$/graphene (black) and $C_{60}F_{48}$/SL-WSe_2/graphene (red) systems [88]. It thereby explains the liquid–solid phase transition observed in Figure 8.8c,d. As the molecular density increases, the intermolecular vdW interactions increase due to the reduced intermolecular separations, while the repulsive intermolecular dipole–dipole interactions probably decrease because of the reduced charge transfer per molecule. Thus, the molecules are likely to be stabilised by a subtle balance between the repulsive dipole–dipole interactions and the attractive vdW force in the dilute liquid phase; and the vdW interactions become dominant in the close-packed solid phase.

In Figure 8.9c,d, it is also worth noting that although the increased molecule-graphite distance by the WSe_2 interlayer reduces the amount of the charge transfer, it enhances the dipole moment formed at the interfaces. The underlying substrates indeed play important roles in tuning the electronic structures of 2D materials.

Experimentally, photoemission spectroscopy (PES) is one of the powerful tools to determine the energy levels, e.g. work function (WF), valence bands (VBs), and core levels, of the semiconducting TMD materials and heterostructures. The shifts of the energy levels before and after the dopant adsorption can be used to estimate the direction and the extent of charge transfer at the heterointerfaces. Figure 8.10A demonstrates the coverage-dependent spectra of the $C_{60}F_{48}$/WSe_2/graphite heterostructure. Upon adsorption of a full $C_{60}F_{48}$ monolayer (8 Å), the WF increases by1.56 eV indicating the formation of a significant interfacial dipole induced by the charge transfer across the heterostructure. Besides, both the WSe_2 valence band and W 4f core level shift upwards to the Fermi level by ~0.7 eV, confirming that the WSe_2 interlayer is efficiently p-doped by the $C_{60}F_{48}$ overlayer.

Furthermore, the doping of the WSe_2 with $C_{60}F_{48}$ can be evaluated by an as-fabricated FET device on SiO_2/Si surface (Figure 8.10B) with the measurements of the transport properties. As shown in Figure 8.10C, the intrinsic as-exfoliated SL-WSe_2 (black curve) demonstrates an ambipolar characteristic with a hole-dominated transport behaviour, i.e. a much faster current increase in the negative bias region compared to the positive side (inset). The threshold voltage V_{th} of the as-made FET was determined to be −51.6 V for the hole transport by linear extrapolation, which shifts to −36.6 V after the deposition of 20 nm $C_{60}F_{48}$ (red curve), indicating effective hole-doping. The hole concentration can be derived from the equation $N_h = -C_i(V_g - V_{th})/e$, where C_i is the gate capacitance per unit area. Using C_i 1.15×10^{-4} Fm^{-2} for the SiO_2/Si substrate, the hole concentration is estimated to be 2.4×10^{11} cm^{-2} for the intrinsic SL-WSe_2, and increases to 1.2×10^{12} cm^{-2} after $C_{60}F_{48}$ adsorption. These observations all suggest that WSe_2 can be efficiently p-doped by the $C_{60}F_{48}$ layers, with or without the graphite substrate, consistent with theoretical modelling [45].

An intrinsic/p-doped homogeneous lateral junction can form in the SL-WSe_2 with a precise control of the molecular coverage, as shown in Figure 8.11a. Using

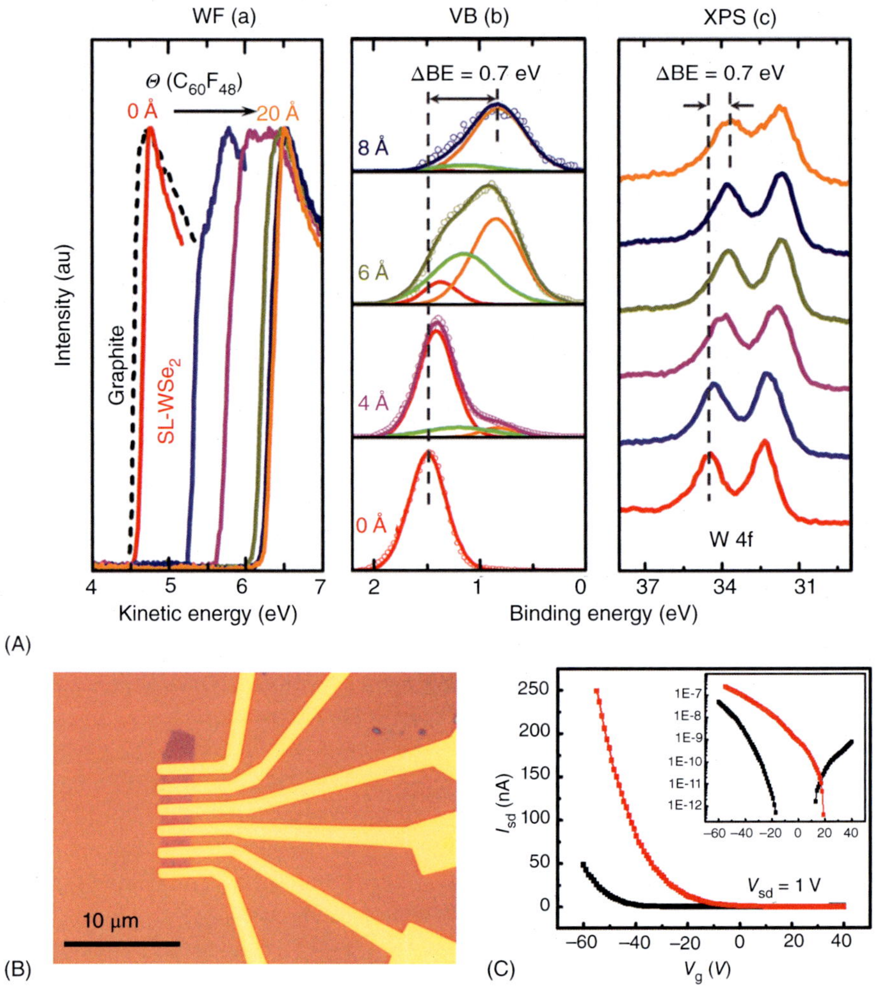

Figure 8.10 Energy levels and transport properties tuned by the molecular dopant. (A) Photoemission spectra of the $C_{60}F_{48}$/SL-WSe_2/graphite heterostructure with the increasing $C_{60}F_{48}$ coverages. (a) The secondary cut-off shifts to higher kinetic energies indicating a work function increasing. (b) Zoom-in of the valence band region after background subtraction and normalisation. Three different contributions with fixed energetic positions are identified by fitting: pristine WSe_2 (red), transition area between the undoped/doped WSe_2 (green) and $C_{60}F_{48}$-doped WSe_2 (orange). (c) A 0.7 eV shift is observed in the W 4f core level, similar to the VB. (B) Optical image of the as-fabricated SL-WSe_2 FET device on 300 nm SiO_2/Si substrate. (C) The transfer curves measured before (black) and after 20 nm $C_{60}F_{48}$ deposition (red) (V_{sd} = 1 V) reveal an efficient p-dope effect. Inset: logarithmic plot of the transfer curve. Source: From Song et al. [45] © 2017 American Chemical Society.

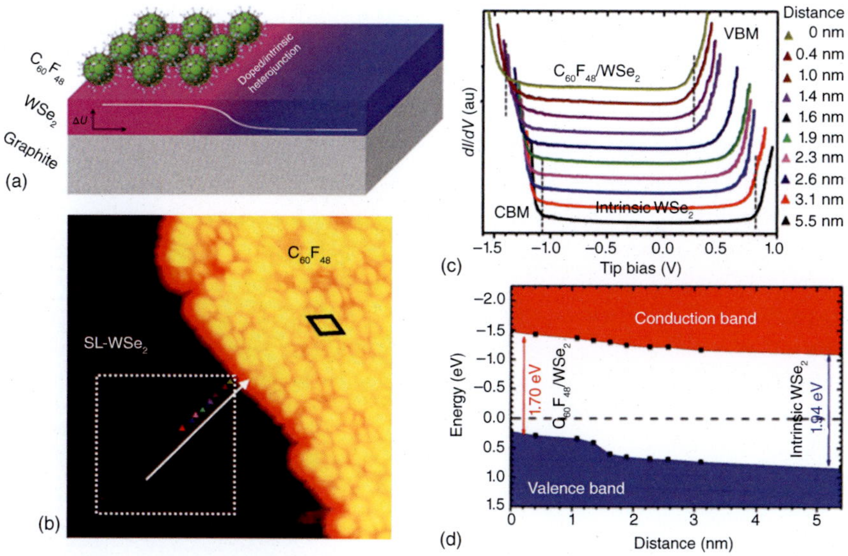

Figure 8.11 Electronic properties of the 1D intrinsic/p-doped interface determined by STM/STS. (a) Schematic model of the lateral undoped/doped heterointerface in the SL-WSe_2. (b) A STM image shows a close-packed $C_{60}F_{48}$ island (0.5 ML) on the SL-WSe_2 surface. (c) The STS spectra recorded in the positions marked by multicoloured triangles in panel (b), reveal the change of the WSe_2 band gap with the distance to the $C_{60}F_{48}$ island, and (d) is the corresponding schematic diagram showing the upward band bending. Source: From Song et al. [45] © 2017 American Chemical Society.

atomically resolved scanning tunnelling microscopy (STM)/scanning tunnelling spectroscopy (STS), we are able to revealed the electronic structure of the atomically sharp 1D interface formed between the undoped (bare) and p-doped (covered with molecule) WSe_2 regions. The STS spectra in Figure 8.11c were obtained at the positions marked by the multicoloured triangles in Figure 8.11b. Obviously, the VBM and the CBM of the SL-WSe_2 vary significantly as the STM tip approached to the $C_{60}F_{48}$ island. The schematic diagram in Figure 8.11d clearly illustrates the gradual upward bending over several nanometres, as a function of the distance to the $C_{60}F_{48}$ island edge. The upward shift of the VBM is ~0.6 eV, consistent with the PES measurements in Figure 8.10A. This transition area with band bending has also been captured in the WSe_2 VB spectra, giving rise to the transient peak (green) in the middle panel of Figure 8.10A. The band bending over several nanometres observed at the lateral heterointerface can be explained by in-plane dielectric screening. Based on nonlinear Thomas–Fermi (TF) theory [92], the change of the potential is a function of distance, $\Delta U \propto e^{-d/L_{TF}}$, where L_{TF} is the TF screening length. By fitting the band bending in Figure 8.11d, the screening length is deduced to be 2.3 ± 0.7 nm from the hole-doped edge, in the intrinsic WSe_2 semiconducting layer. In addition, the carrier density of the intrinsic SL-WSe_2 can be obtained when the screening length is known, using $N_h = kT\varepsilon/e^2L_{TF}^2$. The derived value could reach the range of 10^{11} cm^{-2} at room temperature, which is comparable with that deduced from the transport measurements by the FET device (Figure 8.10B,C) [45].

The understanding of such 1D doped/undoped interface is important for the uses of lateral p–n junctions for future electronic devices.

8.4.3 Screening Effect

The basic concept of electronic screening and its effect on the ELA at the organic-2D TMD heterointerface have been discussed in Section 8.4.1. Here, we present a combined experimental and theoretical study of the heterointerface screening effects between a PTCDA monolayer and a SL-WSe_2 interlayer (Figure 8.12a) on a graphite substrate. Figure 8.12b shows that a sub-monolayer PTCDA assembles into a characteristic herringbone pattern on the SL-WSe_2/graphite surface, which is mainly stabilised by intermolecular hydrogen bonding. Similar herringbone arrangements

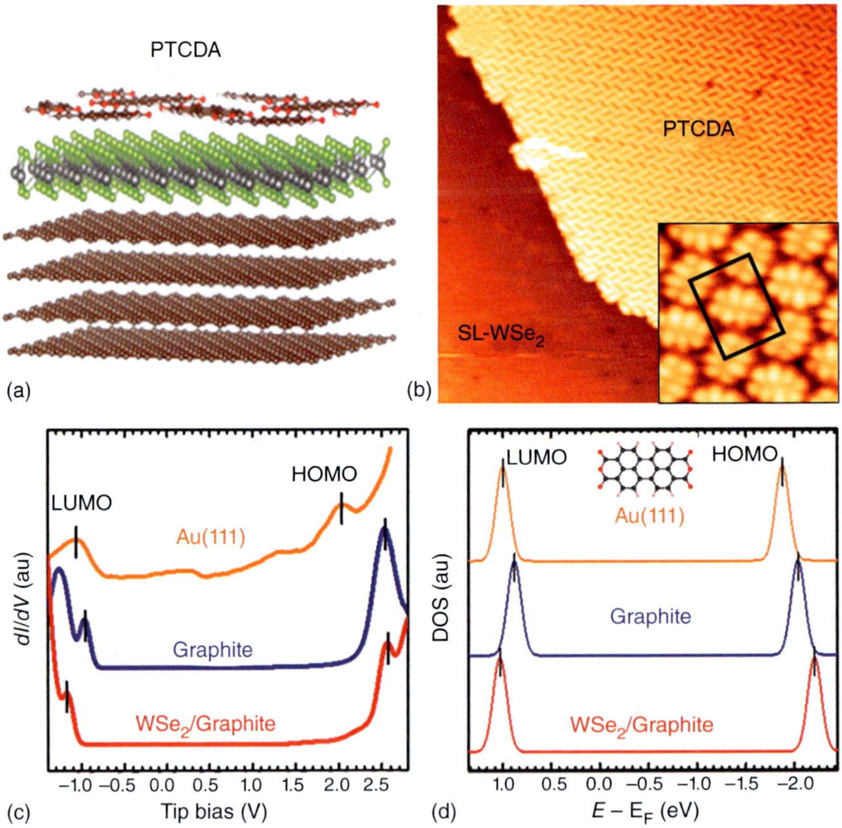

Figure 8.12 Electronic screening at organic/2D TMD interface. (a) Schematic of a PTCDA monolayer on SL-WSe_2/graphite. (b) STM image shows a PTCDA island on SL-WSe_2/graphite. (c) *dI*/*dV* spectra of the PTCDA molecules on Au(111), graphite, and ML-WSe_2/graphite, reveal different HOMO and LUMO positions. (d) Theoretically calculated HOMO and LUMO peaks for PTCDA on the different substrates are consistent well with the experimental measurements. The peaks of the frontier orbitals are artificially broadened with a Gaussian function. Source: From Zheng et al. [44] © 2016 American Chemical Society.

	vdW correction	d (Å)	Z_0 (Å)	Cal. gap (eV)	Exp. gap (eV)
Au(111)	optB86b-vdW	3.35	0.9	2.93	3.10
	PBE-D2	3.09		2.79	
Graphite	optB86b-vdW	3.27	0.7	2.99	3.49
	PBE-D2	3.24		2.97	
		10.18		4.29	
WSe_2/graphite	optB86b-vdW	3.40	0.1	3.29	3.73
	PBE-D2	3.36		3.28	

Figure 8.13 Summary of the calculated PTCDA-substrate distances (d) and HOMO–LUMO gaps for the different substrates based on different calculation methods. The experimental values are given for a better comparison. Particularly, on graphite, $d = 10.18$ Å refers to the case that keeping the molecule and graphite in their original positions in the PTCDA/SL-WSe_2/graphite system while the SL-WSe_2 is removed. z_0 is the computed distance of the image plane from the substrate surface. Source: Zheng et al. [44].

have been reported for the adsorption of PTCDA molecules on other chemically inert surfaces, including Au(111) [93] and graphite [94]. The relatively weak molecule–substrate interactions can be further supported by DFT calculations [44]. As summarised in Figure 8.13, the calculated molecule–substrate distances are all in the range of 3.0–3.4 Å for Au(111), graphite and SL-WSe_2/graphite surfaces, indicating that the PTCDA layered are bound to the underlying substrates mainly by vdW forces.

For these three substrates considered here, however, the HOMO–LUMO gap of PTCDA shows a clear substrate-dependent behaviour. For instance, the PTCDA HOMO–LUMO gap on SL-WSe_2/graphite is determined to be 3.73 eV by STS spectra (Figure 8.12c), which is much smaller than the theoretically predicted value of 5.0 eV for an isolated PTCDA layer [44]. The HOMO–LUMO gap further reduces when the PTCDA molecules adsorb on the semi-metallic graphite substrate (3.49 eV), and on the metallic Au substrate (3.10 eV) (Figure 8.12c,d). Due to the absent of interfacial charge transfer and chemical bonding, these decreases in HOMO–LUMO gaps can be explained only after electronic screening effects from the substrate are taken into account explicitly. More important, the WSe_2 monolayer is not an inert spacer material but participates substantially in the electronic screening. Keeping the molecule and graphite in their original positions in the PTCDA/SL-WSe_2/graphite system while removing the SL-WSe_2 interlayer, i.e. $d = 10.18$ Å on graphite, the HOMO–LUMO gap is predicted to be 4.29 eV, which is of ~1 eV higher than that with the SL-WSe_2 interlayer. Therefore, it suggests that the monolayer TMDs provide substantial – but not complete – screening on the organic–inorganic heterointerface.

These findings highlight the importance of interface screening in determining the final ELAs of hybrid vdW heterostructures. In real applications, the organic-2D TMD heterostructures need to be integrated with dielectric substrates/gates, metal electrodes, and other functional layers. The ELA at the organic-2D TMD heterointerface will be further complicated by the influence from the other component layers, which cannot be screened out due to its atomic thickness.

8.5 Applications in Devices: Organic-2D TMD p–n Heterojunctions

Many recent studies have revealed the great potential of organic-2D TMD hybrid structures for future device applications [20, 70, 95–99]. In particular, p–n junctions with type II alignment have been constructed from the integration of organic molecules with 2D TMD semiconductors, and demonstrated promising properties for next generation flexible devices. The sheer number of available organics provide a wide range of choice and greatly expand the possibilities of the nanostructures. The ultimate goal is to fabricate hybrid organics-2D TMD flexible nanodevices with desired optical and electrical properties, which benefit from both advantages, e.g. the relatively high carrier transport mobility of 2D TMDs and the high optical absorption efficiency of organics [19].

Figure 8.14 demonstrates an elegant organic/2D TMD photovoltaic device fabricated from pentacene and MoS_2 [98]. The formation of the type-II heterojunction is illustrated in Figure 8.14a–c, where the mechanically exfoliated MoS_2 (2-layer) flake was used as the n-type material, and the thermally deposited pentacene (40 nm) film was the p-type component. The built p–n heterojunction shows a rectifying $I_D - V_D$ (output) behaviour at different gate voltage V_G, resulting from the type-II alignment [98]. The transfer I_D–V_G curve in Figure 8.14d further reveals an anti-ambipolar response, where a highly rectifying output current was obtained in an intermediate V_G range, from ~−70 to ~40 V. The anti-ambipolar asymmetry of the pentacene–MoS_2 heterojunction indicates different transconductance in each side, which can be controlled by the ratio of the channel resistances of the two semiconductors. More important, the pentacene–MoS_2 heterojunction also exhibits interesting photovoltaic effect upon optical illumination. In Figure 8.14e, an open circuit voltage V_{OC} as large as ~0.3 V and a short circuit current I_{SC} of ~3 nA was observed at $V_G = 0$ [98]. This V_{OC} value is comparable to that for the optimised pentacene–C_{60} heterojunction reported in the literature ($V_{OC} \leq 0.4$ V) [100, 101], indicating such organic-2D TMD hybrid structures are very promising for applications in solar cells with careful design and optimisation.

Similar transfer behaviours and photovoltaic effects have been observed in various organic-2D TMD hybrid heterostructures. Other organic materials that have been used include dioctylbenzothienobenzothiophene (C8-BTBT) [102], C_{60} [103], rubrene single crystal [104], and so on, which all exhibit excellent rectifying behaviours as well as strong photovoltaic responses after being combined with various 2D TMDs to form p–n heterojunctions. Compared to 2D–2D p–n heterojunctions such as MoS_2–WSe_2 junction [69, 105–108], the organic/2D-TMD heterojunctions exhibit better rectifying ratio and larger open circuit voltages. While these type II organic/2D-TMD heterojunctions have exhibited many advantages

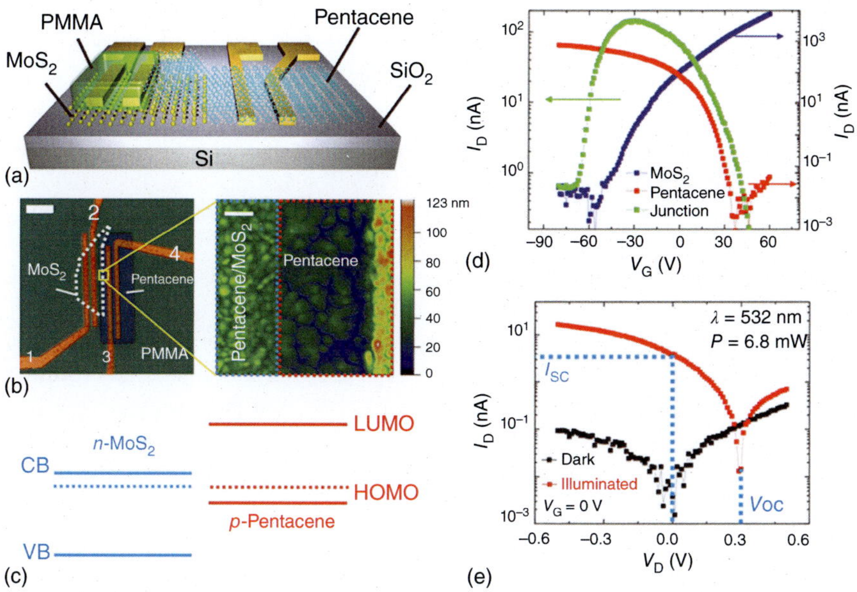

Figure 8.14 A pentacene/MoS_2 photovoltaic device. (a) Schematic structure of the pentacene/MoS_2 p–n heterojunction. (b) The optical image of the device (left) and the corresponding atomic force microscopy (AFM) topography image (right). (c) The pentacene/MoS_2 heterojunction with a type-II alignment. (d) Semilog transfer curves of the pentacene (red, $V_D = 10$ V), MoS_2 (blue, $V_D = 1$ V) and pentacene/MoS_2 FETs (green, $V_D = 10$ V). The pentacene/MoS_2 p–n junction exhibit an asymmetric antiambipolar response. (e) Semilog I_D–V_D curves for the p–n junction ($V_G = 0$ V) without (black) and with illumination (red) reveal a photovoltaic effect. Source: From Jariwala et al. [98] © 2016 American Chemical Society.

for applications in photovoltaics and other optoelectronic devices, further effort is needed to match their performance of industrial bulk photovoltaic devices.

8.6 Conclusion

In this chapter, we provide an overview of the current understanding of the organic/2D-TMD hybrid structures and interfaces. According to the interactions between organic molecules and 2D TMD layers, the fabricated structures can be divided into two categories: non-covalent (vdW) and covalent heterostructures. The vdW forces usually predominate in the epitaxial self-assembly of organic molecules on pristine semiconducting 2D TMDs; while covalently binding interactions are usually observed with the participations of TMD surface defects or metastable metallic phases. The fundamental electronic properties of the interfaces are discussed in detail, where special attention is focused on the two key factors that determine the ELAs: interfacial charge transfer and electronic screening effects. The integration of organics with 2D TMD crystals can benefit from the advantages in both materials, and even enable better functionalities that are not present in either material alone. It is also a promising method to improve the physical properties of TMDs. Device applications based on organic/2D-TMD p–n heterojunctions are highlighted, which show promising and interesting device characteristics such as excellent rectifying behaviours and strong photovoltaic responses.

The fields of organic-2D hybrid nanostructure and devices have immense potential for further developments. The low dimensional organic molecules and 2D materials brings new perspectives and challenges to the well-established interface chemistry and physics. As discussed in Sections 8.3 and 8.4, a clear fundamental understanding of the effects of chemical bond formation, charge transfer effects and electronic screening on the ELAs at the 2D interfaces is still lacking at present. More extensive studies are required for the exploration on the interfacial properties at atomic-scale. Moreover, the fundamental science and technology developed for the organic/2D TMD heterointerfaces can be utilised more generally to other 2D materials including black phosphorus (BP) [109–114] and X-enes [115]. For instance, many recent research works have revealed that the organic functionalisation of BP is also a promising approach to passivate, dope and even modify the electronic band structure of BP [109–114]. Finally, other developing properties that have not been covered in this review include enhanced photoluminescence properties [29], photocurrent response [69, 70], Raman intensities [116–118], as well as valleytronics phenomena at the interfaces, where significant breakthroughs are possible with the combination of various advantages of both organics and 2D materials. We believe that the field of organic/2D hybrid structures is very much in its infancy, and will progress rapidly in the coming years.

Acknowledgements

The authors acknowledge support from MOE grant R-144-000-321-112 and A-STAR SERC grant for the 2D growth project under the 2D pharos program (SERC 1527000012).

References

1 Novoselov, K.S., Jiang, D., Schedin, F. et al. (2005). *Proc. Natl. Acad. Sci. U.S.A.* 102: 10451–10453.
2 Tan, C., Cao, X., Wu, X.J. et al. (2017). *Chem. Rev.* 117: 6225–6331.
3 Bhimanapati, G.R., Lin, Z., Meunier, V. et al. (2015). *ACS Nano* 9: 11509–11539.
4 Novoselov, K.S., Geim, A.K., Morozov, S.V. et al. (2004). *Science* 306: 666–669.
5 Geim, A.K. (2009). *Science* 324: 1530–1534.
6 Castro Neto, A.H., Guinea, F., Peres, N.M.R. et al. (2009). *Rev. Mod. Phys.* 81: 109–162.
7 Elias, D.C., Gorbachev, R.V., Mayorov, A.S. et al. (2011). *Nat. Phys.* 7: 701–704.
8 Mayorov, A.S., Gorbachev, R.V., Morozov, S.V. et al. (2011). *Nano Lett.* 11: 2396–2399.
9 Schwierz, F. (2010). *Nat. Nanotechnol.* 5: 487–496.

10 Butler, S.Z., Hollen, S.M., Cao, L.Y. et al. (2013). *ACS Nano* 7: 2898–2926.

11 Liu, G.B., Xiao, D., Yao, Y.G. et al. (2015). *Chem. Soc. Rev.* 44: 2643–2663.

12 Wang, Q.H., Kalantar-Zadeh, K., Kis, A. et al. (2012). *Nat. Nanotechnol.* 7: 699–712.

13 Wang, H.T., Yuan, H.T., Hong, S.S. et al. (2015). *Chem. Soc. Rev.* 44: 2664–2680.

14 Muccini, M. (2006). *Nat. Mater.* 5: 605–613.

15 Sirringhaus, H. (2014). *Adv. Mater.* 26: 1319–1335.

16 Podzorov, V., Menard, E., Borissov, A. et al. (2004). *Phys. Rev. Lett.* 93: 086602.

17 Sundar, V.C., Zaumseil, J., Podzorov, V. et al. (2004). *Science* 303: 1644–1646.

18 Jariwala, D., Marks, T.J., and Hersam, M.C. (2017). *Nat. Mater.* 16: 170–181.

19 Huang, Y.L., Zheng, Y.J., Song, Z. et al. (2018). *Chem. Soc. Rev.* 47: 3241–3264.

20 Huang, Y., Zheng, W., Qiu, Y., and Hu, P. (2016). *ACS Appl. Mater. Interfaces* 8: 23362–23370.

21 Park, J.H., Sanne, A., Guo, Y.Z. et al. (2017). *Sci. Adv.* 3: e1701661.

22 Yu, Z., Pan, Y., Shen, Y. et al. (2014). *Nat. Commun.* 5: 5290.

23 Cho, K., Min, M., Kim, T. et al. (2015). *ACS Nano* 9: 8044–8053.

24 Ding, Q., Czech, K.J., Zhao, Y. et al. (2017). *ACS Appl. Mater. Interfaces* 9: 12734–12742.

25 Cai, Y., Zhou, H., Zhang, G., and Zhang, Y.-W. (2016). *Chem. Mater.* 28: 8611–8621.

26 Jing, Y., Tan, X., Zhou, Z., and Shen, P.W. (2014). *J. Mater. Chem. A* 2: 16892–16897.

27 Du, Y.C., Liu, H., Neal, A.T. et al. (2013). *IEEE Electron Device Lett.* 34: 1328–1330.

28 Kiriya, D., Tosun, M., Zhao, P.D. et al. (2014). *J. Am. Chem. Soc.* 136: 7853–7856.

29 Amani, M., Lien, D.H., Kiriya, D. et al. (2015). *Science* 350: 1065–1068.

30 Han, H.V., Lu, A.Y., Lu, L.S. et al. (2016). *ACS Nano* 10: 1454–1461.

31 Luo, C.Y., Huang, W.Q., Hu, W. et al. (2016). *Dalton Trans.* 45: 13383–13391.

32 Mouri, S., Miyauchi, Y., and Matsuda, K. (2013). *Nano Lett.* 13: 5944–5948.

33 Zhou, J., Lin, J., Huang, X. et al. (2018). *Nature* 556: 355–359.

34 Eda, G., Fujita, T., Yamaguchi, H. et al. (2012). *ACS Nano* 6: 7311–7317.

35 Chhowalla, M., Shin, H.S., Eda, G. et al. (2013). *Nat. Chem.* 5: 263–275.

36 Wilson, J.A. and Yoffe, A.D. (1969). *Adv. Phys.* 18: 193–335.

37 Kuc, A. and Heine, T. (2015). *Chem. Soc. Rev.* 44: 2603–2614.

38 Sorkin, V., Pan, H., Shi, H. et al. (2014). *Crit. Rev. Solid State Mater. Sci.* 39: 319–367.

39 Lv, R., Robinson, J.A., Schaak, R.E. et al. (2015). *Acc. Chem. Res.* 48: 56–64.

40 Wee, A.T.S. and Chen, W. (2012). *Phys. Scr.* T146: 014007.

41 Cui, X., Troadec, C., Wee, A.T.S., and Huang, Y.L. (2018). *ACS Omega* 3: 3285–3293.

42 Matkovic, A., Genser, J., Luftner, D. et al. (2016). *Sci. Rep.* 6: 38519.

43 Kumar, A., Banerjee, K., and Liljeroth, P. (2017). *Nanotechnology* 28: 082001.

44 Zheng, Y.J., Huang, Y.L., Chenp, Y. et al. (2016). *ACS Nano* 10: 2476–2484.

45 Song, Z., Schultz, T., Ding, Z. et al. (2017). *ACS Nano* 11: 9128–9135.

46 Voiry, D., Goswami, A., Kappera, R. et al. (2015). *Nat. Chem.* 7: 45–49.
47 Lin, X., Lu, J.C., Shao, Y. et al. (2017). *Nat. Mater.* 16: 717–721.
48 Wang, Y.L., Li, L.F., Yao, W. et al. (2015). *Nano Lett.* 15: 4013–4018.
49 Shen, N. and Tao, G. (2017). *Adv. Mater. Interf.* 4: 1601083.
50 Tuxen, A., Kibsgaard, J., Gobel, H. et al. (2010). *ACS Nano* 4: 4677–4682.
51 Helveg, S., Lauritsen, J.V., Lægsgaard, E. et al. (2000). *Phys. Rev. Lett.* 84: 951–954.
52 Lauritsen, J.V., Kibsgaard, J., Helveg, S. et al. (2007). *Nat. Nanotechnol.* 2: 53–58.
53 Chou, S.S., De, M., Kim, J. et al. (2013). *J. Am. Chem. Soc.* 135: 4584–4587.
54 Kim, J.S., Yoo, H.W., Choi, H.O., and Jung, H.T. (2014). *Nano Lett.* 14: 5941–5947.
55 Nguyen, E.P., Carey, B.J., Ou, J.Z. et al. (2015). *Adv. Mater.* 27: 6225–6229.
56 Sim, D.M., Kim, M., Yim, S. et al. (2015). *ACS Nano* 9: 12115–12123.
57 Makarova, M., Okawa, Y., and Aono, M. (2012). *J. Phys. Chem. C* 116: 22411–22416.
58 Li, Q., Zhao, Y., Ling, C. et al. (2017). *Angew. Chem. Int. Ed.* 56: 10501–10505.
59 Chou, S.S., Huang, Y.K., Kim, J. et al. (2015). *J. Am. Chem. Soc.* 137: 1742–1745.
60 Chen, X. and McDonald, A.R. (2016). *Adv. Mater.* 28: 5738–5746.
61 Presolski, S., Wang, L., Loo, A.H. et al. (2017). *Chem. Mater.* 29: 2066–2073.
62 Vishnoi, P., Sampath, A., Waghmare, U.V., and Rao, C.N.R. (2017). *Chem. Eur. J.* 23: 886–895.
63 Knirsch, K.C., Berner, N.C., Nerl, H.C. et al. (2015). *ACS Nano* 9: 6018–6030.
64 Sim, D.M., Han, H.J., Yim, S. et al. (2017). *ACS Omega* 2: 4678–4687.
65 Benson, E.E., Zhang, H., Schuman, S.A. et al. (2018). *J. Am. Chem. Soc.* 140: 441–450.
66 Kroemer, H. (2000). *Nobel Lectures, Physics 1996–2000*, 449–469. Singapore: World Scientific Publishing Co. Pte. Ltd.
67 Chen, W., Qi, D.-C., Huang, H. et al. (2011). *Adv. Funct. Mater.* 21: 410–424.
68 Opitz, A. (2017). *J. Phys. Condens. Matter* 29: 133001.
69 Bettis Homan, S., Sangwan, V.K., Balla, I. et al. (2017). *Nano Lett.* 17: 164–169.
70 Liu, X., Gu, J., Ding, K. et al. (2017). *Nano Lett.* 17: 3176–3181.
71 Quek, S.Y. and Khoo, K.H. (2014). *Acc. Chem. Res.* 47: 3250–3257.
72 Ishii, H., Sugiyama, i., Ito, E., and Seki, K. (1999). *Adv. Mater.* 11: 605–625.
73 Neaton, J.B., Hybertsen, M.S., and Louie, S.G. (2006). *Phys. Rev. Lett.* 97: 216405.
74 Egger, D.A., Liu, Z.F., Neaton, J.B., and Kronik, L. (2015). *Nano Lett.* 15: 2448–2455.
75 Garcia-Lastra, J.M., Rostgaard, C., Rubio, A., and Thygesen, K.S. (2009). *Phys. Rev. B* 80: 245427.
76 Chen, Y.F., Tamblyn, I., and Quek, S.Y. (2017). *J. Phys. Chem. C* 121: 13125–13134.
77 Sau, J.D., Neaton, J.B., Choi, H.J. et al. (2008). *Phys. Rev. Lett.* 101: 026804.
78 Ugeda, M.M., Bradley, A.J., Shi, S.-F. et al. (2014). *Nat. Mater.* 13: 1091–1095.
79 Wang, L., Wang, Z., Wang, H.Y. et al. (2017). *Nat. Commun.* 8: 13906.

80 Chen, W., Qi, D., Gao, X., and Wee, A.T.S. (2009). *Prog. Surf. Sci.* 84: 279–321.
81 Chen, W., Chen, S., Qi, D.C. et al. (2007). *J. Am. Chem. Soc.* 129: 10418–10422.
82 Mao, H.Y., Lu, Y.H., Lin, J.D. et al. (2013). *Prog. Surf. Sci.* 88: 132–159.
83 Lu, W.C.Y.H. and Feng, Y.P. (2009). *J. Phys. Chem. B* 113: 2–5.
84 Garnica, M., Stradi, D., Barja, S. et al. (2013). *Nat. Phys.* 9: 368–374.
85 He, D.W., Pan, Y.M., Nan, H.Y. et al. (2015). *Appl. Phys. Lett.* 107: 5.
86 Yu, W.J., Liao, L., Chae, S.H. et al. (2011). *Nano Lett.* 11: 4759–4763.
87 Voggu, R., Das, B., Rout, C.S., and Rao, C.N.R. (2008). *J. Phys. Condens. Matter* 20: 472204.
88 Song, Z., Wang, Q., Li, M.-Y. et al. (2018). *Phys. Rev. B* 97: 134102.
89 Huang, Y.L., Ding, Z.J., Zhang, W.J. et al. (2016). *Nano Lett.* 16: 3682–3688.
90 Yokoyama, T., Takahashi, T., Shinozaki, K., and Okamoto, M. (2007). *Phys. Rev. Lett.* 98: 206102.
91 Renard, J., Lundeberg, M.B., Folk, J.A., and Pennec, Y. (2011). *Phys. Rev. Lett.* 106: 156101.
92 Resta, R. (1977). *Phys. Rev. B* 16: 2717–2722.
93 Nicoara, N., Roman, E., Gomez-Rodriguez, J.M. et al. (2006). *Org. Electron.* 7: 287–294.
94 Ludwig, C., Gompf, B., Petersen, J. et al. (1994). *Z. Phys. B: Condens. Matter* 93: 365–373.
95 Ren, Q., Xu, Q., Xia, H. et al. (2017). *Org. Electron.* 51: 142–148.
96 Velusamy, D.B., Haque, M.A., Parida, M.R. et al. (2017). *Adv. Funct. Mater.* 27: 1605554.
97 Shastry, T.A., Balla, I., Bergeron, H. et al. (2016). *ACS Nano* 10: 10573–10579.
98 Jariwala, D., Howell, S.L., Chen, K.S. et al. (2016). *Nano Lett.* 16: 497–503.
99 Li, H.M., Lee, D., Qu, D. et al. (2015). *Nat. Commun.* 6: 6564.
100 Yang, J. and Nguyen, T.-Q. (2007). *Org. Electron.* 8: 566–574.
101 Cheyns, D., Gommans, H., Odijk, M. et al. (2007). *Sol. Energy Mater. Sol. Cells* 91: 399–404.
102 He, D., Pan, Y., Nan, H. et al. (2015). *Appl. Phys. Lett.* 107: 183103.
103 Chen, R., Lin, C., Yu, H. et al. (2016). *Chem. Mater.* 28: 4300–4306.
104 Liu, F., Chow, W.L., He, X. et al. (2015). *Adv. Funct. Mater.* 25: 5865–5871.
105 Lee, C.H., Lee, G.H., van der Zande, A.M. et al. (2014). *Nat. Nanotechnol.* 9: 676–681.
106 Furchi, M.M., Pospischil, A., Libisch, F. et al. (2014). *Nano Lett.* 14: 4785–4791.
107 Ceballos, F., Bellus, M.Z., Chiu, H.-Y., and Zhao, H. (2014). *ACS Nano* 8: 12717–12724.
108 Hong, X., Kim, J., Shi, S.F. et al. (2014). *Nat. Nanotechnol.* 9: 682–686.
109 Ryder, C.R., Wood, J.D., Wells, S.A. et al. (2016). *Nat. Chem.* 8: 597–602.
110 Sofer, Z., Luxa, J., Bousa, D. et al. (2017). *Angew. Chem. Int. Ed.* 56: 9891–9896.
111 Cao, Y., Tian, X., Gu, J. et al. (2018). *Angew. Chem. Int. Ed.* 57: 4543–4548.
112 He, Y., Xia, F., Shao, Z. et al. (2015). *J. Phys. Chem. Lett.* 6: 4701–4710.
113 Wang, C., Niu, D., Liu, B. et al. (2017). *J. Phys. Chem. C* 121: 18084–18094.
114 Zhao, Y., Zhou, Q., Li, Q. et al. (2017). *Adv. Mater.* 29: 1603990.
115 Liu, X., Wei, Z., Balla, I. et al. (2017). *Sci. Adv.* 3: e1602356.

116 Ling, X., Fang, W., Lee, Y.H. et al. (2014). *Nano Lett.* 14: 3033–3040.
117 Tan, Y., Ma, L., Gao, Z. et al. (2017). *Nano Lett.* 17: 2621–2626.
118 Lee, Y., Kim, H., Lee, J. et al. (2016). *Chem. Mater.* 28: 180–187.

9

Surface Self-Assembly of Hydrogen-Bonded Frameworks

Nicholas Pearce and Neil R. Champness

University of Birmingham, School of Chemistry, Edgbaston, Birmingham B15 2TT, UK

9.1 Introduction

There is an extensive history of employing hydrogen-bonding interactions to create two-dimensional self-assembled frameworks on surfaces [1–3]. Success in this endeavour results from the rich diversity of chemistry, using hydrogen bonds that has been developed across many fields, in particular, supramolecular chemistry [4]. Hydrogen bonds have been used to develop many complex supramolecular structures, including capsules and cages [5], polymeric and oligomeric constructs [6, 7], and interlocked arrangements [8]. In the solid-state, a plethora of two- and three-dimensional structures have been fabricated using hydrogen-bonding interactions [9]. Many of these structures have been inspired by nature, whether from important biological constructs, e.g. DNA [10] or proteins [11] or from naturally occurring materials as simple as ice [12]. In this chapter, the application of hydrogen bonding to the formation of self-assembled frameworks on surfaces is discussed. We will illustrate that it is possible to design framework structures using the knowledge of reliable and robust hydrogen-bonding interactions and also illustrate the complexity of the design that leads to unexpected and exciting directions of research.

There are many and varied approaches to designing molecules that contain hydrogen-bonding groups and to designing suitable supramolecular synthons for the construction of self-assembled framework structures. Supramolecular synthons have been defined and described as 'spatial arrangements of intermolecular interactions' that 'play the same focusing role in supramolecular synthesis that conventional synthons do in molecular synthesis' [13], a principle that facilitates a design strategy that can be adapted for making complex structures. The supramolecular synthon approach is particularly useful in the creation of hydrogen-bonded frameworks on surfaces and has been used to create many diverse molecular architectures.

Supramolecular Chemistry on Surfaces: 2D Networks and 2D Structures, First Edition.
Edited by Neil R. Champness.

9.2 Carboxylic Acid Supramolecular Synthons

Perhaps the simplest supramolecular synthon that employs hydrogen bonds is the carboxylic acid homodimer, which adopts mutual hydrogen bonds between two carboxylic acids, each containing both a hydrogen-bond donor (OH) and a hydrogen-bond acceptor (C=O). The compatibly between two carboxylic acid groups leads to the formation of the classic $R_2^2(8)$ intermolecular arrangement (Figure 9.1a) [15]. The use of this interaction to create extended surface-based frameworks goes back to the origins of the field with studies of trimesic acid on highly-oriented pyrolytic graphite (HOPG) substrate by Griessl et al. [14]. Imaging

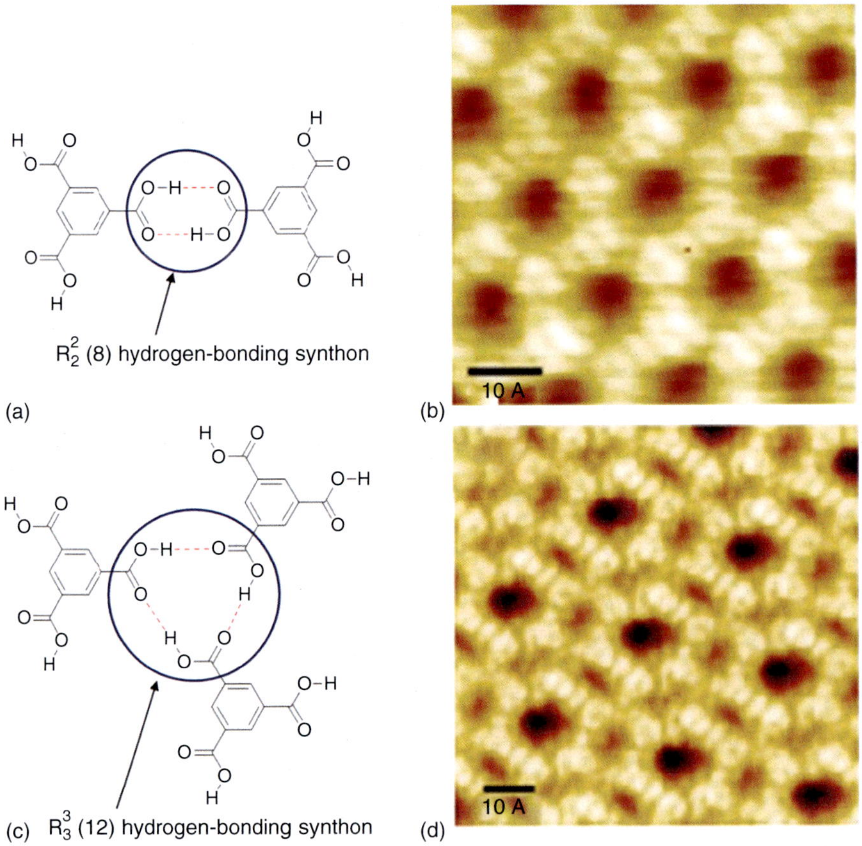

Figure 9.1 (a) and (c) Schematic representations of trimesic acid hydrogen bonding synthons: (a) the $R_2^2(8)$ hydrogen-bonding synthon (highlighted by the blue ring) found in the hexagonal 'chicken wire' arrangement and (c) the $R_3^3(12)$ hydrogen-bonding synthon (highlighted by the blue ring) found in the 'flower' arrangement. (b) and (d) STM images of networks of trimesic acid on HOPG showing (b) the open 'chicken-wire' network formed by using the $R_2^2(8)$ hydrogen-bonding synthon [14] and (d) the 'flower' arrangement formed by the same molecule using $R_3^3(12)$ hydrogen-bonding synthons [14]. Source: (b) Griessl et al. [14]/ with permission from John Wiley & Sons, Inc.

using scanning tunnelling microscopy (STM) reveals details of the self-assembled molecular structure (Figure 9.1b,c). Two open structures, in preference to a close-packed arrangement, are observed: a 'chicken-wire' or honeycomb arrangement (Figure 9.1b) and a secondary arrangement called a 'flower' structure by the authors (Figure 9.1c). The latter structure arises due to the formation of $R_3^3(12)$ supramolecular synthons formed by three carboxylic acid moieties from three separate trimesic acid molecules (Figure 9.1d).

Subsequently, a study of the trimesic acid analogue 1,3,5-tris(4-carboxyphenyl) benzene showed the formation of a honeycomb framework when studied at low temperatures, but at higher temperatures, a more densely packed phase is observed [16]. In contrast to the trimesic acid study, which was investigated under ultra-high vacuum (UHV) conditions, the studies of the larger analogue, 1,3,5-tris(4-carboxyphenyl)benzene, were conducted using solvent deposition from a range of aliphatic acids onto the HOPG substrate. The study evaluated the conditions that influence the phase transition, concluding that both concentration of the tricarboxylic acid and nature of the solvent used for deposition play roles in determining which phase is formed. It was suggested that co-adsorption of solvent molecules at low temperatures plays an important role in stabilising the more open honeycomb structure.

The carboxylic acid hydrogen-bonded homodimer can be used as a supramolecular synthon in a wide variety of molecules, and due to the ability to image the resulting structures at both the molecular and sub-molecular levels [17, 18], it is possible to observe some unique examples of self-assembled supramolecular structures. For example, it is possible to study tetracarboxylic acid molecules, which comprise isophthalic acid (*meta*-benzene dicarboxylic acid) groups at either end of a rigid backbone. One such example is the study of the self-assembly of *p*-terphenyl-3,3″,5,5″-tetracarboxylic acid (TPTC) [19]. Once adsorbed onto a HOPG substrate, TPTC forms a two-dimensional hydrogen-bonded structure that, as with trimesic acid discussed above, uses the $R_2^2(8)$ intermolecular carboxylic acid–carboxylic acid synthon. STM imaging allows direct visualisation of the resulting structure, which is best described as non-ordered. In a self-assembled monolayer of TPTC, the relative positions of molecules within the array are random and similar in structure to a dynamically arrested system, such as a glass (Figure 9.2a). This behaviour arises as a result of the dimensions of the molecule; the length between the centres of the terminal phenyl groups of the molecule and the separation between the centres of the same groups separated by the $R_2^2(8)$ hydrogen-bonded synthon are highly similar. The structure forms an extremely rare example of a random, entropically stabilised rhombus tiling in which intermolecular hydrogen bonding leads to the adoption of hexagonal junctions formed from three, four, five, or six molecules (Figure 9.2b). Due to the molecular level imaging of the STM experiment, it is possible to determine the degree of randomness of the resulting rhombus tiling through a detailed analysis of each STM image. It was found that both the solvent used for deposition and the temperature of the experiment affect the degree of random ordering observed in the tiling structure [22]. The importance of the molecular dimensions and rhomboidal shape on the formation of the

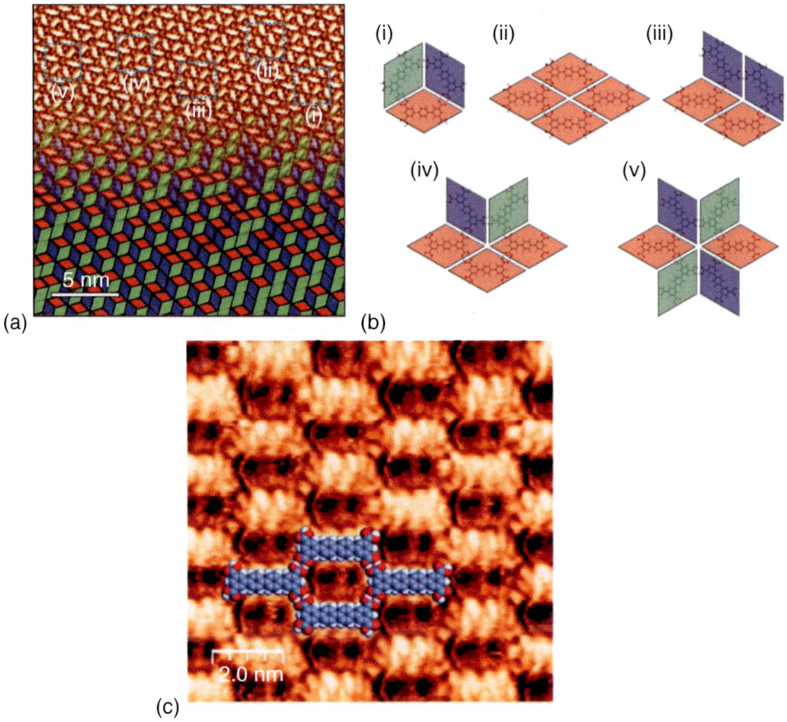

Figure 9.2 (a) STM image of a typical area of TPTC network at the nonanoic acid/HOPG interface. The group of three phenyl rings of the molecule backbone appear as bright features in the image. In the lower half of the image, each molecule has been represented using a rhombus to represent the tiling arrangement observed for this system; (b) the five different arrangements (i–v) of TPTC around a network pore with accompanying rhombus tiles representing the orientation of each component. The locations of the magnified regions are marked in (a) by blue dashed squares; (c) STM image of QPTC adsorbed on HOPG. Showing the 'parallel' arrangement adopted by this longer molecule in contrast to TPTC. Source: (a) Steed (2013) [20] / with permission from John Wiley & Sons, Inc. (c) Blunt et al. [21] / with permission from Royal Society of Chemistry.

random tiling structure is demonstrated by structures of related molecules such as *p*-quaterphenyl-3,3‴,5,5‴-tetracarboxylic acid (QPTC), which is longer and forms regular 'parallel' two-dimensional structures (Figure 9.2c) [21].

Interestingly, the random rhombus tiling framework formed by TPTC on HOPG contains pores that are able to act as hosts to trap guest C_{60} molecules (Figure 9.3) [23]. C_{60} was deposited onto the already formed rhombus tiling array, which led to preferential adsorption into one of the five possible pores (Figure 9.3a, described above). The STM images reveal a strong preference for adsorption of C_{60} into pores of type A (Figure 9.3b), which make up around 40% of the pores in the framework but approximately 76% of pores that trap a C_{60} molecule. This observation was supported by calculations that confirm that adsorption in pore type A is favoured because its boundary contains the highest proportion of phenyl

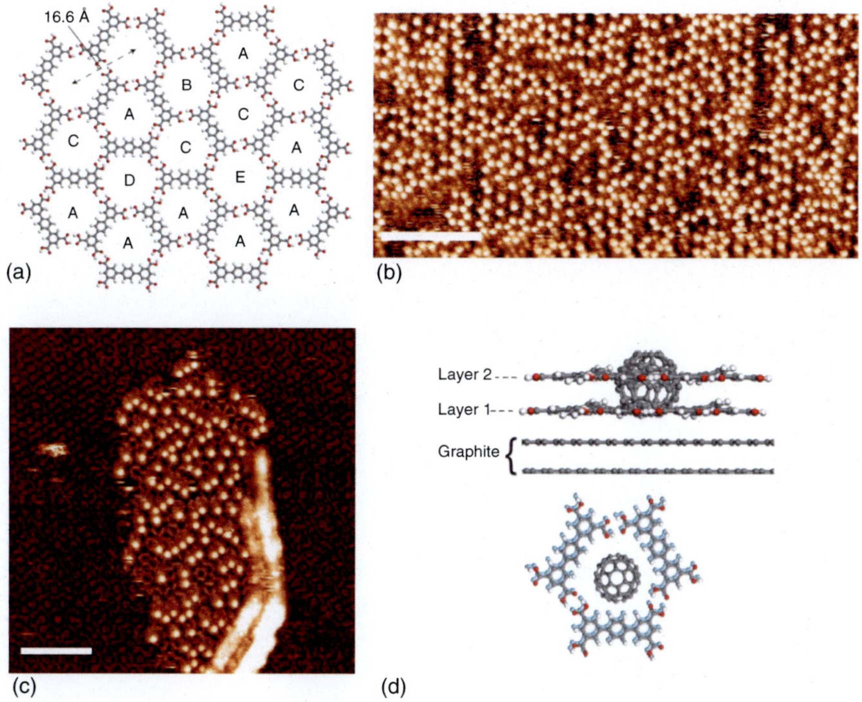

Figure 9.3 (a) Schematic representation of a section of the random rhombus tiling adopted by the TPTC framework that highlights the hexagonally ordered network of pores, A–E; (b) STM image of an area of TPTC framework c. 24 hours after deposition of C_{60}. The locations of C_{60} are clearly visible as the bright spots in the image; the underlying TPTC framework structure is not visible. Scale bar = 160 Å; (c) STM image of TPTC framework immediately after C_{60} deposition. An island of C_{60} and bilayer TPTC framework grows away from a surface defect. The initial layer of TPTC framework is visible with an altered contrast and the TPTC molecules in the second layer appear with the long axis of the molecules as bright, rod-like features. Scale bar = 110 Å. (d) Side-view of the C_{60}–bilayer network that consists of two overlying pores of type A and a view parallel to the surface plane with the C_{60} placed at its minimum-energy position for both the first- (light blue) and second-layer pores (grey). The two layers are displaced slightly with respect to each other to aid clarity, but are expected based on calculations. Source: (b,c) Blunt et al. [23] / with permission from Nature Publishing Group.

ring edges which interact more strongly with the guest C_{60} molecules. Perhaps the most surprising observation of the network is that in areas where adsorption of C_{60} molecules is seen, a second TPTC framework layer assembles and sits over the initial layer resulting in the formation of a bilayer structure (Figure 9.3c). The observation of the growth of a secondary layer had not been reported previously for such arrays, and it was suggested that the presence of the spherical C_{60} molecule allowed anchoring and hence the formation of the second layer. The significance of the trapped C_{60} molecules for the formation of the second rhombus tiling layer is shown by the addition of coronene to the bilayer samples. The entrapped C_{60} was displaced by the flat coronene molecules, which are more strongly adsorbed on the HOPG substrate, resulting in the disassembly of the second supramolecular layer, which is no longer anchored by the protruding C_{60} molecules.

The influence of even small amounts of co-adsorbents during the self-assembly process can have a dramatic effect on the framework formed. Such influence is observed in the case of the random rhombus tiling seen with TPTC [24]. When a small amount of QPTC is present in the deposited TPTC sample, an exclusively parallel self-assembled phase results. The QPTC molecule is known to form solely the parallel phase [21], as discussed above (Figure 9.2c), and is shown to template the same framework arrangement for TPTC (Figure 9.4). Indeed, commercial samples of TPTC were found to contain trace amounts of both QPTC and the longer *p*-quinquephenyl-3,3‴,5,5‴-tetracarboxylic acid (QQPTC), both of which favour the formation of the ordered parallel arrangement.

The random rhombus tiling of TPTC is related to the mathematical concept of Penrose tiles [25]. Penrose tiles do not exhibit translational symmetry, like the TPTC rhombus tiling but do exhibit rotational or reflectional symmetry. The first observation of a structure related to Penrose tiling achieved using surface-based self-assembly was reported for ferrocenecarboxylic acid Fc(COOH) on a Au(111) substrate [26]. This structure is described in more detail elsewhere in this book, but briefly, Fc(COOH) assembles through intermolecular carboxylic acid· · ·carboxylic acid O—H· · ·O interactions, additionally stabilised by C—H· · ·O hydrogen bonds. This hydrogen bonding leads to the formation of cyclic pentagonal arrangements of molecules which were found to be more stable by DFT calculations than other possible hydrogen-bonding arrangements, including the $R_2^2(8)$ hydrogen-bonded synthon [15].

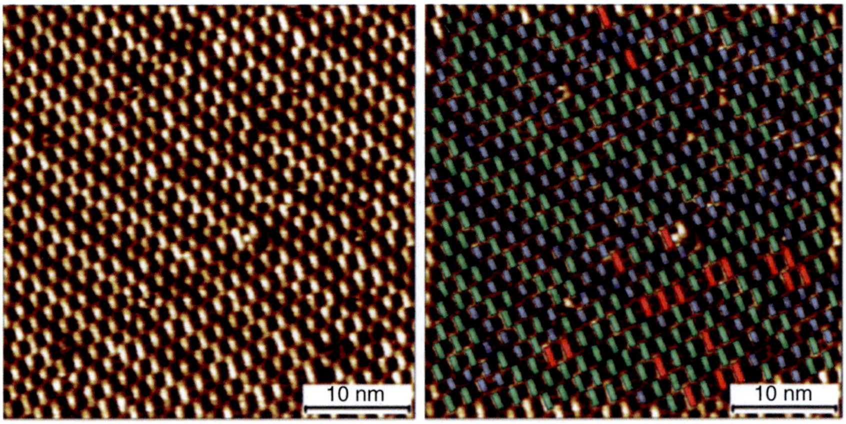

(a) (b)

Figure 9.4 (a) STM image of a TPTC framework self-assembled using commercially obtained sample at the nonanoic acid/graphite interface. Note the ordered structure in comparison to random tilings seen in Figure 9.2. (b) The same STM image as shown in (a) but with overlaid colours indicating the presence of longer molecules as impurities which template the parallel arrangement previously seen for QPTC. TPTC: blue, QPTC: green, QQPTC: red. Source: Steeno et al. [24] / with permission from Royal Society of Chemistry.

Carboxylic acid groups are a valuable supramolecular synthon in the formation of hydrogen-bonded networks, particularly due to the $R_2^2(8)$ homo-dimerisation, which can program edge-like properties into two-dimensional molecular tiles, as seen for TPTC and QPTC. Examples, where alternative arrangements of carboxylic acid hydrogen-bonding patterns occur, such as the $R_3^3(12)$ trimer in the flower arrangement of trimesic acid or the unique combination of $R_5^5(20)$ and $R_2^2(7)$ motifs in the Penrose tiling of Fc(COOH), mean that the hydrogen bonding of carboxylic acid groups may not always behave quite as expected. Alternative bonding possibilities should be considered when employing this synthon in the formation of self-assembled frameworks.

9.3 Imide-Melamine Supramolecular Synthons

Moving beyond carboxylic acids, many other potential hydrogen-bonding supramolecular synthons are available to those wishing to employ hydrogen-bonds to surface self-assembly. One family of systems that has been explored in depth is that formed by imides. A simple example of imide-driven unimolecular self-assembly is demonstrated by naphthalene-1,4:5,8-tetracarboxylic diimide (NTCDI) [27]. NTCDI contains imide moieties at opposing termini of the rod-shaped molecule which can adopt imide···imide hydrogen bonds. As with carboxylic acids, the imide···imide hydrogen bonds adopt an $R_2^2(8)$ intermolecular interaction (Figure 9.5c), and this in turn leads to the formation of self-assembled chains of molecules when the molecule is adsorbed onto a Ag:Si(111)-($\sqrt{3}\times\sqrt{3}$) R30° substrate, heretofore termed a Ag/Si(111) surface (Figure 9.5a). As discussed elsewhere in this volume, non-contact atomic force microscopy (ncAFM) [28] has been used to image these chains with sub-molecular resolution on a variety of surfaces (Figure 9.5b) [18, 29, 30].

Imide groups are capable of forming robust heteromolecular, triply hydrogen-bonded, supramolecular synthons with 2,6-di(acetylamino)pyridine moieties, a synthon that has been widely studied in supramolecular chemistry [31, 32]. This synthon has been developed and employed widely to form bimolecular structures on surfaces [2]. For example, perylene-3,4:9,10-tetracarboxylicdiimide (PTCDI), adopts hydrogen-bonding interactions with melamine to form a heteromolecular honeycomb network on a Ag/Si(111) substrate under UHV conditions (Figure 9.6) [33]. Triple hydrogen bonds are formed between the PTCDI imide groups and each edge of the triangular melamine (Figure 9.6a). The resulting framework (Figure 9.6b) contains large hexagonal pores with 2.4 nm diameter. These pores are capable of acting as hosts for several large guest molecules such as C_{60} (Figure 9.6c) [33]. Indeed, heptameric C_{60} clusters are formed within the pores of the network with a close-packed hexagonal arrangement aligned parallel to the principal axes of the substrate, in contrast to close-packed islands of C_{60} on the same substrate which do not align with the axes of the Ag/Si(111). These results not only show the utility of porous frameworks on surfaces but also demonstrate the stabilising effects of the network. Indeed, the PTCDI-melamine family of arrays has been shown to

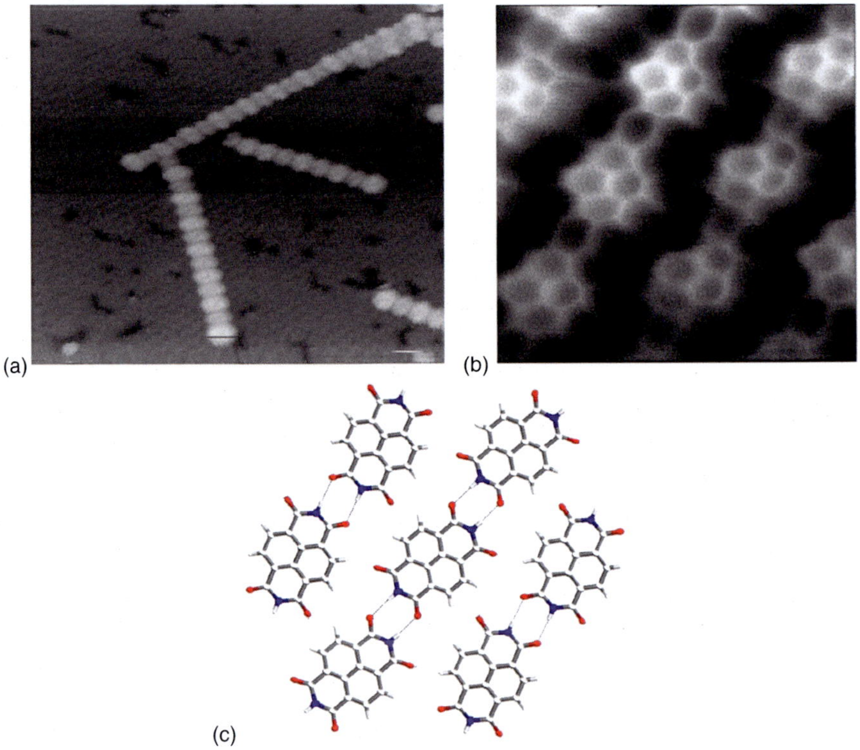

Figure 9.5 (a) STM image of three molecular chains of NTCDI adsorbed on Ag/Si(111) [27]; (b) Constant height DFM image of NTCDI adsorbed on Ag/Si(111) at 77 K (2.1 × 2.0 nm) [18]; (c) View of hydrogen-bonded chains in the single crystal X-ray structure of NTCDI [27]. Source: (a) Keeling et al. [27] / with permission from American Chemical Society. (b) Sweetman et al. [18] / Springer Nature / CC BY 3.0.

host other fullerenes, including C_{84} [34], Lu@C_{82} [35], and a range of molecules discussed further below.

The PTCDI-melamine network [33] can also be prepared on a Au(111) surface [36], generating a honeycomb arrangement analogous to that observed on a Ag/Si(111) substrate. However, when using a Au(111) substrate, a parallelogram phase was also observed following heating to approximately 90 °C, in addition to and ultimately replacing the honeycomb phase. This parallelogram phase has the same stoichiometric ratio as the honeycomb structure but is more densely packed [37]. The restricted cavity size in the parallelogram phase allows trapping of only two C_{60} molecules rather than the seven observed in the honeycomb phase. Related studies have shown Lu@C_{82} entrapment in the parallelogram phase, but due to the larger size of this fullerene, only a single molecule can be accommodated [35]. The potential versatility of the PTCDI-melamine systems as hosts is further demonstrated by the use of the parallelogram phase to trap two decanethiol molecules within its pores. It is important to appreciate that adsorption of decanethiol onto the PTCDI-melamine arrays led to decomposition of the honeycomb phase

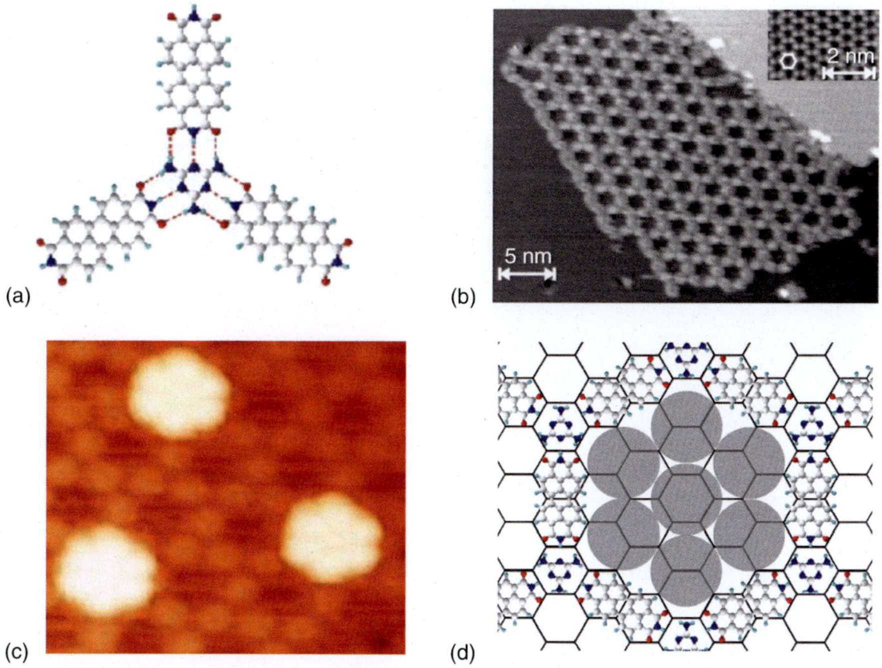

Figure 9.6 (a) A schematic of the PTCDI-melamine junction, showing the triple hydrogen-bonds that make up the supramolecular synthon using dashed red lines. (b) STM image of the PTCDI-melamine framework on Ag/Si(111); inset, high resolution view of the Ag/Si(111) surface. (c) STM image of fullerenes trapped within the pores of the hexagonal network, seen as bright white features; (d) A schematic diagram of a C_{60} heptamer sitting within a pore [33]. Source: (b) Theobald et al. [33] / with permission from Nature Publishing Group. (c) Slater et al. [2] / with permission from American Chemical Society.

but not the parallelogram phase, suggesting distinct relative stability of the two phases [38].

Buck et al. made significant advances in the use of the PTCDI-melamine network by preparing the framework from the solution phase onto a Au(111) substrate [39, 40]. As the framework can be prepared from the solution, it is readily possible to employ traditional Au-thiolate self-assembled monolayer (SAM) synthesis in combination with the PTCDI-melamine template. Thus, the adsorption of adamantanethiol within the network pores has been demonstrated with the thiolates sitting perpendicular to the substrate, which contrasts with related UHV studies [38]. The trapped thiols can be considered as confined SAMs, and in the case of adamantanethiol, they were shown to act as guides for the further deposition of Cu atoms within the pores using underpotential deposition (UPD) methods. Thus, copper metal was deposited only on areas of the surface that were covered by thiols and not areas masked by the PTCDI-melamine framework [39], and subsequently, it was shown that the framework acts as a diffusion barrier for Cu adatoms [41].

The concept of using the PTCDI-melamine network as a diffusion barrier was further extended by Saywell et al. to control reaction pathways [42].

1,3,5-Tris(4-iodophenyl)benzene (TIPB) was deposited into the PTCDI-melamine pores, and by so doing, it was possible to inhibit diffusion of this reactive molecule across the substrate. Iodo-aryl species are widely studied in the pursuit of covalently coupled arrays on surfaces, with reactions proceeding via an Ullman-like pathway using surface adatoms as catalytic sites [43]. Such reactions inside the network's pores, however, are controlled due to the restricted space available, which precludes polymerisation processes and the stabilisation of different reactive intermediates following C—I bond scission, even with the identification of iodine atoms within the pores (Figure 9.7).

Other molecules can be co-adsorbed onto the PTCDI-melamine framework and trapped within the pores of the structure. Typically, this guest deposition is achieved by solution methods or by sublimation into UHV. Although these approaches are not always possible depending on the nature and stability of the target, an alternative approach is to employ electrospray deposition [44, 45]. Using the latter method, it has been shown that magnetically interesting $Mn_{12}O_{12}(O_2CCH_3)_{16}(H_2O)_4$ clusters can be deposited onto the PTCDI-melamine array on a Au(111) substrate [45]. Some of the clusters sit within the pores of the PTCDI-melamine array, but others are positioned on top of the array (Figure 9.8). The inefficient entrapment of the $Mn_{12}O_{12}(O_2CCH_3)_{16}(H_2O)_4$ clusters by the array is attributed to a mismatch of dimensions between the cluster (1.6 nm) and pore (2.5 nm).

It is also possible to covalently modify either of the components of the bimolecular array. In the case of the PTCDI-melamine framework, it is feasible to functionalise the PTCDI component [46–48] prior to network self-assembly. This approach can be used to modify the pore dimensions of the framework and hence

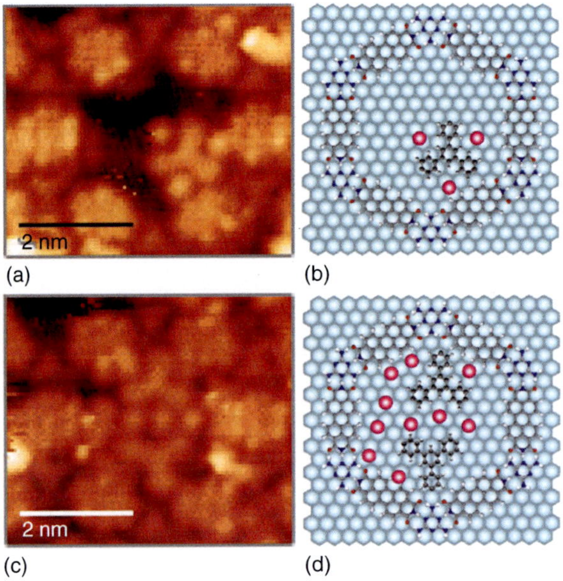

Figure 9.7 (a) and (c) STM images of 1,3,5-tris(4-iodophenyl)benzene (TIPB) deposited into the pores of the PTCDI-melamine framework showing the entrapment of both iodine atoms and dissociated TIPB in the pores; (b) and (d) models of the trapped species corresponding to STM images (a) and (c) respectively [42]. Source: (a) Judd et al. [42] / with permission from John Wiley & Sons, Inc.

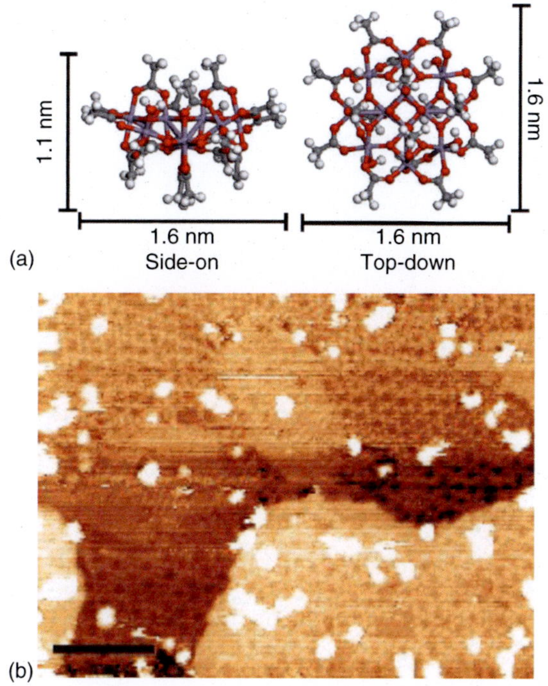

Figure 9.8 (a) Views of the molecular structure of $Mn_{12}O_{12}(O_2CCH_3)_{16}(H_2O)_4$; (b) STM image of $Mn_{12}O_{12}(O_2CCH_3)_{16}(H_2O)_4$ molecules deposited onto a PTCDI-melamine framework on a Au(111) surface. Scale bar = 200 Å [45]. Source: (b) Saywell et al. [45] / with permission from Springer Nature.

influence the array's ability to trap guest molecules. For example, whereas the array formed by replacing PTCDI with the two bromine atom functionalised analogue,1,7-dibromo-PTCDI, still hosts C_{60} heptamers (similar to the parent PTCDI-melamine array); using a thiopropyl-functionalised PTCDI, $(SPr)_2$-PTCDI, to form the honeycomb network does not trap heptamers or even hexamers [46]. In the latter case, C_{60} molecules are positioned in an irregular fashion within the pores and on top of the honeycomb array in a similar manner to the case of $Mn_{12}O_{12}(O_2CCH_3)_{16}(H_2O)_4$ clusters. This observation suggests that the thiopropyl groups inhibit fullerene adsorption, most likely by sterically hindering the C_{60}-surface interaction by the alkyl chains of the thiopropyl groups, which extend into the pores of the network.

Similarly, self-assembly of a related thio-adamantyl functionalised PTCDI, $(SAdam)_2$-PTCDI, and melamine leads to the functionalised PTCDI-melamine honeycomb framework [47]. The thio-adamantyl side chains are considerably more rigid and bulky than those of the thiopropyl analogue $(SPr)_2$-PTCDI and were anticipated to limit the available space of the framework pores. In reality, the system exhibits far more complicated behaviour, with some of the thio-adamantyl groups being detached during the formation of the $(SAdam)_2$-PTCDI-melamine network. As this cleavage process is inconsistently applied across the PTCDI molecules, multiple distinct pore sizes and configurations are adopted with varying numbers of thio-adamantyl groups in each individual pore. The distribution of pore types

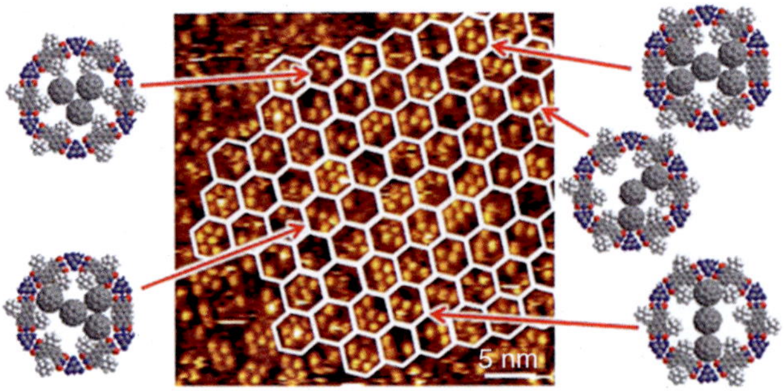

Figure 9.9 C_{60} entrapment in a $(SAdam)_2$-PTCDI-melamine framework on Au(111), as a guide to the eye the honeycomb network is indicated by the hexagonal white grid. The schematic figures illustrate the different C_{60} arrangements within the pores of the array. Different configurations arise as a result of missing thio-adamantyl units [47]. Source: Räisänen et al. [47] / with permission from Royal Chemical Society.

can be assessed by STM imaging of the sample following C_{60} adsorption since it is easier to image individual C_{60} molecules by STM than thio-adamantyl groups (Figure 9.9). The number of C_{60} molecules adsorbed within a pore is controlled by the number of thio-adamantyl groups remaining; C_{60} clusters ranging from dimers to heptamers are observed. Indeed, clusters of greater than five C_{60} molecules can only be explained by the loss of thio-adamantyl groups, and when heptamers are observed, this indicates that no thio-adamantyl groups remain attached to the edges of PTCDI molecules lining that particular pore.

The hydrogen-bonding synthon employed in the PTCDI-melamine frameworks can be readily introduced into other building blocks. The nucleobase thymine, for example, includes the same imide group as PTCDI and can be subject to a vast range of further synthetic modifications. Thus, it is possible to form other related arrays using this supramolecular synthon. The simplest related example is the self-assembled array formed by melamine and cyanuric acid, which has been extensively studied [49, 50]. Cyanuric acid contains three imide groups on each edge of the molecule and therefore provides the complementary configuration of hydrogen-bonding acceptors and donors to melamine. Beton and coworkers have studied this cyanuric acid-melamine (CAM) system and used it to develop a bilayer structure with two distinct hydrogen-bonding layers [51]. In particular, a monolayer of the CAM array is deposited onto a hexagonal boron nitride (hBN) substrate. As noted above, such CAM networks have been studied previously on other substrates by STM [49, 50], but growth on hBN allows imaging by atomic force microscopy (AFM), which directly confirms the anticipated framework arrangement. Subsequently, trimesic acid, which has also been studied previously as described above [14], was deposited onto the CAM monolayer, leading to the formation of a secondary hydrogen-bonded framework atop the initial layer (Figure 9.10). This seminal study indicates that it is feasible to generate multilayer structures, including layered arrangements of different types of molecule, opening many possible future research directions.

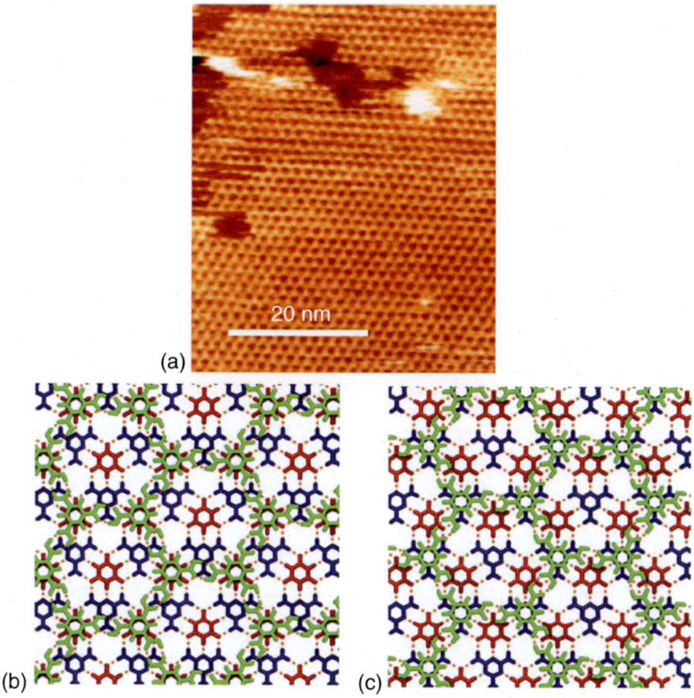

Figure 9.10 (a) AFM image of a honeycomb arrangement of trimesic acid framework (bright contrast) adsorbed on a cyanuric acid-melamine framework (dark contrast) on an hBN substrate. (b and c) Show two models of a trimesic acid network (green molecules) adsorbed on a cyanuric acid-melamine network (cyanuric acid in red/melamine in blue). The two orientations maximise van der Waals contributions [51]. Source: (a) Korolkov et al. [51] / with permission from Springer Nature.

9.4 From Hydrogen-bonding Synthons to Covalently-organic Frameworks

One interesting direction that research in this area has taken is to use hydrogen-bonding synthons to organise molecules into arrays that can be subsequently locked in place by covalent bond formation. Covalently-bonded frameworks are more commonly known as covalent-organic frameworks or COFs, as a result of their structural relationship to metal–organic frameworks, or MOFs. The preparation of such systems on surfaces represents a growing field of study, and surface-based COFs have recently been reviewed [52]. To study the formation of COFs in a stepwise fashion, it is necessary to prepare assemblies in which the components of the final framework interact in a supramolecular fashion on the timescale of the imaging process prior to covalent coupling. Indeed, it is possible to achieve the assembly of suitable building blocks from analogues of PTCDI, namely perylene-3,4:9,10-tetracarboxylic dianhydride (PTCDA), and linking amines [53, 54], including the reaction between the two components [55]. However, herein, we focus on a system that is reminiscent of one of the most widely studied COF synthons, those based on boroxine rings. Boroxine-based COFs are prepared by the dehydration of boronic acids, stimulating the formation of boron–oxygen covalent bonds between neighbouring molecules. Interestingly, when 1,4-benzenediboronic acid (BDBA) is deposited on a Ag(100) substrate [56], the BDBA initially forms a hydrogen-bonded array in which parallel rows of molecules are assembled (Figure 9.11a). This network can be imaged by STM, but upon appropriate stimulation using the STM tip, it is possible to initiate polymerisation, dehydration, and boroxine ring formation, which leads to the creation of a more open, porous COF on the surface (Figure 9.11b–e). Impressively, since the polymerisation is induced by the tip, it is possible to choose exactly where to initiate the polymerisation reaction. The preparation of COFs on surfaces is

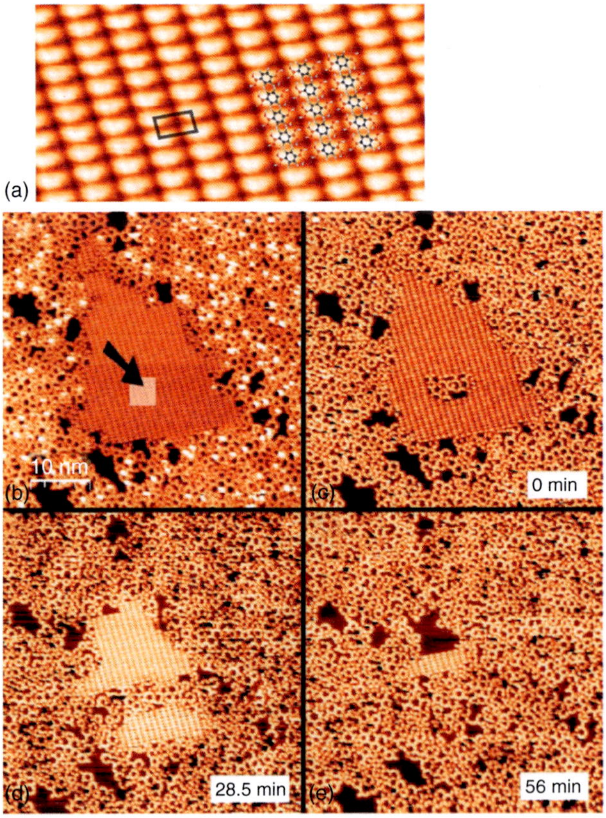

Figure 9.11 (a) STM image showing the hydrogen-bonded array formed by 1,4-benzenediboronic acid (BDBA) with a superimposed model showing parallel rows of BDBA molecules [56]; (b–e) STM images following tip-induced polymerisation of BDBA. In (b) the STM tip is brought into close-proximity and scanned over the hydrogen-bonded array in the location identified by the arrow. In (c–e) it can be seen that the polymerisation reaction initiates immediately and then slowly propagates until the hydrogen-bonded arrangement disappears [56]. Source: Clair et al. [56] / with permission from Royal Chemical Society.

clearly a developing field and highly promising for the formation of both stable porous frameworks and exotic structures, including the recently reported Sierpiński triangles prepared from the simple *meta*-isomer of BDBA, 1,3-benzenediboronic acid [57].

9.5 Heteromolecular Hydrogen-bonding Synthons

The intricacy exhibited by these Sierpiński triangles and the rhombus tiling described above are indicative of a wider field studying complex self-assembly using molecular imaging [58]. Supramolecular synthons based on heteromolecular arrangements lend themselves to the assembly of multiple molecules and therefore are an intriguing target for the assembly of complex systems. Some of the more obvious heteromolecular supramolecular synthons are those based on the nucleobases of the genetic alphabet: adenine (A), cytosine (C), guanine (G), thymine (T), and also uracil (U). The 2D self-assembly of DNA bases has been studied over a number of years [59], and the variety of hydrogen bonds observed for such systems are many and varied. For example, surface-based self-assembly of guanine has received prominent attention [60–63]. STM imaging of assemblies of guanine on a Au(111) surface [60, 63] reveals the formation of guanine quartets via Hoogsteen-style hydrogen bonding, which are in turn associated through further N—H⋯N hydrogen bonds to give two-dimensional arrays. Thus, it can be seen that the self-assembly of even a relatively simple molecule such as guanine can give rise to highly distinct arrays due to the wide variety of potential hydrogen-bonding motifs that can be adopted by molecules that possess a large number of different hydrogen-bond donors and acceptors.

Despite this, it is possible to introduce a degree of control over the variety of hydrogen-bonding interactions that can be adopted by nucleobases through their attachment to molecular scaffolds, which block certain hydrogen-bonding sites, thereby limiting the range of possible supramolecular synthons that can be formed. An example of this strategy is the study of nucleobase-functionalised porphyrin molecules (tetra-TP and tetra-AP) in which the porphyrin core is functionalised in each meso-position by a phenylthymine, or phenyladenine moiety (Figure 9.12a). In these molecules, or tectons, each nucleobase presents a hydrogen-bonding group exo to the porphyrin core [64, 65]. Either tetra-TP or tetra-AP can self-assemble on a HOPG substrate to give rise to two-dimensional lattice arrays. In the case of tetra-TP, the molecules interact through $R_2^2(8)$ intermolecular thymine⋯thymine hydrogen-bonds (see Figure 9.12b,c for the directly analogous Zn-tetra-TP assembly in which the porphyrin is metalated with Zn(II) cations), similar in structure to the carboxylic acid⋯carboxylic acid synthon discussed extensively above. The two-dimensional array of tetra-TP molecules adopts a chiral network due to the asymmetric arrangement of the thymine groups. Indeed, the almost perfectly square unit cell observed for the array suggests that all the thymine groups within an individual tetra-TP molecule adopt the same orientation and that individual domains contain only molecules of the same handedness [64, 65]. Tetra-AP also

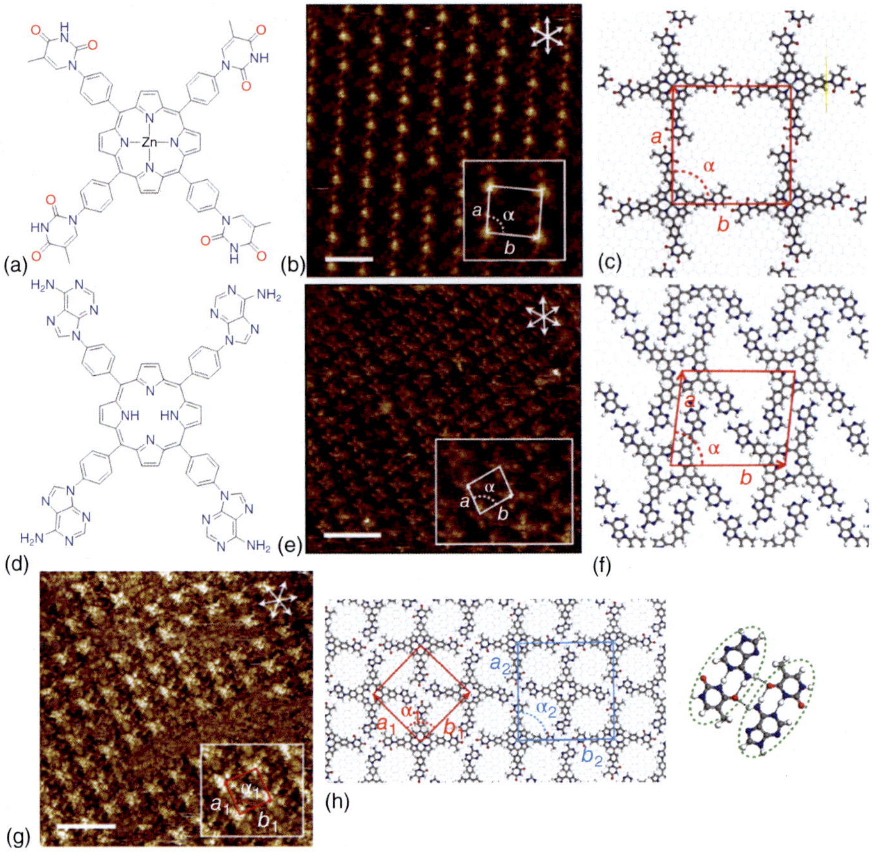

Figure 9.12 (a) and (d) Zn-tetra-TP and tetra-AP tectons used in two-dimensional array assembly; STM images at the liquid–HOPG interface for (b) Zn-tetra-TP; (e) Tetra-AP; (g) the Zn-tetra-TP and tetra-AP network accompanied by molecular models, (c), (f), and (h) respectively, of the resulting assemblies from molecular mechanics simulations. In each case the unit cell is shown. In the case of (h) the ATAT quartet formed by the Zn-tetra-TP and tetra-AP network is highlighted in the red square [64]. Source: (b,g) Blunt et al. [64] / with permission from American Chemical Society.

forms a two-dimensional array [64], although alternating orientations of the adenine appendages are observed in this case. This arrangement facilitates each adenine group to adopt three hydrogen bonds, including an $R_2^2(8)$ interaction between the termini of the adenine groups (Figure 9.12d,e). Co-assembly of tetra-TP and tetra-AP leads to the formation of ATAT adenine$\cdots$thymine quartets (Figure 9.12f,g,h) in which pairs of A$\cdots$T groups adopt Watson–Crick base-pairing, and then each of these pairs also interacts with an adjacent A$\cdots$T couple to form the quartet structure. STM imaging of this complex structure was only achieved when the tetra-TP was metalated with a Zn(II) cation, Zn-tetra-TP, providing sufficient contrast between the two different porphyrin molecules. The assembly process through the ATAT quartets ultimately generates an alternating chessboard pattern of Zn-tetra-TP and tetra-AP molecules across the substrate (Figure 9.12e). Although the alternating nucleobase-porphyrin arrangement is anticipated, the nature of the hydrogen-bonding configuration is substantially less so and therefore presents an ongoing challenge to the predictable design of two-dimensional assemblies on surfaces.

González-Rodríguez and coworkers have developed a related approach [66], which had been established previously in solution-phase studies [67–69]. In their case, rod-shaped molecules appended by the RNA nucleobases G, C, A, and U are used as the building blocks for the formation of self-assembled arrays. The strategy developed places complementary hydrogen-bonding groups (G· · ·C or A· · ·U) at opposite termini of the rods such that hydrogen-bonding interactions lead to the formation of cyclic structures. This approach also leads to cyclic arrangements on HOPG substrates. It is interesting to note that the A· · ·U system forms an AUAU quartet structure similar to the ATAT quartet observed for Zn-tetra-TP· · ·tetra-AP discussed above [64].

9.6 Conclusions

In this chapter, we have sought to demonstrate the versatility of employing hydrogen bonds to create surface-based two-dimensional supramolecular arrays and frameworks. The supramolecular synthon strategy provides a pathway to attempt the rational design of such systems, but it is apparent that, just as in crystal engineering, the expected supramolecular synthons are not always successfully formed due to other subtle energetic balances in the self-assembly process. However, successful strategies have been developed for the synthesis of surface-based frameworks, and some of these approaches have led to the discovery of highly unusual tiling processes and frameworks that are unlikely to be discovered using other strategies or environments. It is clear that significant progress has already been made in the field, and in some cases, such as the PTCDI-melamine arrays, the first steps towards functional materials are already well underway.

References

1 Mali, K.S., Pearce, N., De Feyter, S., and Champness, N.R. (2017). *Chem. Soc. Rev.* 46: 2520–2542.
2 Slater, A.G., Perdigao, L.M.A., Beton, P.H., and Champness, N.R. (2014). *Acc. Chem. Res.* 47: 3417–3427.
3 De Feyter, S., Gesquière, A., Abdel-Mottaleb, M.M. et al. (2000). *Acc. Chem. Res.* 33: 520–531.
4 Seto, C.T. and Whitesides, G.M. (1993). *J. Am. Chem. Soc.* 115: 905–916.
5 Liu, Y., Hu, C., Comotti, A., and Ward, M.D. (2011). *Science* 333: 436–440.
6 Yang, S.K. and Zimmerman, S.C. (2013). *Isr. J. Chem.* 53: 511–520.

7 Goor, O.J.G.M., Hendrikse, S.I.S., Dankers, P.Y.W., and Meijer, E.W. (2017). *Chem. Soc. Rev.* 46: 6621–6637.

8 Gil-Ramirez, G., Leigh, D.A., and Stephens, A.J. (2015). *Angew. Chem. Int. Ed.* 54: 6110–6150.

9 Wuest, J.D. (2005). *Chem. Commun.* 47: 5830–5837.

10 Watson, J.D. and Crick, F.H.C. (1953). *Nature* 171: 737–738.

11 Fersht, A.R. (1987). *Trends Biochem. Sci.* 12: 301–304.

12 Li, J. and Ross, D.K. (1993). *Nature* 365: 327–329.

13 Desiraju, G.R. (1995). *Angew. Chem. Int. Ed. Engl.* 34: 2311–2327.

14 Griessl, S., Lackinger, M., Edelwirth, M. et al. (2002). *Single Mol.* 3: 25–31.

15 Etter, M.C. (1990). *Acc. Chem. Res.* 23: 120–126.

16 Gutzler, R., Sirtl, T., Dienstmaier, J.F. et al. (2010). *J. Am. Chem. Soc.* 132: 5084–5090.

17 Gross, L., Mohn, F., Moll, N. et al. (2009). *Science* 325: 1110–1114.

18 Sweetman, A.M., Jarvis, S., Sang, H. et al. (2014). *Nat. Commun.* 5: 3931.

19 Blunt, M.O., Russell, J., Giménez-López, M.C. et al. (2008). *Science* 322: 1077–1081.

20 Steed, J.W., Atwood, J.L. (2021). *Supramolecular Chemistry*, 3e.Wiley.

21 Blunt, M., Lin, X., Giménez-López, M.C. et al. (2008). *Chem. Commun.* 20: 2304–2306.

22 Stannard, A., Russell, J.C., Blunt, M.O. et al. (2012). *Nat. Chem.* 4: 112–117.

23 Blunt, M.O., Russell, J.C., Giménez-López, M.C. et al. (2011). *Nat. Chem.* 3: 74–78.

24 Steeno, R., Minoia, A., Giménez-López, M.C. et al. (2021). *Chem. Commun.* 57: 1454–1457.

25 Penrose, R. (1978). *Eureka* 39: 16–32.

26 Wasio, N.A., Quardokus, R.C., Forrest, R.P. et al. (2014). *Nature* 507: 86–89.

27 Keeling, D.L., Oxtoby, N.S., Wilson, C. et al. (2003). *Nano Lett.* 3: 9–12.

28 Sweetman, A., Champness, N.R., and Saywell, A. (2020). *Chem. Soc. Rev.* 49: 4189–4202.

29 Sweetman, A.M., Jarvis, S.P., Champness, N.R. et al. (2014). *Phys. Rev. B.* 90: 165425.

30 Jarvis, S.P., Sweetman, A.M., Lekkas, I. et al. (2015). *J. Phys. Condens. Matter* 27: 054004.

31 Whitesides, G.M., Simanek, E.E., Mathias, J.P. et al. (1995). *Acc. Chem. Res.* 28: 37–44.

32 Zerkowski, J.A., Mathias, J.P., and Whitesides, G.M. (1994). *J. Am. Chem. Soc.* 116: 4305–4315.

33 Theobald, J.A., Oxtoby, N.S., Phillips, M.A. et al. (2003). *Nature* 424: 1029–1031.

34 Theobald, J.A., Oxtoby, N.S., Champness, N.R. et al. (2005). *Langmuir* 21: 2038–2041.

35 Silly, F., Shaw, A.Q., Porfyrakis, K. et al. (2008). *Chem. Commun.* 38: 4616–4618.

36 Perdigão, L.M.A., Perkins, E.W., Ma, J. et al. (2006). *J. Phys. Chem. B* 110: 12539–12542.

37 Staniec, P.A., Perdigão, L.M.A., Saywell, A. et al. (2007). *ChemPhysChem* 8: 2177–2181.

38 Perdigão, L.M.A., Staniec, P.A., Champness, N.R., and Beton, P.H. (2009). *Langmuir* 25: 2278–2281.

39 Madueno, R., Räisänen, M.T., Silien, C., and Buck, M. (2008). *Nature* 454: 618–621.

40 Silien, C., Räisänen, M.T., and Buck, M. (2010). *Small* 6: 391–394.

41 Silien, C., Raisanen, M.T., and Buck, M. (2009). *Angew. Chem. Int. Ed.* 48: 3349–3352.

42 Judd, C.J., Champness, N.R., and Saywell, A. (2018). *Chem. Eur. J.* 24: 56–61.

43 Judd, C.J., Haddow, S.L., Champness, N.R., and Saywell, A. (2017). *Sci. Rep.* 7: 14541.

44 Saywell, A., Magnano, G., Satterley, C.J. et al. (2008). *J. Phys. Chem. C* 112: 7706–7709.

45 Saywell, A., Magnano, G., Satterley, C.J. et al. (2010). *Nat. Commun.* 1: 75.

46 Perdigão, L.M.A., Saywell, A., Fontes, G.N. et al. (2008). *Chem. Eur. J.* 14: 7600–7607.

47 Räisänen, M.T., Slater, A.G. (née Phillips), Champness, N.R., and Buck, M. (2012). *Chem. Sci.* 3: 84–92.

48 Phillips, A.G., Perdigão, L.M.A., Beton, P.H., and Champness, N.R. (2010). *Chem. Commun.* 46: 2775–2777.

49 Perdigão, L.M.A., Champness, N.R., and Beton, P.H. (2006). *Chem. Commun.*: 538–540.

50 Staniec, P.A., Perdigão, L.M.A., Rogers, B.L. et al. (2007). *J. Phys. Chem. C* 111: 886–893.

51 Korolkov, V.V., Baldoni, M., Watanabe, K. et al. (2017). *Nat. Chem.* 9: 1191–1197.

52 Cui, D., Perepichka, D.F., MacLeod, J.M., and Rosei, F. (2020). *Chem. Soc. Rev.* 49: 2020–2038.

53 Ma, J., Rogers, B.L., Humphry, M.J. et al. (2006). *J. Phys. Chem. B* 110: 12207–12210.

54 Swarbrick, J.C., Rogers, B.L., Champness, N.R., and Beton, P.H. (2006). *J. Phys. Chem. B* 110: 6110–6114.

55 Treier, M., Fasel, R., Champness, N.R. et al. (2009). *Phys. Chem. Chem. Phys.* 11: 1209–1214.

56 Clair, S., Ourdjini, O., Abel, M., and Porte, L. (2011). *Chem. Commun.* 47: 8028–8030.

57 Cui, D., Fang, Y., MacLean, O. et al. (2019). *Chem. Commun.* 55: 13586–13589.

58 Pfeiffer, C.R., Pearce, N., and Champness, N.R. (2017). *Chem. Commun.* 53: 11528–11539.

59 Ciesielski, A., El Garah, M., Masiero, S., and Samorì, P. (2016). *Small* 12: 83–95.

60 Otero, R., Schock, M., Molina, L.M. et al. (2005). *Angew. Chem. Int. Ed.* 44: 2270–2275.

61 Ciesielski, A., Perone, R., Pieraccini, S. et al. (2010). *Chem. Commun.* 46: 4493–4495.

62 Ciesielski, A., Haar, S., El Garah, M. et al. (2015). *L'Actualité Chimique* 399: 31–36.

63 El Garah, M., Perone, R.C., Santana Bonilla, A. et al. (2015). *Chem. Commun.* 51: 11677–11680.

64 Blunt, M.O., Hu, Y., Toft, C.W. et al. (2018). *J. Phys. Chem. C* 122: 26070–26079.

65 Slater, A.G., Hu, Y., Yang, L. et al. (2015). *Chem. Sci.* 6: 1562–1569.

66 Bilbao, N., Destoop, I., De Feyter, S., and González-Rodríguez, D. (2016). *Angew. Chem. Int. Ed.* 55: 659–663.

67 Montoro-García, C., Camacho-García, J., Lýpez-Põrez, A.M. et al. (2015). *Angew. Chem. Int. Ed.* 54: 6780–6784.

68 Montoro-García, C., Bilbao, N., Tsagri, I.M. et al. (2018). *Chem. Eur. J.* 24: 11983–11991.

69 Mayoral, M.J., Bilbao, N., and González-Rodríguez, D. (2016). *ChemistryOpen* 5: 10–32.

Index

Supramolecular Chemistry on Surfaces: 2D Networks and 2D Structures, First Edition.
Edited by Neil R. Champness.

p

q